# SOLAR ENERGY
# HANDBOOK SECOND EDITION

# SOLAR ENERGY
# HANDBOOK SECOND EDITION

## AMETEK, Inc.

### Theory and Applications

## CHILTON BOOK COMPANY
Radnor, Pennsylvania

## NOTE ON FORMULAS AND EQUATIONS

The formulas and equations in this handbook have been organized in a manner that is convenient for use with handheld programmable calculators. The primary computational features of each of the chapters may be incorporated by the reader into a magnetic card program for the Texas Instruments, Hewlett-Packard, or equivalent programmable calculators. This may be of particular value where numerous repetitive calculations occur in the worksheets for determining collector area in Chapter 7.

---

## NOTE ON UNITS AND NOTATION

Both English and metric (SI) units are used in this handbook. English units appear in places where engineering or practical formulas result in order to conform to prevailing construction practices in the United States. Metric units are used in the theoretical sections and the unit notation is indicated for each formula and equation.

Copyright © 1984, 1979 by AMETEK, Inc.
Second Edition All Rights Reserved
Published in Radnor, Pennsylvania 19089, by Chilton Book Company
Designed by Arlene Putterman
Manufactured in the United States of America

Library of Congress Cataloging in Publication Data
Main entry under title:
Solar energy handbook.
Bibliography: p. 259
Includes index.
1. Solar energy—Handbooks, manuals, etc.
I. Ametek Inc.
TJ810.S62445   1983        621.47        83-71091
ISBN 0-8019-7154-3

1 2 3 4 5 6 7 8 9 0   3 2 1 0 9 8 7 6 5 4

# CONTENTS

# PART I
# Sun and Climate

**PART III
Solar Electrical
Technology**

# LIST OF TABLES

# SOLAR ENERGY
# HANDBOOK SECOND EDITION

# 1　Solar Energy Markets

The past decade of the seventies has seen major changes in the sources of domestic energy consumed in the United States. The availability of oil and gas from our nation's wells has declined. The use of imported fuels and domestic coal resources has increased. In addition, the development of new sources of energy has begun—sources which will grow in importance not only in this decade but in future decades as well. Figure 1–1 displays the estimated structure of these changes for the prior and future decades. The figure shows that as we move forward toward the year 2000, the proportionate share of energy supplied by coal will increase until it becomes the predominant fossil fuel, and alternative energy sources, including nuclear and solar, will grow from a minor share of the pre-1970 sources to nearly one-quarter of the total by the year 2000.

## TRADITIONAL RESOURCES

The employment of oil from foreign sources as a major component of our energy supply is totally unsatisfactory. It limits our freedom with the burdens of the international *quid pro quo* and strains our economy as the balance of payments is afflicted with increasing deficits. Moreover, even without unilateral action like the OPEC consortium initiated in 1973, world competition for remaining reserves of petroleum will have the effect of worsening price disadvantage and could cause conflict between otherwise peaceful nations that must compete in the same market for the lifeblood of their industries and commerce. Although the abundance of petroleum-based energy may vary up and down, we must strive to narrow the gap between supply and demand. We would be well advised to do it not with more imported oil and gas, but with equivalent domestic substitutes.

The obvious choice is to expand the supply of other traditional energy resources from domestic reserves, but the outlook is not bright. Domestic petroleum reserves are now declining as wellhead production exceeds the rate of new discoveries. There are new sources ahead, potential offshore reserves for example, but these will provide only partial and temporary relief. Natural gas is also a limited resource (although there are prospects for the development of new but costly reserves). Even allowing for the unexpected discovery of sufficient oil and natural gas reserves to diminish the need for foreign oil, these resources are inherently finite. We cannot produce more than the earth has left for us and we are already well advanced in the depletion of our economically recoverable reserves.

Coal is our most abundant energy resource; in fact, a large part of the total

1

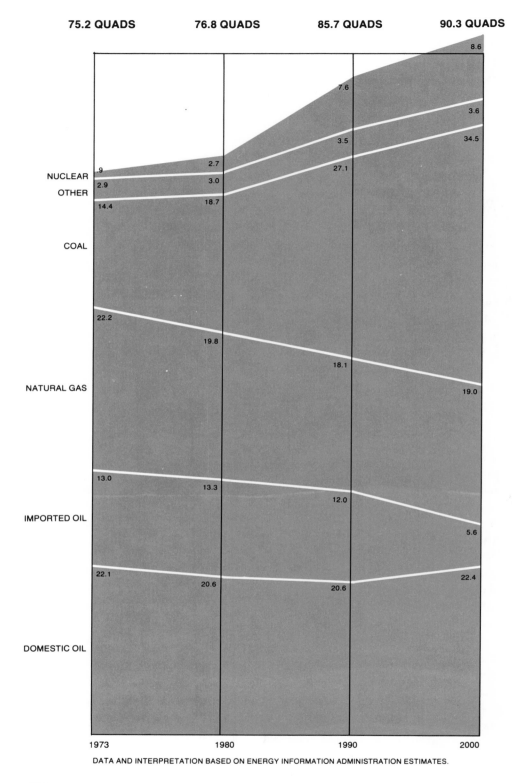

**FIGURE 1–1** Future estimates show less dependence on imports and an increase in innovative energy sources.

world reserves are within the United States. In addition to the safety hazards of mining, the use of coal presents many environmental risks. We might proceed with the utilization of coal wherever possible, with due concern for the potential damage of strip mining, the danger of mining, and environmental pollution caused by its combustion. But coal is a less than ideal source. It must be mined, processed, loaded, transported, and unloaded before being put to use. It is bulky and unsuited to modern pollution-control requirements. And it will become more expensive in the presence of greater demand and as the costs of its production rise.

## ALTERNATIVE ENERGY SOURCES

The search for alternative sources of energy, supported by federal and private funds, has yielded a variety of possibilities with differing degrees of promise. They include coal gasification, coal liquification, shale oil extraction, nuclear energy (both fusion and fission), and various forms of solar energy. Many of these possibilities are in early stages of technological development or have economic or environmental consequences similar to oil, natural gas, and coal. Nuclear energy production utilizing radioactive fuels can be expected to increase its contribution as experience and wider use make it more cost effective. But there are problems here as well; the questions of nuclear waste disposal and nuclear safety engineering are of such gravity that a prolonged period of research and testing must precede wider use. Most importantly, nuclear energy, like modern coal-based energy, is fundamentally limited to large-scale central production. Of all the alternative sources, only solar energy combines the advantages of present-day technology, potential for significant future technical and economic gains, and minimal impact on the environment.

## THE UNIQUE RESOURCE

Solar energy possesses characteristics which make it highly attractive as a primary energy source. It is based upon a continuously renewable resource which cannot be depleted and which is not subject to political control. Of all energy sources, it is the least encumbered by environmental and safety hazards. And, most significantly, it is possible to collect, convert, and store solar energy with present technology. Future advances can be expected to improve and reduce the cost of solar energy equipment.

There are unique problems, to be sure, which can affect the rate of growth. Energy can only be collected during daylight hours and then most efficiently on sunny days. Thus, some provision must be made for storage to accommodate nighttime and cloudy-day demands. The energy in sunlight is distributed through space in such a way that the amount collected, and therefore the energy yield of the system, is dependent upon the area intercepted. In addition, the sun is uncooperative enough to persist in its apparent motion in daily and seasonal ranges of considerable expanse across the sky with a consequent variation in the amount of power available. But all these disadvantages are outweighed by the twin benefits of the universality of the source and the long-term economy of operating a solar energy system.

Long-term operating economy is another attractive characteristic of solar energy systems. Unlike traditional energy sources, however, the initial costs of tapping and converting the resource of the sun are relatively large per unit of power produced. Furthermore, the costs of those systems which are built to serve a single house or building must be borne by individual users who often lack the financial capabilities of large utilities and other central power generating authorities. For this reason, governmental incentives for the development of solar energy and its financing at the local level may play a significant role in the market's development.

## THE GROWTH OF SOLAR ENERGY

The proportion of the total United States energy requirement supplied by solar energy will increase as economic and technical advances are made. Today the solar thermal system (that which produces heat) is the most economically accessible technology. Rapidly advancing is the availability of low cost solar electricity as the technology of photovoltaic systems is further developed.

A probable timetable for the initial and commercial availability of a variety of solar technologies is illustrated in Figure 1–2. The graph illustrates the approximate time period during which the technology is initially introduced, used in limited or selective applications, and enters general commercial usage. These would correspond respectively to the economic periods when usage would require support or subsidization, when the economic benefits are favorable with respect to other sources in selective or unique applications (such as remote or isolated locations), and when the technology is generally competitive with conventional energy.

Irrespective of the anticipated future introduction of new solar technologies, the rapid growth solar equipment has experienced in the past several years is impressive enough. Table 1–1 displays the production history of solar

**TABLE 1–1**   Solar Collector Production

| | Millions of Square Feet | | | | | | | | | |
|---|---|---|---|---|---|---|---|---|---|---|
| | 1974 | 1975 | 1976 | 1977 | 1978 | 1979 | 1980 | 1981 | 1982 | 1983 |
| Solar Thermal | | | | | | | | | | |
| Low temperature | 1.14 | 3.03 | 3.88 | 4.74 | 5.87 | 8.40 | 12.23 | 8.56 | 7.47 | 4.85 |
| Medium & high temperature | .13 | .71 | 1.92 | 5.57 | 4.99 | 5.85 | 7.17 | 11.39 | 11.14 | 11.98 |
| Total production (in millions of Sq. Ft.) | 1.27 | 3.74 | 5.80 | 10.31 | 10.86 | 14.25 | 19.40 | 19.95 | 18.61 | 16.83 |
| Number of manufacturers | 45 | 131 | 186 | 321 | 340 | 349 | 276 | 217 | 274 | 224 |
| Solar Electric | | | | | | | | | | |
| Peak kilowatts of module | NA | NA | NA | 450 | 950 | 1450 | 4204 | 6748 | 6897 | 12620 |
| Number of manufacturers | NA | NA | NA | NA | NA | NA | 13 | 13 | 19 | 18 |

NA: Not Available.
* Annualized based on first six months of production.
Source: DOE/EIA-0174 (811.)

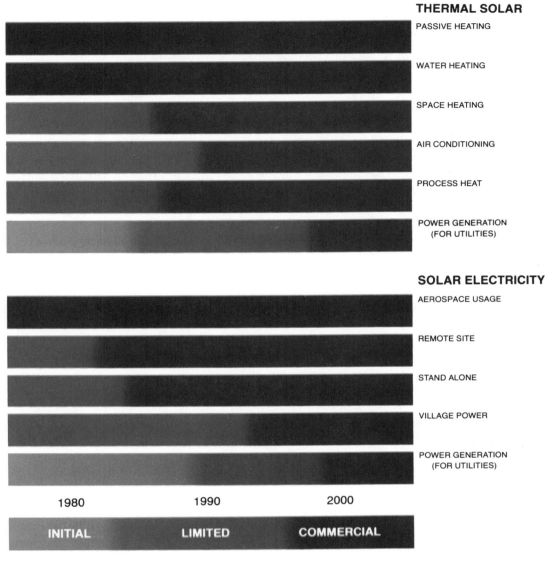

**THERMAL SOLAR**

PASSIVE HEATING

WATER HEATING

SPACE HEATING

AIR CONDITIONING

PROCESS HEAT

POWER GENERATION
(FOR UTILITIES)

**SOLAR ELECTRICITY**

AEROSPACE USAGE

REMOTE SITE

STAND ALONE

VILLAGE POWER

POWER GENERATION
(FOR UTILITIES)

1980    1990    2000

INITIAL    LIMITED    COMMERCIAL

SOURCE: AMETEK ESTIMATES

**FIGURE 1–2**  Estimated timetable of current solar technologies.

thermal collectors and photovoltaic modules. The information highlights several meaningful trends. First, the average yearly growth rate of total solar collector production was in excess of 50 percent per year for the period shown. It can also be seen that in the earlier years of the decade, the low temperature collector (used primarily for swimming pools) was lower in cost and greater in volume. By the end of the decade major growth swung to the medium and/ or higher temperature collectors which are used primarily for domestic water heating. Also of note is the peaking of the numbers of manufacturers and its current decline. This is characteristic of many growth industries which initially attract many participants, followed by the inevitable maturing when technical, manufacturing, and marketing skills determine the successful companies.

The history of photovoltaic module production is much briefer, but exhibits the same rapid growth characteristics. Of note is the considerably smaller number of manufacturers. This reflects the substantial technical and financial investment required to become an active participant in this field.

## SOLAR ENERGY APPLICATIONS

An attempt to list all present and potential solar energy applications would probably be forever incomplete since the breadth of present technology and the versatility of the process allow a wide range of innovation. The major current applications for solar *thermal* energy are swimming pool heating, domestic hot water heating, space heating, industrial process heat, space cooling, and agricultural process heat.

A ranked breakdown of applications for solar *electrical* energy is not readily available. However, if the purely experimental residential demonstration installations are excluded, almost all the usage will be found to be supplied by "stand-alone" equipment. These are applications in which the photovoltaic module replaces gasoline- or diesel-engine powered generators, thermoelectric generators, or storage batteries. Their use is typically restricted to remote sites where no utility grid exists or it is inaccessible. Their functions include: supplying electrical power for communication equipment, agricultural water pumping, cathodic protection, navigational and avionic aids, as well as such conventional uses as lighting and refrigeration.

The following section is devoted to a description of these applications, their growth pattern and future expectations.

### SOLAR THERMAL

*Swimming Pool Heating.* The major application for solar heating, based on square feet of collector, is the heating of swimming pools. Figure 1–3 indicates the rate of growth of usage of swimming pool heaters. Although there are many solar-heated public or institutional swimming pools, the majority of the usage is for pools of privately-owned, single-family dwellings.

The primary motivation for the installation of a solar-heated swimming pool is to reduce the cost of maintaining the pool at a comfortable temperature and to extend the length of the swimming season. Since the temperature difference between the pool water and the ambient air is usually small during the normal usage season, it is possible to use solar collectors for this application that would not be effective for other applications. This includes unglazed collectors or collectors made from extruded black plastic.

The data reported for the first half of 1981 indicates that more than 80 percent of the square footage utilized in this application was of non-metallic construction. Since this type of construction is inherently lower in cost than glazed, enclosed, metallic collectors, this application can be economically attractive, especially when tax credits or incentives are available.

It is estimated that in 1945 there were 2500 privately-owned, residential swimming pools in the United States. According to the National Spa and Pool Institute the current number is 2,000,000 pools, of which half are heated. Of the

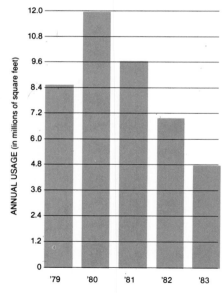

**SWIMMING POOL HEATING**

ANNUAL USAGE (in millions of square feet)

**FIGURE 1–3** Growth of solar swimming pool heating.

1,000,000 heated pools only 8 percent or 80,000 are solar heated. This relatively small degree of penetration and the continuing increases in the cost of electricity or natural gas (the predominant energy source for pool heating) should provide strong support for the continuing growth of this important solar application.

*Domestic Water Heating.* Solar water heaters have become a major solar application for several reasons. Hot water requirements are year round, so the user's investment in a solar water heater will yield the maximum return. The necessary equipment is not complex and is small in size, thereby lending itself to retrofitting in existing installations. Furthermore, during the early years of solar market development, while the total number of solar installations is still modest and long-term operating experience at a minimum, the smaller investment necessary for a solar water heater, compared with solar space-heating units, will lead many users to limit their risk by keeping expense at a minimum.

The growth of solar domestic water heating for the past several years is shown in Figure 1–4. Underlying this steady growth has been the development of the critical business structure of solar installers having the technical, marketing, and financial strengths to transfer the developed solar technology to the ultimate user. It is only recently that national merchandising organizations have engaged in this facet of the solar business. They will have a strong and positive impact on future growth.

Initial penetration of this market, including installations for schools, hospitals, office buildings, laundries, and so on, is expected to be slower than in the case of the residential market. This is a function of the diverse nature of the market and the fact that many of the individual applications are highly specialized. The ultimate size of the market will not be diminished, however, since a number of influences will promote greater acceptance. The inevitability of the end of traditional fuel sources has already been noted; another factor is the effect of regulator philosophy. Experience has shown that, during periods of fuel shortage, the energy consumption of the commercial sector is more subject to curtailment or allocation than is residential consumption. Substitution of the solar alternative in industrial and commercial establishments could well become vital to uninterrupted operations.

*Space Heating.* The growth of solar space heating is shown in Figure 1–5. For the period covered, about one third of the total square footage used air as the fluid medium and the balance used water or other liquids as the heat transfer medium. Although the volume of solar space heating is considerably less than that of domestic water heating, it would seem (without consideration of the associated minimum costs which make the simpler water heating systems more attractive) to be the wiser choice, since the vast majority of these systems would also inherently satisfy all the domestic hot water needs. In any case, space heating will widen the solar market since such systems are especially suited to new residential construction.

Because increased capacity of each installation is accompanied with an increase in costs, non-residential space-heating markets, like those of non-residential water heating, may proceed at a somewhat slower pace. It is a reasonable assumption, though, that as the supply pressures of traditional energy materials increase and further research yields a unit cost decrease in solar equipment, solar energy applications in non-residential use will achieve success.

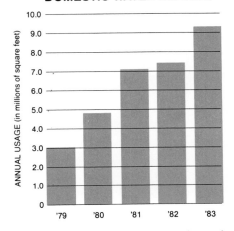

**DOMESTIC WATER HEATING**

**FIGURE 1–4**  Growth of solar domestic water heating.

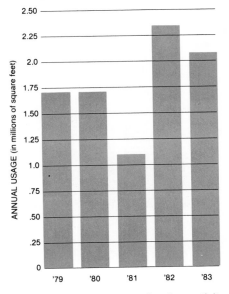

**SPACE HEATING**

**FIGURE 1–5**  Solar space-heating activity.

## AIR CONDITIONING

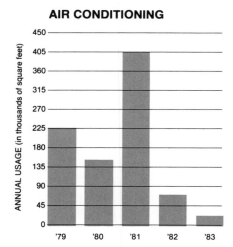

**FIGURE 1–6** Solar air-conditioning activity.

*Air Conditioning.* Most practical methods of air conditioning require high-temperature sources of heat energy for efficient operation. This means that the collectors in a solar-powered system must be of higher efficiency than those required for space or water heating. Thus, though technically suitable solar air-conditioning equipment is available, it is only cost effective in a limited range of applications. Even so, the rising costs of conventional energy sources will progressively widen the solar air-cooling market. Figure 1–6 displays the erratic level of activity in solar cooling.

This market presents a contrast to the water and space-heating markets in that non-residential air-conditioning applications are expected to grow more rapidly than residential ones. The reasons are inherent in the characteristics of the demand load and the configurations of new and existing equipment. The cooling season for commercial buildings, stores, offices, and the like is longer than that for private dwellings; in the southern United States, it is very nearly a year-round demand. Therefore, solar air-conditioning equipment would be better utilized and more cost effective. In addition, the configurations of equipment in existing conventional systems make the adaption to solar power doubly attractive. Since the majority of non-residential systems yield chilled water as their product, they can be retrofitted with a solar unit in which the collectors and solar heat-driven chiller supplement the existing equipment. Those systems already incorporating a heat-actuated chiller (and the number is substantial) need only add a solar collection system. Finally, the typical solar-driven, heat-actuated chiller requires the use of a cooling tower. Cooling towers are now widely used for heat rejection in commercial air-conditioning installations but are not generally used in residences. The operating complexity imposed by the cooling tower is therefore not a material deterrent to the use of solar air-conditioning in non-residential operations.

*Process Heat.* While water heating, space heating, and air conditioning are expected to provide the highest initial demand for solar equipment, the direct thermal-energy requirements of many process industries make them well suited for solar technologies. The volume of activity in solar-powered process heat, both for industrial and agricultural processing, is shown in Figure 1–7.

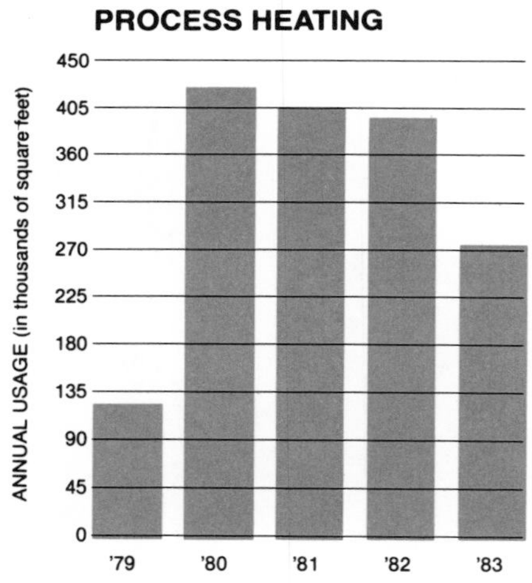

Although the feasibility of using solar energy for many processes can and has been adequately demonstrated, many barriers exist to its adaption. These barriers include the following:

## PROCESS HEATING

**FIGURE 1–7** Solar-powered process heating activity.

1. Many processes require higher temperatures than those attainable with flat plate collectors and would require the use of the more expensive and sophisticated concentrating collectors.
2. The lower temperature process heat requirements can be supplied by waste-heat recovery techniques, creating a competitive technology.
3. There are smaller economic incentives for industry to use solar energy than those provided for residential applications.

## SOLAR ELECTRICAL

*Photovoltaic Modules.* Electricity is a high quality form of energy which is suitable for almost every type of energy-consuming activity. Its uses range from such highly technical tasks as powering computers or electron-beam welders to domestic tasks such as water heating or powering the kitchen toaster.

Electricity can be used to generate temperatures approaching those at the surface of the sun or as cold as those found in the empty reaches of space. Electricity can be used to generate pressures high enough to transform ordinary carbon into diamonds, or vacuums as low as one billionth of a billionth of an atmosphere. It is the almost limitless utility of electrical energy that has resulted in such great interest in the direct conversion of sunlight into electricity.

The applications for such a technology seem endless and the opportunities for market growth stagger the imagination. However despite its superiority as an energy source, electricity is a commercial commodity and the factors affecting the growth and utilization of new sources are economic as well as functional.

The economic measure used by consumers of electricity is the cost per kilowatt-hour, that is, the actual or computed charge for the utilization of 1000 watts of electricity for a period of one hour.

Residential and commercial customers of electrical utilities receive a notification of charge for regular time periods based on the metered consumption of kilowatt-hours of electricity. Users of electricity generated by gasoline- or diesel-powered generators must compute the cost of the electricity they consume. The computation consists basically of summing all the costs associated with the operation of the generating equipment during some defined time period and dividing this cost by the total number of kilowatt-hours of electricity generated by the equipment which is utilized. Note the emphasis on the term "utilized." Generating equipment which is operating but not being used to operate electrical appliances or devices is generating electricity having an infinite cost. This is because the operating costs during this period are divided by zero kilowatt-hours of electricity utilized. Similarly, when the generating equipment is being operated at its full capacity on a continuous basis and its output is fully utilized, it is generating electricity at its lowest possible kilowatt-hour cost. Figure 1–8, displays this relationship for a 10 kilowatt diesel-powered generator. The cost which varies with the total amount of the capacity utilized is that of the diesel fuel. The fixed and recurring costs which must be paid irrespective of the amount of capacity utilized are primarily the equipment amortization costs and the operator and maintenance costs. As the graph indicates, these costs rapidly drive up the generated electricity cost at low capacity utilization, while at the higher utilizations the fuel cost establishes an almost constant relationship between usage and cost per kilowatt-hour. Although highly simplified this type of relationship between cost and utilization is descriptive of all fuel-powered generators.

In the case of photovoltaic generators of electricity, some major differences should be noted. First is the fact that there is no fuel cost as such; it is only the fixed and recurring costs which contribute to the cost of the electricity generated. A second major difference is the variability in the intensity or availability of the powering sunlight from day to night, season to season, and geographic location to location. This means that the cost of electricity generated by a photovoltaic module or array is not only a function of the percentage of

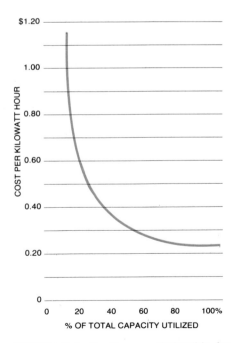

**FIGURE 1–8** Cost/usage relationship for 10-kilowatt diesel-powered generator.

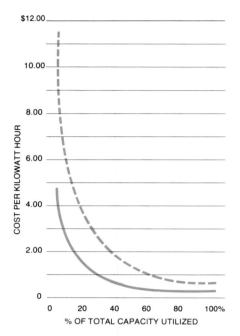

**FIGURE 1–9** Computed cost per kilowatt-hour for photovoltaic array with 10 percent efficiency at module cost of $12.50 per square foot.

maximum rated capacity used, but also a function of the particular time it is used and the geographic location at which it is used.

This difference is displayed in Figure 1–9, which indicates the change in computed cost per kilowatt-hour for a photovoltaic array having an efficiency of 10 percent and a module cost of $12.50 per square foot. The solid line is for the southwest area of the United States where sunlight is abundant and the dotted line is representative of a cloudier, more northerly area (such as upper New England or Alaska).

Here, unlike the fuel-powered generators, the calculated electrical cost increases in direct proportion to the reduction in percentage of capacity used. The figure indicates several important points. The first is that a careful matching of the size of the electrical generator to electrical capacity required is more important with photovoltaic than conventional fuel-powered generators. The second important consequence is the desirability of adding storage capacity to the photovoltaic system. This storage effectively transfers the utility of the photovoltaic system from one time period to another, permitting more constant utilization, a smaller nameplate capacity, and reduced costs.

Although batteries are normally thought to be the typical storage medium, there are many other forms of energy storage which can be used. Among the more unusual are flywheel storage or the electrochemical generation of hydrogen gas. In a more conventional system such as one designed for irrigation, energy can be stored by pumping water to an elevated tank or in the case of a photovoltaic-powered refrigeration system, the production of ice for cooling during the nighttime.

Perhaps the most unique storage procedure of all would be to have excess electricity from the photovoltaic system put to use producing new photovoltaic cells by means of the electrodeposition of a semiconductor, such as the cadmium-telluride photovoltaic cell recently developed by AMETEK. In this case, the term ''growth'' rather than ''storage'' might be more descriptive.

Photovoltaic systems offer the potential for unique functional advantages as well as economic advantages, and in many cases this may be of the greater importance. For example, they require no supply of fuel, they are quiet and can operate with minimal or no attendance. Perhaps most importantly, being non-mechanical in nature, they do not require the periodic services of a skilled technician or mechanic—an individual whose services may be nonexistent in many parts of the world.

As noted previously, the applications for photovoltaic-generated electricity are almost limitless. It is therefore perhaps more appropriate to describe their usage in terms of the existing types of electrical generating equipment. Table 1–2 lists these sources in groups according to the typical computed cost of electricity generated by them. As the cost of photovoltaic equipment and photovoltaic electricity continues to drop, more and more of the conventional generating equipment will be displaced or replaced by photovoltaic devices.

*Small or Integral Units.* This group consists primarily of primary storage batteries or thermoelectric generators. Usage ranges from providing power for automotive starter motors, to battery-powered lift trucks or mine locomotives, to standby power for critical communication systems or navigational aids.

At the present time, photovoltaic generators can and have displaced such

**TABLE 1-2**  Current Applications and Computed Costs for Photovoltaic-Generated Electricity

| Electrical Source | Typical Capacity and Cost Per Kilowatt-Hour |
|---|---|
| Small or Integral Units<br>—Primary Batteries<br>—Thermoelectric Generators | Typically in excess of $1.00 per kilowatt hour. Capacity to several kW. |
| Small Stand-Alone Units<br>—Gasoline-Powered Generators<br>—Small Diesel-Powered Generators | Typically ranging from $0.50 to $1.00 per kilowatt hour. Capacity to 10 kW. |
| Large Stand-Alone Units<br>—Diesel-Powered Generators<br>—Extension of Utility Grid to Remote Sites | Typically ranging from $0.75 to $0.50 per kilowatt hour. Capacity to 100 kW or more. |
| Utilities<br>—Direct Connection to Primary Grid | Less than $0.25 per kilowatt hour. Capacity unlimited. |

equipment even at the relatively high cost of the current available silicon cell modules.

The use of photovoltaics is limited in many of the applications due to the small space available e.g. a lighted buoy, or the environment, e.g. indoor or mine equipment.

*Small Stand-Alone Units.* This group, which is comprised of the many small gasoline- or diesel-powered generators, finds use in a wide variety of remote locations such as construction sites, agricultural locations, military installations, and small remote village or residential sites.

The intermittent use of many of these units raises their generating cost to where even the higher priced silicon modules can displace them. Even when the economics are not favorable, a photovoltaic generation unit may be preferred because it eliminates the need for fuel or the requirement for trained mechanics. The market served by this group will represent the initial major market for photovoltaics.

*Large Stand-Alone Units.* This group basically satisfies the same requirements as utility generated power would where no utility grid exists. Because electricity generated by these sources is much higher in cost than utility generated electricity, it is not utilized for expendable residential functions that could be more economically served by other means.

This is a common form of electrical generation in remote areas of the world and developing countries. Large stand-alone generators, being of a static nature, represent an area well suited for photovoltaic displacement. But it may require the availability of low-cost photovoltaic modules based on thin film semiconductors such as electro-deposited cadmium telluride or vacuum-deposited amorphous silicon.

*Utilities.* Electricity delivered from very large generating plants via the grid is the lowest in cost and most widely utilized type in developed nations. It is the primary source for industry, commercial, and residential requirements.

The entry of photovoltaics into this area (except for demonstration projects) will require the existence of a photovoltaic industry having a mammoth capacity and capable of producing photovoltaic units at costs of less than $1.00 per peak watt, as compared to the current cost of in excess of $10.00 per peak watt for the silicon cell module.

As this point in time approaches, the availability of photovoltaic units of ever decreasing cost will lead to the investigation, evaluation, and establishment of new applications for solar derived electricity. As confidence in and knowledge of the unsurmountable potential of solar energy develops, the world may very well enter what can be called—with due respect to the ultimate source—a new golden age.

# PART I

## Sun and Climate

# 2 Solar Physics and Geometry

Science is a fragmented and accumulative process. Its practitioners often work independently, theorizing, experimenting, proving; but pile the resultant conclusions in a vast common storehouse. Discoveries are sometimes triumphs, but triumphs of the moment, for each unveiled fact quickly assumes the anonymity of data, history to be drawn from the common fund when needed, by discoverer or stranger alike, as support for future theories and further questioning. And, while each discipline in science is separate and distinct, a community concerned in the main with its own factual past and theoretical future, it is not independent. It is benefited by, and often depends upon, the work performed in other disciplines; physics, for example, reaches into chemistry for guidance and information while moving toward new knowledge.

This intramural exchange as the basis for progress is most evident in today's practicality of solar power. The debts are many, for the achievement of solar energy utility has drawn on the knowledge gained in physics, chemistry, thermodynamics, and a hundred others; similarly, its application equipment is a mixture of metallurgy, semiconductor technology, fluid mechanics and other contributors. Because of this, a complete study of solar energy entails the complexities of optics, structural engineering, heat exchange, electrical engineering, and many other disciplines, treatments which are far beyond the scope of this handbook since the object here is practical knowledge useful in today's applications. It will be valuable, however, when considering specific applications in later chapters, to have an understanding of a few basic theoretical concepts that, since they are fundamental, are concerned with the most fundamental part of all solar energy study: the sun.

The sun, like all functioning stars, is a gaseous sphere in a state of such complete and violent agitation that "turbulent" seems a pale description. Columns of gases, heated by core temperatures of many millions of degrees, fume up to the surface, lose some of their heat, and descend to be heated again. Giant magnetic storms, sunspots, split the seething cauldron of solar gases to add their own definition of bedlam. And, as if to extend the violence past imagination, immense solar flares, cataclysmic eruptions of the surface, occur periodically to spew a bewildering array of radiation outward into space—radio waves, gamma waves, physical fluxes, visible light, ultraviolet light, infrared light, and x rays—energy enough in a single solar flare to satisfy earth's energy needs for 100,000 years. All of it is raw, naked power, the prime physical force of the universe, and all of it originates within the tiny simplicity of the hydrogen atom.

For 5 billion years the sun has gathered the elements from the vast reaches of interstellar space by gravitational attraction until, today, 1 percent of its mass is composed of traces of all elements, 24 percent is helium, and 74 percent is hydrogen. This is a necessary predominance, for hydrogen is the fuel for the sun's thermonuclear furnace. The energy-producing process has been duplicated on earth only in the hydrogen bomb where hydrogen transmuted into helium and the higher elements leaves a surplus of energy derived from the intranuclear "glue" which binds the particles of all nuclei together. The effects of this transformation are gargantuan, but the process itself and the vessel which contains it, the sun, are stable. When viewed on the mammoth scale of solar time and place, it must end at some point either in a white dwarf star shriveling in the absence of hydrogen fuel or in a supernova of incomprehensible explosive force. But, when framed in the lilliputian scale of human interests, the sun can be safely regarded as in eternal existence.

The sun's stability is a product of transference phenomena within the solar sphere. Energy is conveyed to the surface, the photosphere, where it is released into space to balance the stream erupting from the interior furnace. The photosphere consists of a plasma of ionized gases which dispose of energy in several ways. The solar wind, cosmic rays, and radiated neutrinos expel some of the vast output, but the majority of radiated energy is in the form of electromagnetic waves characteristic of the plasma's elemental composition. These waves are combined into the phenomena we know as sunlight.

## THE SOLAR SPECTRUM

The sun's electromagnetic radiation occurs in a large number of pure waves of different wavelengths combined into the "extra-terrestrial solar radiation" depicted in Figure 2–1. Also shown, in order of increasing magnitude of wavelength, are its components: x-rays, ultraviolet light, visible light, infrared light, radiant heat, and radio waves. Most of the sun's energy which can be recovered in a useful way is in the form of ultraviolet, visible, and infrared light.

All pure electromagnetic waves are characterized by wavelength, $\lambda$, and frequency, $f$, through the relationship:

$$\lambda f = c \tag{2–1}$$

where $c$ is the speed of light in a vacuum: $2.998 \times 10^8$ meters per second.

The unit of energy intensity used in Figure 2–1 is the Langley, named for Samuel Pierpont Langley (1834–1906), an American scientist and aeronautical pioneer who made major contributions to methods of measuring solar energy. One Langley (Ly) is one calorie (cal) of energy incident normal to one square centimeter of area (cal cm$^{-2}$); one Langley per minute (Ly min$^{-1}$) is equivalent to 221.2 Btu per hour per square foot (Btu/hr-ft$^2$). The spectral energy, $S$, in Figure 2–1 is expressed in Langleys per minute per micron ($\mu$) of wavelength on a logarithmic scale. The unit for wavelength, $\lambda$, is the micron, $\mu$, or one millionth of a meter, also on a logarithmic scale. The logarithmic scales are used so that a wide range of conditions can be displayed.

In an ideal world one would seek to capture as much of this prodigious outpouring as possible, but the width of the solar spectrum in general (and the longer radiant-heat wavelengths in particular) are not ideal. In many cases, as will become apparent in the section on selective coatings, it is better to utilize a

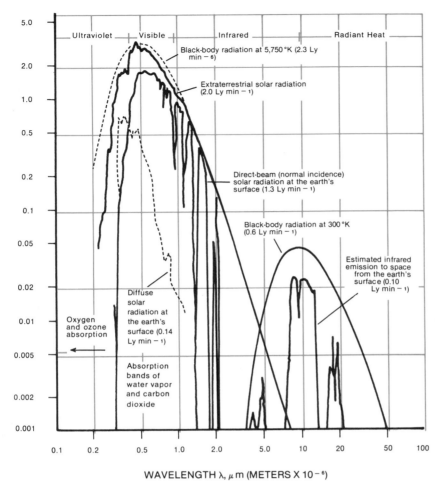

**FIGURE 2–1** The solar spectrum. After Kreider and Kreith, *Solar Heating and Cooling,* Scripta Book Co., 1975, page 30.

collector which absorbs sunlight and reflects the longer wavelengths. Similarly, collector covers should be constructed of materials that transmit radiation best within the solar spectrum while reflecting radiant heat waves back into the collector.

Figure 2–1 represents a general view of the way the earth's atmosphere changes the extra-terrestrial radiation. However, each hour of the day and each season produces a different air mass, water vapor, and turbidity (amount of moisture and particles).

For clearly commercial solar products, two new standards have been written by SERI (Solar Energy Research Institute) and have been accepted by ASTM (American Society for Testing Materials):

ASTM E892-82 "Terrestrial Solar Spectral Irradiance Tables at Air Mass 1.5 for a 37° tilted surface."

ASTM E891-82 "Terrestrial Direct Normal Solar Spectral Irradiance Tables for Air Mass 1.5."

In addition, recent work by a number of researchers strongly suggests that a revised extraterrestrial spectrum by Neckel and Labs[1] is a significantly better representation of the solar extraterrestrial spectrum. Comparisons have shown that approximately a five percent difference can exist in narrow band widths of the spectrum, but for the integrated total little difference is apparent. It is very probable that in the near future this standard will be updated using Neckel and Labs revised extraterrestrial spectrum.

A SERI document "Terrestrial Solar Spectral Data Sets" by Richard E. Bird (SERI/TR-215-1149) compares the two and shows the comparison between the model and recent data.

## THE SOLAR CONSTANT

The solar constant is the reference point to determine the amount of solar energy available for utilization. It is defined as the total energy emitted by the sun per unit of area perpendicular to the sun's rays in near-earth space at an average earth distance from the sun per unit of time. As might be suspected, this does not represent power available to earthbound collectors, since the changing geometry of the sun as it moves across the sky and the interposition of atmosphere and climate all have an effect. In addition, though the solar constant represents a useful reference, its stated value differs slightly in various source publications. This is a consequence of the measurement techniques used, spacecraft instrumentation or terrestrial astronomical data (both of which contain sources of inaccuracy). Other variances in absolute value result from natural variations in the sun's intensity, sunspots, and temporary spectral abnormalities, but the greatest natural variation results from the elliptical configuration of the earth's orbit. In early January the earth occupies its nearest approach to the sun, thereby increasing the solar constant by about 3 percent. In early July the average value declines by about 3 percent when the earth reaches its greatest distance from the sun. Table 2–1 expresses the average value of the solar constant in different systems of units.

**TABLE 2–1**  The Solar Constant, $I_0$

| | | |
|---|---|---|
| $I_0 =$ | 1354 | Watts per square meter |
| $=$ | 1.354 | Kilowatts per square meter |
| $=$ | 429 | Btu per hour per square foot |
| $=$ | 1.94 | Langleys per minute = calories per square centimeter per minute |
| $=$ | 4870 | Kilojoules per hour per square meter |
| $= 4.87 \times 10^6$ | | Joules per hour per square meter |
| $=$ | 1.52 | Horsepower per square yard |

1. Neckel, H., and Labs, D. "Improved Data of Solar Spectral Irradiance from 0.33 to 1.25 $\mu$m." *Solar Physics* 74:1981, pages 231–249.

# BLACKBODY RADIATION LAWS

As is true in many disciplines, the foundation for understanding real events is an understanding of an ideal case which exists only in theory. Economists, for example, use the impossibility of perfect competition as their laboratory case, then proceed to study the ways in which reality approximates or departs from that ideal.

The theoretical case underlying the reality of solar energy utilization is that of the blackbody. A blackbody, so called because of the resulting color, is defined as an object that absorbs all, and reflects none, of the radiation incident upon it. No perfect blackbody exists, but the concept is important because the radiation laws derived from that perfect case can then be used to yield a relationship between the properties of real objects in actual radiative environments. Reality here, of course, is the energy exchange between the originating sun and the receiving object on earth.

A detailed discussion of blackbody radiation laws and the theoretical derivation of relevant equations is contained in Appendix A. Though complete familiarity with this theoretical background is not necessary to understanding solar energy utilization, the material does form the basis for practical application. As such, it is well worth study.

Real objects are characterized by their absorption and emission compared to those of a blackbody at the same temperature. The absorption and emission of an object are expressed as coefficients, $\alpha$ and $\varepsilon$ respectively, ranging from zero (for no absorption and no emission) to one (for absorbing and emitting as well as a blackbody at the same temperature). Both the absorption and emission coefficient depend upon the temperature of the object, the material of which it's made, and surface characteristics such as roughness, coatings, corrosion, and features. In general, the absorption and emission of an object vary with temperature, wavelength, and angle of incidence, with the absorption and emission coefficients usually expressed as average values over a range of these variables.

When radiation is incident upon an object that does not completely absorb it, the portion that is not absorbed is either reflected by, or transmitted through, the object. The coefficient of transmittance, $\tau$, represents the fraction of incident radiation that passes through a transparent or translucent object. Transmission can be direct (as through clear glass), diffuse (frosted glass), or in multiple refracted beams (certain crystals). The coefficient of reflectance, $\rho$, represents the fraction of incident radiation that is reflected from an object. Reflectance can be diffuse (as from snow), or specular (as from a mirror). In general, transmittance, $\tau$, and reflectance, $\rho$, vary with temperature, wavelength, angle of incidence, and polarization.

The coefficient of transmittance and reflectance, $\tau$ and $\rho$, and the ratio of $\alpha$ to $\varepsilon$ (see equation A–8) are significant characteristics of any solar absorber. The $\alpha/\varepsilon$ ratio is often regarded as an important characteristic, since it relates to the temperature that the absorber can achieve if all heat losses other than radiative can be eliminated. It would be misleading, however, to assume that a high value of $\alpha/\varepsilon$ implies greater energy collection by the object, since the amount of energy collected increases with increases in $\alpha$ even if the ratio $\alpha/\varepsilon$ remains fixed.

## APPARENT MOTION OF THE SUN

The relation between a particular site on this planet and the sun must be expressed with terms which include accommodation for the geographical location on the earth, the status of the earth's rotation (usually expressed as the time of day), and the relationship between the positions of the earth and sun in space. Each has its effect on the absolute value of solar energy available for utilization.

The first, geographical location, is the easiest with which to deal. Any position on the earth can be designated by its latitude, longitude, and elevation above sea level. The position of the sun as observed from this site is designated by the angle of elevation *above* the horizon and its azimuth (the angle formed by the sun's projection to the horizon and due south.)

The position of the earth during its annual journey around the sun can be expressed by the angle of declination, D, which represents the amount by which the earth's north polar axis is tilted toward the sun. An approximation for *D* is:

$$D = 23.44 \sin (360 \ d/365) \text{ degrees} \tag{2-2}$$

where:  $D$  is the angle of declination in degrees

$d$  is the number of days following the vernal equinox which usually falls near March 21 (or day 80)

and where the number 23.44 is the angle, in decimal degrees, between the earth's axis and the plane of the ecliptic (earth's orbital plane).

The time of day, for solar purposes is expressed in terms of the hour angle, $H$, which represents one 24-hour day as 360 degrees of angle.

$$H = 15 \ t \text{ degrees} \tag{2-3}$$

where:  $t$  is the time in hours (decimally) from solar noon

$H$  is the hour angles in decimal degrees.

Each time period of 4 minutes advances the hour angle by one degree.

The local time, or clock time, at which solar noon occurs will differ from 12:00 noon as a result of several effects which are incorporated into the following equation:

$$t_{sn} = 12{:}00 + 4 \ (Y - Y_0) + t_c, \text{ minutes} \\ + 1 \text{ hour if daylight savings time is in effect} \tag{2-4}$$

where:  $Y$  is the longitude of the site in decimal degrees

$Y_0$  is the longitude of the center of the local time zone (see Table 2–2)

$t_c$  is the correction from the equation of time (see Figure 2–5 or Equation 2–5)

$t_{sn}$  is the time at which solar noon occurs.

Within a given time zone, local time is established as a discrete multiple of one hour from the time at the zero meridian at Greenwich, England; thus, a location west of the zone's center will experience its solar noon at a later time than at the center and a location east of the center will experience an earlier

**TABLE 2–2**  Time Zones and Longitudes

| Time Zone | $Y_0$, Longitude at Center | Representative Cities |
|-----------|---------------------------|----------------------|
| Eastern | 75° | Camden, New Jersey |
| Central | 90° | Memphis, Tennessee |
| Mountain | 105° | Denver, Colorado |
| Pacific | 120° | South Lake Tahoe, California |

solar noon. In specific terms, solar noon will occur four minutes later for each degree of longitude *west* of the center of a time zone.

The longitude of the centers of U.S. time zones and representative cities are shown in Table 2–2. This table provides the value of $Y_0$ for use in Equation 2–4. If the longitude of the site is not known, it may be estimated by determining the distance in miles east or west of the center and multiplying by 69 times the cosine of the latitude of the site. This will yield the difference in longitude between the site and the representative city in degrees. Multiplication by four will yield the number of minutes that solar noon will follow 12:00 noon if the site is west of the representative city or the number of minutes by which solar noon will precede 12:00 noon if the site is east of the representative city.

The final correction, $t_c$, is needed to establish the local reckoning of solar time. It is due to two different effects of the earth's motion around the sun. The first is the cumulative advance and retardation of solar time with seasonal changes in declination; the other is the cumulative advance and retardation of solar time with variations in the earth's orbital speed resulting from its slightly elliptical orbit around the sun. These factors are both taken into account by the

**FIGURE 2–2**  The earth's axis remains tilted in the same direction as the earth revolves annually around the sun, giving rise to the seasons.

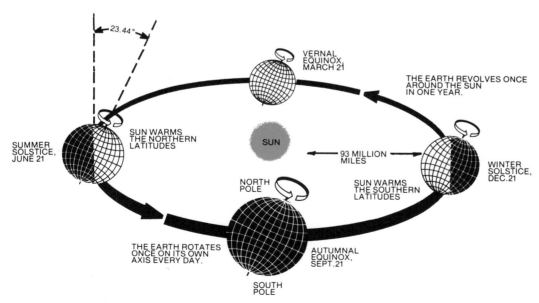

APPARENT DAILY ORBIT OF THE SUN

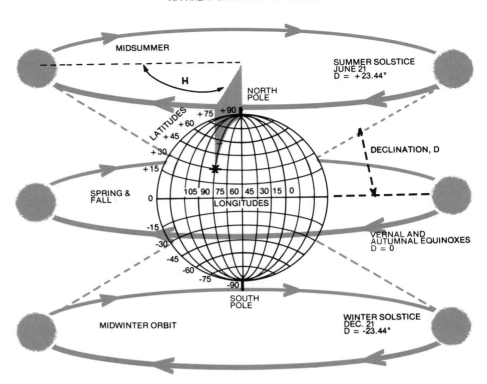

**FIGURE 2–3** From an earth regarded as stationary, the apparent motion of the sun is described by the hour angle, H, which depicts the time of day, and the declination, D, which depicts the time of year. A site on the earth is specified by its latitude, L, and longitude, Y. The hour angle increases uniformly with 15 degrees corresponding to each hour from solar noon, or one degree every four minutes.

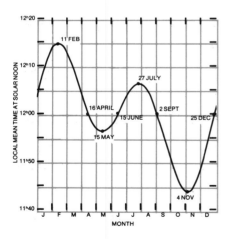

**FIGURE 2–5** The equation of time.

equation of time correction, $t_c$, shown in Figure 2–5 and applied in Equation 2–4. The curve shown in Figure 2–5 is known as the "equation of time" and is represented graphically on many globes as a distorted figure-eight known as an *analemma*. The equation of time may also be represented to within an accuracy of one minute by the formula:

$$t_c = 9.9 \sin (2y) + 7.3 \sin (y - 13°) \text{ minutes} \qquad (2–5)$$

where $y = \dfrac{360}{365} (d + 80)$ is the "year angle" in degrees, and $d$ is the day number defined following Equation 2–2.

The time of solar noon, $t_{sn}$, is the time during the day when the sun achieves its highest elevation angle above the horizon, and when its azimuth angle is zero. Thus, it is located in the sky in the direction of due south. One method of determining directions is to note the direction of the shadow of a plumb vertical at solar noon. It will be oriented in a north–south direction.

The local times at which sunrise and sunset occur can be found by adding and subtracting an hour angle, $H_s$, from that of solar noon and converting to time units. The hour angle, $H_s$, is determined by:

$$\cos H_s = \tan D \tan L \qquad (2–6)$$

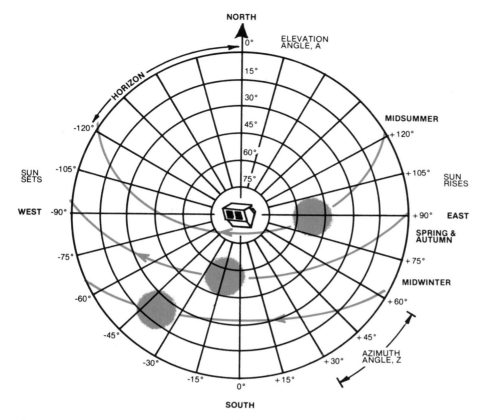

**FIGURE 2–4** In a plan view of a solar site from above, the elevation and azimuth angles of the sun may be plotted on a normal projection of a celestial hemisphere. The site in the center is located at 40° North latitude.

where: $H_s$ is the hour angle between solar noon and sunset or sunrise

$D$   is the declination angle from Equation 2–2

$L$   is the latitude of the site.

The geocentric time of sunrise, $t_{sr}$, and the time of sunset, $t_{ss}$, are given by:

$$t_{sr} = t_{sn} - \frac{H_s}{15} \quad \text{Hours (decimal)}$$

$$t_{ss} = t_{sn} - \frac{H_s}{15} \quad \text{Hours (decimal)}$$

**(2–7)**

where: $t_{sn}$ is the time of solar noon given by Equation 2–4. (The actual time of sunrise and sunset are earlier and later, respectively, than that predicted by Equation 2–7 because of the refractory distortion of the atmosphere and slight parallactic advantage due to the elevation of an observer.)

The astronomical definition of sunrise and sunset is based on the appearance and disappearance, respectively, of the upper limb (edge) of the sun as it moves across a level horizon. The limb of the sun is 16 minutes of arc from its center and atmospheric refraction accounts for about 34 feet of arc in a vertical direction. The correction to be applied to Equation 2–7 to compute the astronomical sunrise and sunset (as distinct from the geocentric) requires an addition, $\Delta H_s$, to the value of $H_s$, given by:

$$\Delta H_s = \frac{1}{0.3 \, (\cos L) \, (\cos D) \, (\sin H_s)} \tag{2-8}$$

where $\Delta H_s$ will be expressed in degrees of angle.

Equations 2–2 through 2–8 can be programmed into card programmable calculators for convenient application.

The position of the sun can be determined approximately by the following equations for its elevation *above* the horizon, $A$, and its azimuth angle with respect to due south, $Z$. All angles are in degrees.

$$\sin A = (\cos D) \, (\cos H) \, (\cos L) + (\sin D) \, (\sin L) \tag{2-9}$$

$$\sin Z = - \frac{\cos D \sin H}{\cos A} \tag{2-10}$$

where:   $A$  is the angle of elevation of the sun, positive above the horizon and negative below the horizon

$Z$  is the azimuth angle of the sun, positive to the east and negative to the west if measured from the south

$D$  is the declination angle from Equation 2–2

$H$  is the hour angle from Equation 2–3, negative in the morning and positive in the afternoon

$L$  is the latitude of the site.

Equation 2–10 represents an interesting geometrical fact. It states that the product $(\sin Z)(\cos A)$ is independent of the latitude from which the sun is observed.

The declination, $D$, is the elevation of the sun to an observer at the North Pole and the hour angle, $H$, is the negative value of its azimuth there. Equations 2–9 and 2–10 constitute a transformation of solar elevation and azimuth angles as reckoned from the North Pole to any desired location on the Earth.

In the application of Equation 2–10 for determining the azimuth angle of the sun, there is an important precaution to observe. That is when the azimuth angle, $Z$, exceeds 90°, the value of $\sin Z$ will be the same as for an angle less than 90° by an amount equal to that by which the solar azimuth exceeds 90°. Thus, once the value of $\sin Z$ is determined by Equation 2–10, the determination of $Z$ is sometimes ambiguous. The ambiguity is resolved if the azimuth angle is redefined as being referred from the north if the sun is in the northern sky and from the south if the sun is in the southern sky.

Another method of resolving the ambiguity is to use the following equation to obtain the azimuth, $Z$:

$$\tan \frac{Z}{2} = \frac{(\cos D)(\sin H)}{(\cos L)(\sin D) - (\sin L)(\cos D)(\cos H) - (\cos A)} \tag{2-11}$$

This equation, when solved for $Z$ (using the principal value of the inverse tangent function) provides the correct azimuth angle under all conditions. In the southern hemisphere, however, the azimuth is referred to south, just as in the northern hemisphere. The usual convention in the southern hemisphere is to refer azimuth angles to the north, which is automatically accomplished in Equation 2–10.

In the southern hemisphere, the latitude, $L$, should be entered into Equations 2–9 and 2–11 as a negative number. If Equation 2–10 is used, the azimuth should be referred to north when the sun is in the north and referred to south when the sun is in the south; however, the sign of Equation 2–10 should be reversed from negative (−) to positive (+). If, instead, Equation 2–11 is used to determine the sun's azimuth, $Z$, then it will be referred to south at all times when either northern latitudes or southern latitudes are used.

The accuracy of these formulas (Equations 2–9 through 2–11) is based upon the assumption that the earth is a sphere and that daily variation in declination, $D$, may be ignored. In fact, the earth is slightly pear shaped, the declination changes slightly every day, there is some parallax in the direction of the sun as seen from different points on the earth, the earth's motion is affected by the moon, and there are long-term changes in the obliquity of the equinox. There are, additionally, totally unexplained changes in the earth's speed of rotation that some astronomers relate to changes in its internal structure. Fortunately, all these effects are small and can safely be ignored in most solar energy applications. The formulas presented here can be considered accurate to within the angular diameter of the sun. The sun's average angular size is 16′0″ of arc from its center to its edge, or slightly more than one-half degree from side to side as viewed from the earth.

A flat plane surface with normal (perpendicular to the surface) directed at an elevation angle, $\psi$, from the horizontal and at an azimuth angle, $\phi$, from due south forms an angle, $\theta$, with the sun. The angle, $\theta$, is the angle between the sun's rays and the normal to the plane surface and is given by:

$$\cos \theta = (\sin A)(\cos \Sigma) + (\cos A)(\sin \Sigma)[\cos(Z - \phi)] \qquad \textbf{(2–12)}$$

where the geometry is shown in Figure 2–6.

The angle, $\Sigma$, is the tilt angle of the plane as measured up from a horizontal surface. By definition, it is the complement to $\psi$, or $\Sigma = 90° - \psi$. The angle, $\phi$, is the direction in which the plane is tilted measured from $\phi = 0$ at due south and increasing *positive* as it faces a more easterly direction, and *negative* to the west. The angle, $\theta$, between the sun's rays and the normal to a plane surface is known as the incident angle.

## SOLAR IRRADIANCE

The incident angle, $\theta$, is important because it determines the magnitude of solar irradiance. Solar irradiance is defined as the solar intensity that is incident perpendicular (normal) to one unit area of the plane surface. The direct beam of solar energy incident to a tilted plane surface is given by:

$$I_b = I_n \cos \theta \qquad \textbf{(2–13)}$$

where:   $I_b$ is the direct beam solar irradiance on the tilted plane surface in Btu/hr-ft$^2$

$I_n$ is the solar irradiance or direct normal intensity of the sun's radiation in Btu/hr-ft$^2$ (Equation 2–14)

$\theta$ is the incident angle given by equation 2–12.

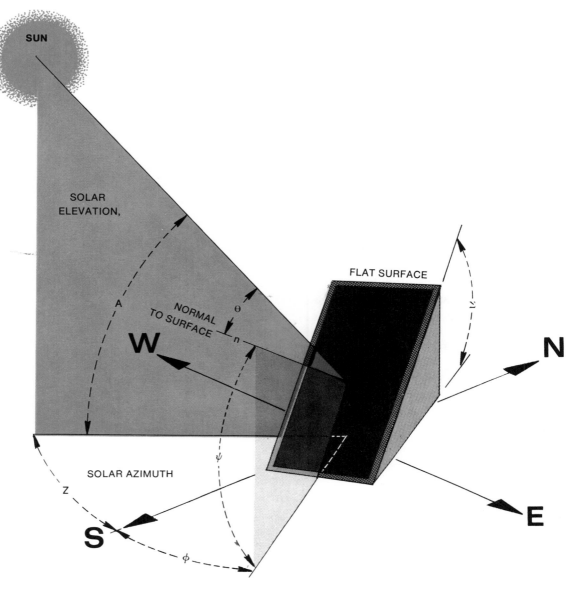

**FIGURE 2–6** Solar geometry.

The direct normal intensity of the sun's radiation, $I_n$, on a clear day is given by:

$$I_n = I_r e^{-mB} \text{ Btu/hr-ft}^2 \tag{2–14}$$

where $e$ is the base of the natural logarithm with value of 2.71828, and where $I_r$ and $B$ are constants given in Table 2–3 for each month of the year, and $m$ is the air mass given by:

$$m = \frac{1}{\sin A} \tag{2–15}$$

where $A$ is the elevation angle of the sun given by Equation 2–9. The values of

26

**TABLE 2-3**  Values of $I_r$ and $B$ for Equation 2-14

| Month | $I_r$ (Btu/hr-ft$^2$) | $B$ |
|-------|------|------|
| Jan. | 390 | 0.142 |
| Feb. | 385 | 0.144 |
| Mar. | 376 | 0.156 |
| Apr. | 360 | 0.180 |
| May | 350 | 0.196 |
| June | 345 | 0.205 |
| July | 344 | 0.207 |
| Aug. | 351 | 0.201 |
| Sept. | 365 | 0.177 |
| Oct. | 378 | 0.160 |
| Nov. | 387 | 0.149 |
| Dec. | 391 | 0.142 |

$I_r$ = apparent solar irradiation at air mass = 1.
$B$ = atmosphere extinction coefficient.

$B$ in Table 2-3 are only considered accurate in latitudes from 24 to 56 degrees. Some error may occur if these values are used in equatorial or polar latitudes where seasonal changes and atmospheric conditions are different.

Although the values for $I_r$ and $B$ are determined empirically to fit measured radiation data, there is some basis for determining them through calculation. The values of $I_r$ represent the effect of the upper atmosphere in removing a certain spectral portion of solar energy as the sun's rays pass through, and take into account the variation in distance between the earth and the sun. An approximate formula for $I_r$ is given by:

$$I_r = 0.862\,[I_0 + 27 \cos (y - 8°)] \text{ Btu/hr-ft}^2 \qquad \textbf{(2-16)}$$

Where $y$ is the year angle defined following Equation 2-5, and $I_0 = 429$ Btu/hr-ft$^2$ is the solar constant from Table 2-1.

The values of $B$ correspond to seasonal changes in the total air mass as a result of temperature, humidity, and biological debris which affect the transmission of solar energy through the atmosphere. An approximate formula for $B$ is given by:

$$B = 0.172 - 0.033 \cos (y - 11°) \qquad \textbf{(2-17)}$$

## EFFECTS OF ELEVATION ABOVE SEA LEVEL

The availability of direct-beam solar energy tends to increase with the elevation of the site because of the reduced air mass through which the sun's rays must travel. At Denver, Colorado, which is about one mile above sea level, the direct normal solar insolation is greater than at Wilmington, Delaware, which is at the same latitude as Denver, but only a few feet above sea level.

An approximate method of determining the effect of altitude on solar irradiance can be calculated by dividing the atmosphere into two layers, upper and lower. Certain portions of the sun's spectrum are quickly absorbed in the upper atmosphere, leaving an amount, $I_r$, given in Table 2–3 or Equation 2–16. Thus, the effect of elevation is not reflected in any change in $I_r$. The lower atmosphere, which extends to about 80,000 feet, has an effect on solar irradiance that is proportional to the air mass through which the sun's rays must travel. Since the density of air decreases nearly exponentially with altitude, an approximate formula for the air mass at elevation, $E$, is given by:

$$m' = me^{-E/E_0}$$ (2–18)

where:  $m'$  is the air mass at elevation $E$

$m$  is the air mass at sea level given by Equation 2–15

$E$  is the elevation above sea level of the site, in feet

$E_0$  = 30,000 feet, is the approximate elevation where the atmospheric density is $1/e$ = 37 percent of the standard atmospheric density at sea level.

To account for the effects of site elevation, $m'$ should be substituted for $m$ in Equation 2–14.

There is a limit to the accuracy that can be obtained in formulas for $I_r$, $B$, and $m'$ because of the variability of the atmosphere. Each of these quantities depends on site elevation and latitude to some extent and also on specific features of a local region (in decreasing order of importance): pollution, moisture, pollen count, wind-blown dust, and volcanic dust.

The effect of site elevation is often treated by the use of an atmospheric clearness number which is to be multiplied by the direct normal solar intensity, $I_n$. The advantage of the clearness number is that it also accounts for the presence of high levels of moisture in low-elevation areas and compensates for the variations in the value of $B$ due to changes in latitude. Table 2–4 summarizes the clearness numbers in the various *non-industrial* regions of the United States.

**TABLE 2–4**  Clearness Numbers Used as a Multiplier for $I_n$

| Geographic Region | Clearness Number |
|---|---|
| Gulf Coast and Florida | Winter: 0.95 <br> Summer: 0.90 |
| Deep South | All year: 0.90 |
| Middle South | All year: 0.95 |
| Mid-Atlantic, New England, Midwest, Lower Great Lakes to Texas Panhandle | All year: 1.00 |
| East Canadian Border, Upper Great Lakes, Northern Plains | All year: 1.05 |
| Rocky Mountains | Winter: 1.05 <br> Summer: 1.10 |
| West and Southwest Desert Areas | Winter: 1.00 <br> Summer: 1.10 |
| West Coast | Winter: 0.95 <br> Summer: 1.05 |

## SUMMARY

The determination of solar irradiance at any location requires knowledge of climatic factors, as well as the geometric relationships between the earth and the sun. In this chapter, the geometric relationships were used to determine the amount of available direct-beam solar energy that can be realized on a surface of any orientation. Starting with the time of day, date, and location of a site, Equations 2–9 and 2–10 or Equations 2–9 and 2–11 provide the position of the sun in the sky. From this and the orientation of a plane surface, Equation 2–12 is used to determine the incident angle, $\theta$. The incident angle is used to convert the direct normal solar irradiance, $I_n$, to the value realized on the plane surface by Equation 2–13.

Direct normal solar irradiance can be approximately determined by the methods shown in Equations 2–14 and 2–15, using empirical values for $I_r$, $B$, and clearness number.

Since no formula for clearness number is provided, readers may find that it will suffice to use an appropriately modified value for $E$, the site elevation above sea level, in Equation 2–18. In most areas, the actual altitude is reasonably accurate; however, in the Gulf coastal areas, Florida, and the deep and middle south, the effects of moisture and biological debris can have the effect of producing conditions comparable to an elevation somewhat below sea level. It may be more appropriate to rely on the clearness number in that region and in areas where the clearness number shows a marked seasonal variation.

There is no substitute for measured values of solar irradiance (Chapter 3). Measured values are usually based on a horizontal orientation where the incident angle is the 90° complement to the elevation angle of the sun, $A$. The relationship between the direct normal irradiance, $I_n$, and the horizontal value of beam irradiance is thus:

$$I_n = \frac{I_{Hb}}{\cos (90 - A)} = \frac{I_{Hb}}{\sin A} = mI_{Hb} \qquad \textbf{(2–19)}$$

where:   $I_{Hb}$ is the measured value of beam solar irradiance on a horizontal surface

   $I_n$   is the direct normal solar irradiance

   $A$   is the elevation angle of the sun given by Equation 2–9

   $m$   is the air mass given by Equation 2–15.

This method of determining the direct normal solar irradiance, $I_n$, requires measured values of the beam solar irradiance on a horizontal surface. Most solar irradiance measurements on a horizontal surface, however, include diffuse and reflected radiation from the sky and clouds, so that some accounting must be made for these effects. Diffuse radiation is treated in Chapter 3, since it is primarily dependent on climatic conditions.

## SHADOW SIZES

The sizes of shadows can be important in applications of solar engineering, since shadows cast by trees and buildings can interfere with the performance of solar collectors. For the same reason, linear arrays of solar collectors should be spaced so that their shadows do not interfere with the collection of adjacent rows during the times when optimum performance is desired.

The vertical pole in Figure 2–7 has a height, Y, and casts a shadow on a flat horizontal surface with length $l$.

$$l = Y/\tan A \qquad\qquad (2–20)$$

where: $l$ is the length of the shadow of a vertical pole of height, Y, expressed in the same units as Y.

A is the elevation angle of the sun given by Equation 2–9.

The vertical fence in Figure 2–7 has a height, $\Gamma$, and is oriented along a straight line forming an angle, $\chi$, from a north–south line, casting a shadow on a flat horizontal surface with a width, $w$, given by:

$$w = \Gamma|\sin (Z - \chi)/\tan A| \qquad\qquad (2–21)$$

where: $w$ is the width of the shadow of a fence of height, $\Gamma$, directed at an angle, $\chi$, from a north–south line

Z is the azimuth angle of the sun given by Equation 2–10 or 2–11.

Care must be taken to be sure that the two angles, Z and $\chi$, are determined in the same sense from due south. A change in sign of the sine function in Equation 2–21 occurs as the shadow changes from one side of the fence to the other so the absolute value function is used to restore a positive value to $w$, regardless of which side of the fence is shaded.

The general treatment for finding the length and direction of a shadow cast by a tilted pole upon a tilted surface requires the use of a sequence of formulas (Table 2–5). Consider a pole of length, Y, tilted from the zenith by an angle, $\xi$, in a direction, $\omega$, from due south. If its base is given as a plane surface tilted

**FIGURE 2–7** Shadows cast by vertical objects.

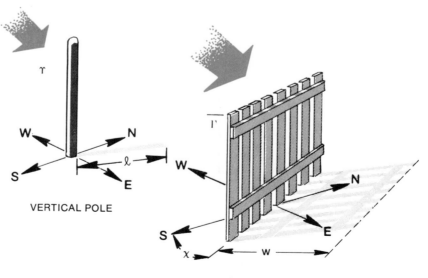

VERTICAL POLE

VERTICAL FENCE

**TABLE 2–5**   The Shadow of a Pole: Length and Direction

| | |
|---|---|
| Vertical pole of height, Y, on horizontal plane: | $l = \dfrac{Y}{\tan A}$ |
| | $\Omega = Z + 180°$ |
| Pole of length, Y, normal to plane tilted by $\Sigma$ from horizontal in direction, $\phi$, from south: | $l = Y \tan \theta$ |
| | $\Omega = \beta + 180°$ |
| | where $\sin \beta = \dfrac{\cos A \sin (Z - \phi)}{\sin \theta}$ |

Vertical pole of height, Y, on plane tilted by $\Sigma$ from horizontal in direction. $\phi$, from south:

$$l = Y \sqrt{\sin^2 \Sigma + \cos^2 \Sigma \tan^2 \theta - 2 \sin \Sigma \cos \Sigma \tan \theta \cos \beta}$$

$$\Omega = \tan^{-1} \left\{ \frac{\cos \Sigma \sin \beta \tan \theta}{\cos \Sigma \tan \theta \cos \beta - \sin \Sigma} \right\} + 180°$$

Pole, tilted by $\xi$ in direction, $\psi$, from south on horizontal surface:

$$l = \gamma \sqrt{\sin^2 \xi + \frac{\cos^2 \xi}{\tan^2 A} - 2 \sin \xi \cos \xi \frac{\cos (Z - \omega)}{\tan A}}$$

$$\Omega = \tan^{-1} \left\{ \frac{\sin \xi \sin \psi \tan A - \cos \xi \sin Z}{\sin \xi \cos \psi \tan A - \cos \xi \cos Z} \right\} + 180°$$

$Y$ = height (length) of pole
$\xi$ = angle by which pole is tilted from vertical
$\psi$ = direction from south by which pole is tilted
$\Sigma$ = angle by which plane is tilted from horizontal
$\phi$ = angle from south in which plane is tilted
$\theta$ = incident angle of sun (Equation 2–24)
$\beta$ = tilted surface azimuth of sun (Equation 2–25)
$A$ = elevation angle of sun
$Z$ = azimuth angle of sun

from the horizontal by an angle, $\Sigma$, in a direction, $\phi$, from south, then the length of its shadow may be determined by using the following procedure:

Determine the angle, $\zeta$, between the pole and the normal to the surface by:

$$\cos \zeta = \cos \Sigma \cos \xi + \sin \Sigma \sin \xi \cos (\psi - \phi) \qquad \textbf{(2–22)}$$

Then, the azimuth angle, $\alpha$, between the projection of the pole on the surface and the line of steepest slope is determined by:

$$\tan \left( \frac{\alpha}{2} \right) = \frac{\sin \xi \sin (\psi - \phi)}{\sin \zeta - \sin \Sigma \cos \xi + \cos \Sigma \sin \xi \cos (\psi - \phi)} \qquad \textbf{(2–23)}$$

Next, determine the incident angle, $\theta$, of the sun to a line normal to the tilted plane surface by:

$$\cos \theta = \cos \Sigma \sin A + \sin \Sigma \cos A \cos (Z - \phi) \qquad \textbf{(2–24)}$$

where: $A$ and $Z$ are the elevation and azimuth, respectively, of the sun from Equations 2–9 and 2–10, or 2–9 and 2–11.

Then, determine the azimuth sun-line angle, $\beta$, on the tilted surface measured from the direction of steepest tilt from:

$$\tan\left(\frac{\beta}{2}\right) = \frac{\cos A \sin (Z - \phi)}{\sin \theta - \sin \Sigma \sin A + \cos \Sigma \cos A \cos (Z - \phi)} \quad (2\text{-}25)$$

Next, determine the orthogonal parameters:

$$M = \sin \zeta \sin \alpha - \cos \zeta \sin \beta \tan \theta \quad (2\text{-}26)$$
$$N = \sin \zeta \cos \alpha - \cos \zeta \cos \beta \tan \theta \quad (2\text{-}27)$$

The length of the shadow is:

$$l = Y\sqrt{M^2 + N^2} \quad (2\text{-}28)$$

and its orientation on the plane will be at an angle, $\Omega$, from the direction of steepest tilt determined by:

$$\Omega = \tan^{-1}\left(\frac{M}{N}\right) + 180° \quad (2\text{-}29)$$

Although this sequence of equations seems complex, it is often surprising how quickly they reduce to simpler forms for the usual geometries which are normally encountered in solar energy applications.

For example, consider the case of a pole of length $Y$, normal to a tilted surface. In that case $\xi = \Sigma$, $\phi = \psi$, and from Equation 2–22, $\zeta = 0$. This causes Equations 2–26 and 2–27 to reduce to $M = -\sin \beta \tan \theta$ and $N = -\cos \beta \tan \theta$ which, in turn, results in the shadow length, $l = Y \tan \theta$, from Equation 2–28. Equation 2–29 provides the angle formed by the shadow: $\Omega = \tan^{-1} (\tan \beta) + 180° = 180° + \beta$, as would be expected where $\beta$ is determined through Equation 2–25.

The general treatment for finding the width of the shadow cast by a tilted fence on a tilted surface can now be considered with the formulas discussed here and repeated in Table 2–6.

Consider a fence of height, $\Gamma$, where "height" is the perpendicular extent of the fence from its base to its upper edge. Suppose it is oriented along a line forming an angle, $\chi$, with the lines of steepest tilt on a tilted plane surface. The plane surface is tilted by an angle, $\Sigma$, from the horizontal in a direction, $\phi$, from south. Furthermore, let the fence be tilted at an angle, $\xi$, from the normal to the plane.

The width of its shadow, measured from the base of the fence along a perpendicular upon the tilted surface is:

$$w = \Gamma|\cos \xi \sin (\beta - \chi)\tan \theta| \pm \Gamma \sin \xi \quad (2\text{-}30)$$

where the "+" sign is used when the shadow is *in* the direction of tilt and a "−" sign is used when the shadow is directed *opposite* to the tilt. The incident angle, $\theta$, and surface azimuth, $\beta$, are found from Equations 2–24 and 2–25.

Equation 2–30 reduces to other convenient forms when special cases are considered. If the fence is normal to the plane, $\xi = 0$, and:

$$w = \Gamma|\sin (\beta - \chi)\tan \theta| \quad (2\text{-}31)$$

If the fence is oriented along the line of steepest tilt and normal to the plane, then $\xi = 0$ and $\chi = 0$ so that $w = \Gamma \sin \beta \tan \theta$, or from Equation 2–25, it

**TABLE 2–6**  The Width of the Shadow of a Fence

| | |
|---|---|
| Vertical fence of height, $\Gamma$, on horizontal surface oriented along line at angle, $\chi$, from south: | $w = \Gamma \left\| \dfrac{\sin (Z - \chi)}{\tan A} \right\|$ |
| Fence normal to tilted plane. Plane tilted by $\Sigma$ from horizontal in direction, $\phi$, from south. Fence directed at angle $\chi$ from steepest slope on tilted plane: | $w = \Gamma \left\| \sin (\beta - \chi) \tan \theta \right\|$ |
| Fence normal to tilted plane along line of steepest slope. Plane tilted by $\Sigma$ in direction, $\phi$: | $w = \Gamma \left\| \dfrac{\cos A \sin (Z - \phi)}{\cos \theta} \right\|$ |
| Fence normal to tilted plane perpendicular to line of steepest slope. Plane tilted by $\Sigma$ in direction, $\phi$: | $w = \Gamma \dfrac{\sqrt{\|\sin^2 \theta - \cos^2 A \sin^2 (Z - \phi)\|}}{\cos \theta}$ |
| Fence tilted from normal by angle $\xi$ on tilted plane. Plane tilted by $\Sigma$ in direction, $\phi$. Fence oriented at angle $\chi$ from steepest slope: | $w = \Gamma \left\|\cos \xi \sin (\beta - \chi) \tan \theta\right\|$ $\pm\ \Gamma \sin \xi$ |

$\Gamma$ = height (extent) of fence
$\chi$ = angle formed between base of fence and steepest slope
$\Sigma$ = tilt angle of plane from horizontal
$\phi$ = angle from south by which plane is tilted
$\theta$ = incident angle (Equation 2–24)
$\beta$ = tilted surface azimuth of sun (Equation 2–25)
$A$ = elevation angle of sun
$Z$ = azimuth angle of sun

reduces to:

$$w = \Gamma \frac{\cos A \sin (Z - \phi)}{\cos \theta} \tag{2-32}$$

Equation 2–32 can also be used to determine the width of the shadow cast by the side wall of a tilted solar collector onto its absorber surface.

If the fence is normal to the surface and oriented perpendicular to the line of steepest tilt on the plane surface, then $\xi = 0$, and $\chi = 90°$ so that $w = \Gamma \cos \beta \tan \theta$, and:

$$w = \Gamma \frac{\sqrt{|\sin^2 \theta - \cos^2 A \sin^2 (Z - \phi)|}}{\cos \theta} \tag{2-33}$$

Equation 2–33 can be used to determine the width of the shadow cast by the end, top, or bottom wall of a tilted solar collector onto its absorber surface.

For a vertical fence on a horizontal surface, then $\xi = 0$, $\Sigma = 0$, $\beta = Z$, and $\theta = 90° - A$ so that $w = \Gamma|\sin (Z - \chi)/\tan A|$ as stated in Equation 2–21. This equation is particularly useful in establishing the minimum spacing between rows of solar collectors.

Typically, solar collectors may be aligned in rows that are in an east–west direction ($\chi = 90°$) and tilted toward the south by an angle, $\Sigma$. If regarded as a tilted fence on a horizontal surface, then $\xi = 90° - \Sigma$, $\theta = 90° - A$, $\beta = Z$, and $\chi = 90°$. For this important case the shadow width is:

$$w = \Gamma|\sin \Sigma \cos Z/\tan A| \pm \Gamma \cos \Sigma \tag{2-34}$$

as measured from the base or foot of the row of collectors.

**TABLE 2–7** Shadows at Solstices and Equinoxes for Collector Spacing Example

| Solar Time | | Sun's Elevation | Sun's Azimuth | Shadow Width (feet) |
|---|---|---|---|---|
| AM | PM | A | Z | w |
| At December 21 (winter solstice) | | | | |
| 8:00 | 4:00 | 5.5° | 53.0° | 37.5 |
| 9:00 | 3:00 | 14.0° | 41.9° | 17.9 |
| 10:00 | 2:00 | 20.7° | 29.4° | 13.8 |
| 11:00 | 1:00 | 25.0° | 15.2° | 12.4 |
| 12:00 noon | | 26.6° | 0.0° | 12.0 |
| At June 21 (summer solstice) | | | | |
| 5:00 | 7:00 | 4.2° | 117.3° | * |
| 6:00 | 6:00 | 14.8° | 108.4° | * |
| 7:00 | 5:00 | 26.0° | 99.7° | * |
| 8:00 | 4:00 | 37.4° | 90.7° | * |
| 9:00 | 3:00 | 48.8° | 80.2° | 0.9 |
| 10:00 | 2:00 | 59.8° | 65.8° | 1.4 |
| 11:00 | 1:00 | 69.2° | 41.9° | 1.7 |
| 12:00 noon | | 73.4° | 0.0° | 1.8 |
| At March 21 and September 21 (equinoxes) | | | | |
| 7:00 | 5:00 | 11.4° | 80.2° | 5.0 |
| 8:00 | 4:00 | 22.5° | 69.6° | 5.0 |
| 9:00 | 3:00 | 32.8° | 57.2° | 5.0 |
| 10:00 | 2:00 | 41.5° | 41.9° | 5.0 |
| 11:00 | 1:00 | 47.7° | 22.6° | 5.0 |
| 12:00 noon | | 50.0° | 0.0° | 5.0 |

* = negative values; the shadow is in front of the collector.

In the winter when the sun's elevation, $A$, is low and its azimuth, $Z$, moves within a low daily range, the shadow width is large. Note, however, that reducing the tilt angle, $\Sigma$, can reduce the shadow width. In the summer when the sun's elevation, $A$, is high and its daily azimuth range is high, the shadow width is reduced and rows of solar collectors could be placed closer together. In designing an array of solar collectors, therefore, the determination of the optimum tilt and spacing between rows requires consideration of the shadows cast by adjacent rows at different times of both the year and the day. This important consideration is illustrated in a sample calculation which follows.

## COLLECTOR SPACING

As has been noted, the width of the shadow cast by a row of solar collectors is needed to determine the minimum spacing between adjacent rows. Because the "sawtooth" array is commonly used for multiple panel installations, calculations of the appropriate spacing between rows can be instructive. Consider a "sawtooth" array at a height of 6 feet above a level rooftop surface, and oriented in an east-west direction so that the collectors will be facing due south. Assume a latitude of 40° north.

In this example, maximum shadow widths will occur on the winter solstice—on or about December 21. The azimuth, $Z$, and elevation, $A$, of the sun on that date, on the summer solstice, and at the equinoxes for various times of the solar day are shown in the following schedule. These values may be calculated from Equations 2–9 and 2–10 or are available from various references. Shadow width calculations use $1 = 6$ feet and $x = 90°$.

From the first formula in Table 2–6:

$$w = 6 \text{ feet} \times \frac{\sin (Z - 90°)}{\tan A} = 6 \frac{\cos Z}{\tan A}$$

From this example, it is clear that at the winter solstice, a minimal spacing of approximately 18 feet between adjacent rows would be necessary to accommodate winter solstice solar angles and shadow widths. Such a spacing would automatically accommodate shadow widths at other times of the year, since the shadow width decreases to 5 feet at the equinoxes (and is constant throughout the day), and is reduced to less than 2 feet on the summer solstice.

If spacing of 18 feet is used, however, a considerable loss of usable rooftop area would occur in the summer months with sunlight falling between adjacent rows. This is one of the unfortunate limitations in making optimal use of horizontal areas for solar energy collection. The most ideal "sawtooth" array geometry for one season cannot be effectively applied for all seasons.

A study of shadow widths also points out other problems inherent in the use of a "sawtooth" array. Ideally, the performance of such a collector array could be improved by tilting the collectors on a seasonal basis. But, if this is done, the height of the collectors becomes higher in the winter, thus requiring even greater spacing between rows since shadow width will increase with increases in height. This had led some collector suppliers to use retro-reflectors on the back (north-facing) side of the rows of a "sawtooth" array. These reflectors serve to increase the energy collection in the summer, since they collect some of the solar energy that would otherwise fall between adjacent rows.

# Climatic Statistics and Solar Irradiance

All the variations affecting the magnitude of solar intensity or, more correctly, solar irradiance at the earth's surface, considered in Chapter 2, are regular in the sense that their recurring effects are predictable. At a given site, seasonal variations of sun and shadow will be reproduced from year to year, the effects of latitude are as unchangeable as latitude itself, and the universal waxing and waning of daily light is like the difference between night and day. But climate is capricious and its effects demand that each installation, even those in locations near enough or similar enough to tempt standardization, be treated individually. Even the smallest planning generalization is unwise because climatological peculiarities can yield surprising results. New York City, for example, is farther south in latitude than Rome, Italy, but because climate is a product of influences more complex than simple latitude, solar energy systems in New York must cope with greater temperature extremes and more cloudy days than those in Rome. Climate is, therefore, the dominant variable to be considered in the design of all solar energy systems. There are three principal concerns: the percentage of sun versus cloud cover, the average ambient temperature, and both peak and average wind speeds.

## SUN VERSUS CLOUD COVER

The percentage of sun versus cloud cover can be expressed in several ways. One might observe the sun over a long period and measure those intervals during which the sun is not obscured versus those when clouds interfere. Another approach, to measure the longest time period between successive appearances of the sun, if done over a long enough time to be statistically valid, could provide data useful in determining minimum solar energy storage requirements. The most complete and accurate approach is to record the average daily solar irradiance (both direct and diffuse components) with a continuously operating solar radiometer. Figures 3–1 through 3–12 depict the results of this method by month for the United States as averaged over a period of twelve years.

The maps in Figures 3–1 through 3–12 were plotted by the National Oceanic and Atmospheric Administration (NOAA), using data from the old network of weather stations in the U.S. Unfortunately, the NOAA network had several deficiencies both in coverage and accuracy of the data. Beginning in 1973, efforts were made to correct the situation with 38 new stations now operational. The NOAA will be releasing new maps based on this more accurate data as they are prepared. In addition, weather tapes for selected sites using rehabilitated data are now being supplied in Fortran magnetic tape.

The maps in Figures 3–1 through 3–12, therefore, are useful for illustration and sample calculations; analysis of a specific application site, and calculations for sizing a solar energy system, should utilize the new, accurate data. The work by Liu and Jordan in correlating diffuse radiation with total radiation is helpful in predicting long-term performance. Whillier, in his doctoral thesis, and Klein have refined this method further.

A numerical representation of the percentage of sun versus cloud cover is given for selected cities in Table 3–1; this specific information will be used in determining load requirements through the solar engineering worksheets. For the moment, it is important to recognize that cloud effects are not limited to blockage of the sun with a consequent diminution in the magnitude of solar irradiance. The presence of clouds not directly obscuring the sun can have effects ranging from zero to a significant increase in the level of solar irradiance resulting from the increase in diffuse radiation by cloud scattering. Accordingly, it is necessary to refine the calculation of total solar irradiation by considering the effects of all its components.

## TOTAL SOLAR IRRADIANCE

The total solar irradiance incident upon a surface consists of three components. The first is direct-beam radiation, following a direct path from the sun to the object and forming an angle $\theta$ with the normal to the surface. The second component is diffuse radiation, sunlight scattered by molecules and particles in the atmosphere (such as the moisture in clouds) and arriving from all directions of the sky. The last component is the radiation of sunlight reflected by the ground and objects on the ground.

Total solar irradiance incident to a flat surface is given by:

$$I_T = I_b + I_d + I_f \qquad \textbf{(3–1)}$$

where: $I_T$ is the total solar irradiance incident on the surface in Btu/hr-ft$^2$

$I_b$ is the direct beam component of solar irradiance given by Equations 2–13 and 2–19 in Btu/hr-ft$^2$

$I_d$ is the diffuse component of solar irradiance (Btu/hr-ft$^2$)

$I_f$ is the reflected component of solar irradiance (Btu/hr-ft$^2$).

On clear days (i.e., cloudless and sunny), the diffuse component, $I_d$, and the reflected component, $I_f$, of solar irradiance are approximated by:

$$I_d = C_d I_n F_s \qquad \textbf{(3–2)}$$
$$I_f = C_f I_n F_g \qquad \textbf{(3–3)}$$

where: $F_s = (1 + \cos \Sigma)/2$
$$\qquad\qquad\qquad\qquad \textbf{(3–4)}$$
$F_g = (1 - \cos \Sigma)/2$

are the weighted view factors for the sky and ground, respectively, $C_d$ and $C_f$ are given in Table 3–2 and the angle, $\Sigma$, is the angle by which the flat surface is

**TABLE 3–1**  Average Percentage of Possible Sunshine

[Airport data, except as noted. For period of record through 1975]

| | State and Station | Length of record (Years) | Jan. | Feb. | Mar. | Apr. | May | June | July | Aug. | Sept. | Oct. | Nov. | Dec. | Annual |
|---|---|---|---|---|---|---|---|---|---|---|---|---|---|---|---|
| AL | Montgomery | 25 | 47 | 53 | 58 | 64 | 66 | 65 | 63 | 65 | 63 | 66 | 57 | 50 | 59 |
| AK | Juneau | 30 | 33 | 32 | 37 | 38 | 38 | 34 | 30 | 30 | 25 | 19 | 23 | 20 | 31 |
| AZ | Phoenix | 80 | 78 | 80 | 83 | 89 | 93 | 94 | 85 | 85 | 89 | 88 | 83 | 77 | 86 |
| AR | Little Rock | 32 | 46 | 54 | 57 | 61 | 68 | 73 | 71 | 73 | 68 | 69 | 56 | 48 | 63 |
| CA | Los Angeles | 32 | 69 | 72 | 73 | 70 | 66 | 65 | 82 | 83 | 79 | 73 | 74 | 71 | 73 |
| | Sacramento | 27 | 45 | 61 | 70 | 80 | 86 | 92 | 97 | 96 | 94 | 84 | 64 | 46 | 79 |
| | San Francisco | 38 | 56 | 62 | 69 | 73 | 72 | 73 | 66 | 65 | 72 | 70 | 62 | 53 | 67 |
| CO | Denver | 26 | 72 | 71 | 69 | 66 | 64 | 70 | 70 | 72 | 75 | 73 | 65 | 68 | 70 |
| CN | Hartford | 21 | 58 | 57 | 56 | 57 | 58 | 58 | 61 | 63 | 59 | 58 | 46 | 48 | 57 |
| DE | Wilmington* | 25 | 50 | 54 | 57 | 57 | 59 | 64 | 63 | 61 | 60 | 60 | 54 | 51 | 53 |
| DC | Washington | 27 | 48 | 51 | 55 | 56 | 58 | 64 | 62 | 62 | 62 | 60 | 53 | 47 | 57 |
| FL | Jacksonville | 25 | 57 | 61 | 66 | 71 | 69 | 61 | 59 | 58 | 53 | 56 | 61 | 56 | 61 |
| | Key West | 17 | 72 | 76 | 81 | 84 | 80 | 71 | 75 | 76 | 69 | 68 | 72 | 74 | 75 |
| GA | Atlanta | 41 | 47 | 52 | 57 | 65 | 69 | 67 | 61 | 65 | 63 | 67 | 60 | 50 | 61 |
| HI | Honolulu | 23 | 63 | 65 | 69 | 67 | 70 | 71 | 74 | 75 | 75 | 67 | 60 | 59 | 68 |
| ID | Boise | 35 | 41 | 52 | 63 | 68 | 71 | 75 | 89 | 85 | 82 | 67 | 45 | 39 | 67 |
| IL | Chicago | 33 | 44 | 47 | 51 | 53 | 61 | 67 | 70 | 68 | 63 | 62 | 41 | 38 | 57 |
| | Peoria | 32 | 45 | 50 | 52 | 55 | 59 | 66 | 68 | 67 | 64 | 63 | 44 | 39 | 57 |
| IN | Indianapolis | 32 | 41 | 51 | 51 | 55 | 61 | 68 | 70 | 71 | 66 | 64 | 42 | 39 | 58 |
| IO | Des Moines | 25 | 51 | 54 | 54 | 55 | 60 | 67 | 71 | 70 | 64 | 64 | 49 | 45 | 59 |
| KS | Wichita | 22 | 59 | 59 | 60 | 62 | 64 | 69 | 74 | 73 | 65 | 66 | 59 | 56 | 65 |
| KY | Louisville | 28 | 41 | 47 | 50 | 55 | 62 | 67 | 66 | 68 | 65 | 63 | 47 | 39 | 57 |
| LA | Shreveport | 23 | 49 | 54 | 56 | 55 | 64 | 71 | 74 | 72 | 68 | 71 | 62 | 53 | 64 |
| ME | Portland | 35 | 55 | 59 | 56 | 56 | 56 | 60 | 64 | 65 | 61 | 58 | 47 | 53 | 58 |
| MD | Baltimore | 25 | 51 | 55 | 55 | 55 | 57 | 62 | 65 | 62 | 60 | 59 | 51 | 48 | 57 |
| MA | Boston | 40 | 54 | 56 | 57 | 56 | 58 | 63 | 66 | 67 | 63 | 61 | 51 | 52 | 59 |
| MI | Detroit | 32 | 32 | 43 | 49 | 52 | 59 | 65 | 70 | 65 | 61 | 56 | 35 | 32 | 54 |
| | Sault Ste. Marie | 34 | 34 | 46 | 55 | 55 | 56 | 57 | 63 | 58 | 46 | 41 | 23 | 28 | 48 |
| MN | Duluth | 25 | 49 | 54 | 56 | 54 | 55 | 58 | 67 | 61 | 52 | 48 | 34 | 39 | 54 |
| | Minneapolis-St. Paul | 37 | 51 | 57 | 54 | 55 | 58 | 63 | 70 | 67 | 61 | 57 | 39 | 40 | 58 |
| MS | Jackson | 11 | 48 | 55 | 61 | 60 | 63 | 67 | 61 | 62 | 58 | 65 | 54 | 45 | 59 |
| MO | Kansas City | 3 | 64 | 54 | 61 | 65 | 67 | 72 | 84 | 69 | 51 | 62 | 46 | 54 | 64 |
| | St. Louis | 16 | 52 | 51 | 54 | 56 | 62 | 69 | 71 | 66 | 63 | 62 | 49 | 41 | 58 |
| MT | Great Falls | 33 | 49 | 57 | 67 | 62 | 64 | 65 | 81 | 78 | 68 | 61 | 46 | 46 | 64 |
| NE | Omaha | 40 | 55 | 55 | 55 | 59 | 62 | 68 | 76 | 72 | 67 | 67 | 52 | 48 | 62 |
| NV | Reno | 33 | 66 | 68 | 74 | 80 | 81 | 85 | 92 | 93 | 92 | 83 | 70 | 63 | 80 |
| NH | Concord | 34 | 52 | 54 | 52 | 53 | 54 | 57 | 62 | 60 | 54 | 54 | 42 | 47 | 54 |
| NJ | Atlantic City | 15 | 49 | 48 | 51 | 53 | 54 | 58 | 60 | 62 | 59 | 57 | 50 | 42 | 54 |
| NM | Albuquerque | 36 | 73 | 73 | 74 | 77 | 80 | 83 | 76 | 76 | 80 | 79 | 78 | 72 | 77 |
| NY | Albany | 37 | 46 | 51 | 52 | 53 | 55 | 59 | 64 | 61 | 56 | 53 | 36 | 38 | 53 |
| | Buffalo | 32 | 34 | 40 | 46 | 52 | 58 | 66 | 69 | 66 | 60 | 53 | 29 | 27 | 53 |
| | New York† | 99 | 50 | 55 | 56 | 59 | 61 | 64 | 65 | 64 | 63 | 61 | 52 | 49 | 59 |
| NC | Charlotte | 25 | 55 | 59 | 63 | 70 | 69 | 71 | 68 | 70 | 68 | 69 | 63 | 58 | 66 |
| | Raleigh | 21 | 55 | 58 | 63 | 64 | 60 | 61 | 61 | 61 | 60 | 63 | 63 | 56 | 60 |
| ND | Bismarck | 36 | 54 | 56 | 60 | 58 | 63 | 64 | 76 | 73 | 65 | 59 | 44 | 47 | 62 |
| OH | Cincinnati | 60 | 41 | 45 | 51 | 55 | 61 | 67 | 68 | 67 | 66 | 59 | 44 | 38 | 57 |
| | Cleveland | 34 | 32 | 37 | 44 | 53 | 59 | 65 | 68 | 64 | 60 | 55 | 31 | 26 | 52 |
| | Columbus | 24 | 37 | 41 | 44 | 52 | 58 | 62 | 64 | 63 | 62 | 58 | 38 | 30 | 53 |
| OK | Oklahoma City | 23 | 59 | 61 | 63 | 63 | 65 | 73 | 75 | 77 | 69 | 68 | 60 | 59 | 67 |
| OR | Portland | 26 | 24 | 35 | 42 | 48 | 54 | 51 | 69 | 64 | 60 | 40 | 27 | 20 | 47 |
| PA | Philadelphia | 33 | 50 | 53 | 56 | 56 | 57 | 63 | 63 | 63 | 60 | 60 | 53 | 49 | 58 |
| | Pittsburgh | 23 | 36 | 38 | 45 | 48 | 53 | 60 | 62 | 60 | 60 | 56 | 40 | 30 | 50 |
| RI | Providence | 22 | 57 | 56 | 55 | 55 | 57 | 57 | 59 | 59 | 58 | 60 | 49 | 51 | 56 |
| SC | Columbia | 22 | 56 | 59 | 64 | 67 | 66 | 65 | 64 | 65 | 65 | 66 | 64 | 60 | 63 |
| SD | Rapid City | 33 | 54 | 59 | 61 | 59 | 57 | 60 | 71 | 73 | 67 | 65 | 56 | 54 | 62 |
| TN | Memphis | 25 | 48 | 54 | 57 | 63 | 69 | 73 | 72 | 75 | 69 | 71 | 58 | 49 | 64 |
| | Nashville | 33 | 40 | 47 | 52 | 59 | 62 | 67 | 64 | 66 | 63 | 64 | 50 | 40 | 57 |
| TX | Amarillo | 34 | 69 | 68 | 71 | 73 | 73 | 77 | 77 | 78 | 74 | 75 | 73 | 67 | 73 |
| | El Paso | 33 | 78 | 82 | 85 | 87 | 89 | 89 | 79 | 80 | 82 | 84 | 83 | 78 | 83 |
| | Houston | 6 | 41 | 54 | 48 | 51 | 57 | 63 | 68 | 61 | 57 | 61 | 58 | 69 | 56 |
| UT | Salt Lake City | 38 | 47 | 55 | 64 | 66 | 73 | 78 | 84 | 83 | 84 | 73 | 54 | 44 | 70 |
| VT | Burlington | 32 | 42 | 48 | 52 | 50 | 56 | 60 | 65 | 62 | 55 | 50 | 30 | 33 | 51 |
| VA | Norfolk | 19 | 57 | 58 | 63 | 66 | 67 | 68 | 65 | 65 | 64 | 60 | 60 | 57 | 63 |
| | Richmond | 25 | 51 | 54 | 59 | 62 | 64 | 67 | 65 | 64 | 63 | 59 | 56 | 51 | 60 |
| WA | Seattle-Tacoma | 10 | 21 | 42 | 49 | 51 | 58 | 54 | 67 | 65 | 61 | 42 | 27 | 17 | 49 |
| | Spokane | 27 | 26 | 41 | 53 | 60 | 63 | 65 | 81 | 78 | 71 | 51 | 28 | 20 | 57 |
| WV | Parkersburg | 78 | 32 | 36 | 43 | 49 | 56 | 59 | 62 | 60 | 59 | 54 | 37 | 29 | 48 |
| WI | Milwaukee | 35 | 44 | 47 | 51 | 54 | 59 | 63 | 70 | 67 | 60 | 57 | 41 | 38 | 56 |
| WY | Cheyenne | 40 | 61 | 65 | 64 | 61 | 59 | 65 | 68 | 68 | 69 | 68 | 60 | 59 | 64 |
| PR | San Juan | 20 | 65 | 69 | 74 | 69 | 61 | 57 | 64 | 65 | 59 | 59 | 57 | 56 | 63 |

Source: U.S. National Oceanic and Atmospheric Administration, *Comparative Climatic Data*.
* Data not available; figures are for a nearby station.
† City office data.

**TABLE 3–2** Coefficients for Diffuse and Reflected Solar Irradiance on a Clear Day

| Month | $C_d$ | Surface | $C_f$ |
|-------|-------|---------|-------|
| Jan. | .058 | Snow | 0.7–0.87 |
| Feb. | .060 | | |
| Mar. | .071 | Bituminous and | |
| Apr. | .097 | gravel roofs | 0.12–0.15 |
| May | .121 | | |
| June | .134 | Bituminous | |
| July | .136 | paving | 0.10 |
| Aug. | .122 | | |
| Sept. | .092 | Concrete | 0.21–0.33 |
| Oct. | .073 | | |
| Nov. | .063 | Grass | 0.20–0.30 |
| Dec. | .057 | | |

tilted from the horizontal, as shown in Figure 2–6. $I_n$ is the direct normal solar irradiance given by Equation 2–14.

The coefficient for diffuse solar irradiance, $C_d$, may also be represented by the formula:

$$C_d = 0.094 - 0.031 \cos y - 0.013 \sin \frac{3y}{2} \qquad \textbf{(3–5)}$$

where:   $y$  is the year angle defined following Equation 2–5.

The maps in Figures 3–1 through 3–12 are used to provide the average daily solar irradiance in Langleys at any selected site within the continental United States. The data is averaged over each day of the month represented by the figure.

The value of solar irradiance in Langleys is converted to Btus per square foot per day by multiplying 3.69. The result is the average total daily solar irradiance available on a *horizontal* surface, $I_H$, in Btu/day-ft². 

Diffuse radiation on clear days can be approximated by Equation 3–2; however, the average solar irradiance data on the maps of Figures 3–1 through 3–12 are based on a composite of clear and cloudy days, such as they naturally occur. The nature and intensity of diffuse radiation on cloudy and partially cloudy days is quite different from that of clear days.

A fully equipped U.S. weather station will supply data on the cloud cover (CC) at the site of the station on an hourly basis for each day of the year. The cloud cover, CC, is usually estimated visually by experienced cloud observers and expressed as a number from 0 to 10. A clear sky with no clouds is designated as CC = 0 and a completely overcast sky is denoted as CC = 10.

A useful expression for hourly diffuse radiation intensity is provided by Kreider and Kreith:

$$I_{Hd} = 0.78 + 1.07A + 6.17CC \qquad \textbf{(3–6)}$$

where: $I_{Hd}$ is the diffuse component of solar irradiance on a horizontal surface in Btu/hr-ft$^2$

$A$ is the sun's angle of elevation in degrees, given by Equation 2–9

$CC$ is the cloud cover (from 0 to 10)

The total daily value of diffuse radiation must be obtained by summing the hourly values of $I_{Hd}$ obtained in Equation 3–6 through the hours of the day. Both $A$ and $CC$ change with each hour of the day.

If cloud cover data are available, the average daily value of diffuse solar irradiance on a horizontal surface $I_{Hd}$ in Btu/day-ft$^2$ can be obtained by a summation of hourly values in Equation 3–6.

The value of $I_H$, the total irradiance on a horizontal surface is:

$$I_H = I_{Hb} + I_{Hd} \text{ (Btu/hr-ft}^2) \tag{3–7}$$

where: $I_{Hd}$ is given on an hourly basis by Equation 3–6

$I_{Hb}$ may be determined from the calculated values of the direct normal solar irradiance and the air mass from Equation 2–19, but should be set $= 0$ when the sun is not visible.

If cloud cover data are not available, the measured daily values of solar irradiance on a horizontal surface, $I_H$ (Btu/day-ft$^2$), may be obtained by multiplying the value in Langleys obtained from the maps of Figures 3–1 through 3–12 by 3.69, and the average daily values of $I_{Hb}$ can be approximated by adding the hourly contributions obtained from Equation 2–19 and multiplying the result by the "percent sun" (in decimal form) value taken from Table 3–1. The average daily diffuse radiation on a horizontal surface, $I_{Hd}$, may then be represented as $I_H - I_{Hb}$ where the units are in Btu/day-ft$^2$.

In summary, the total hourly solar irradiance on a tilted surface consists of direct, diffuse, and reflected components, as indicated in Equation 3–1. The values of the diffuse and reflected radiation for clear days are given by Equations 3–2 and 3–3 in terms of the weighted view factors that depend on the tilt angle, Equation 3–4, the coefficients $C_d$ and $C_f$, and the direct normal solar irradiance, $I_n$, which is provided in Equation 2–14.

On cloudy days the value of diffuse radiation is based on the cloud cover, $CC$, by Equation 3–6, and the value of beam radiation should be set equal to 0 whenever the sun is not visible.

The average daily solar irradiance is shown on the maps in Figures 3–1 through 3–12; the percent sun figures in Table 3–1 can be used in conjunction with these maps to determine the average daily direct and diffuse radiation on a horizontal surface.

In many solar applications, one begins by assuming clear day performance and then multiplies the clear day energy delivered by the percent sun figure to obtain the average daily energy delivered. This is preferable to first assuming performance based on average values of solar irradiance because the performance of a solar system is not linearly dependent upon solar irradiance, with the effect that there is a certain value of solar irradiance below which no net energy can be delivered by most systems. The percent sun figure best represents the proportion of days within any month that a solar system can be expected to be exposed directly to the sun—the most critical figure, since solar

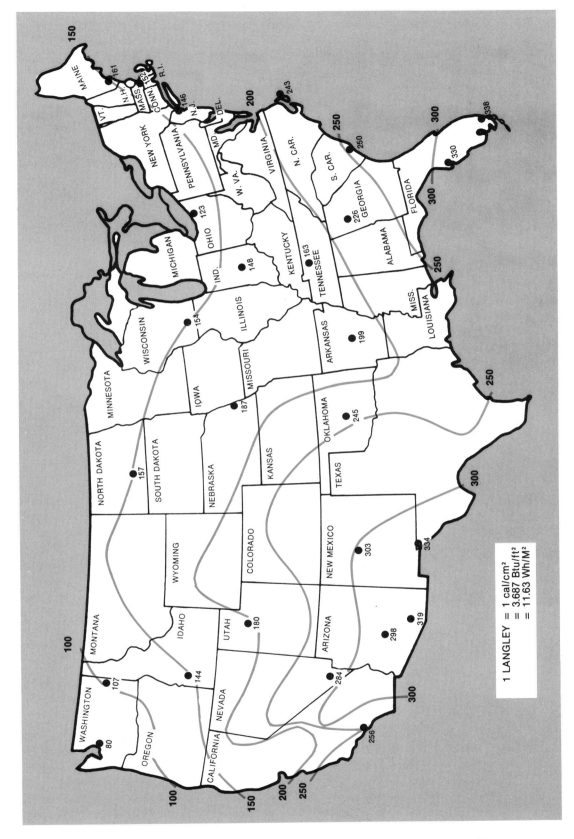

**FIGURE 3–1** Average daily solar radiation (Langleys per day), January.

1 LANGLEY = 1 cal/cm²
= 3.687 Btu/ft²
= 11.63 Wh/M²

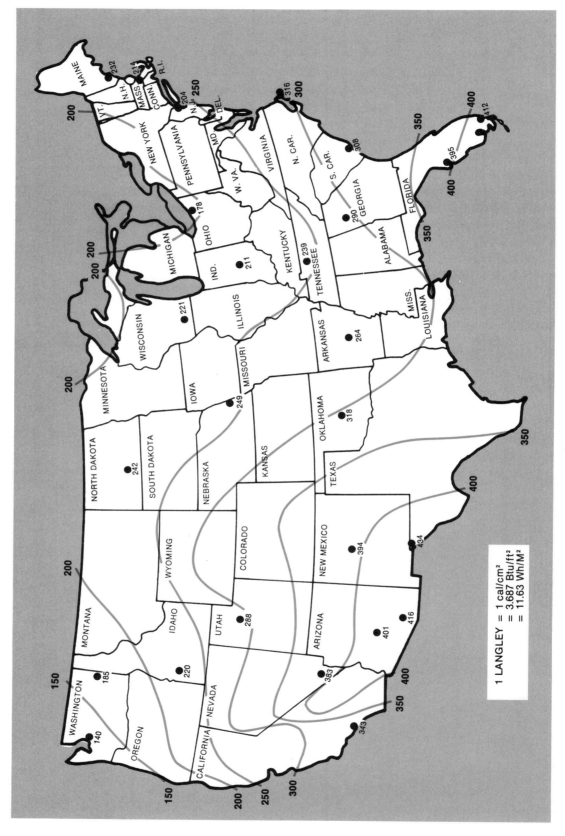

**FIGURE 3-2** Average daily solar radiation (Langleys per day), February.

1 LANGLEY = 1 cal/cm²
= 3.687 Btu/ft²
= 11.63 Wh/M²

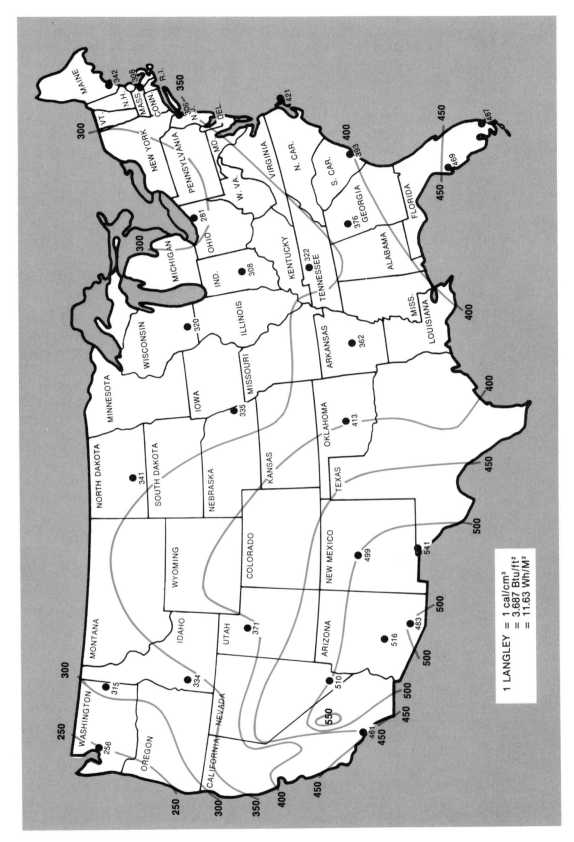

**FIGURE 3-3** Average daily solar radiation (Langleys per day), March.

1 LANGLEY = 1 cal/cm³
= 3.687 Btu/ft²
= 11.63 Wh/M²

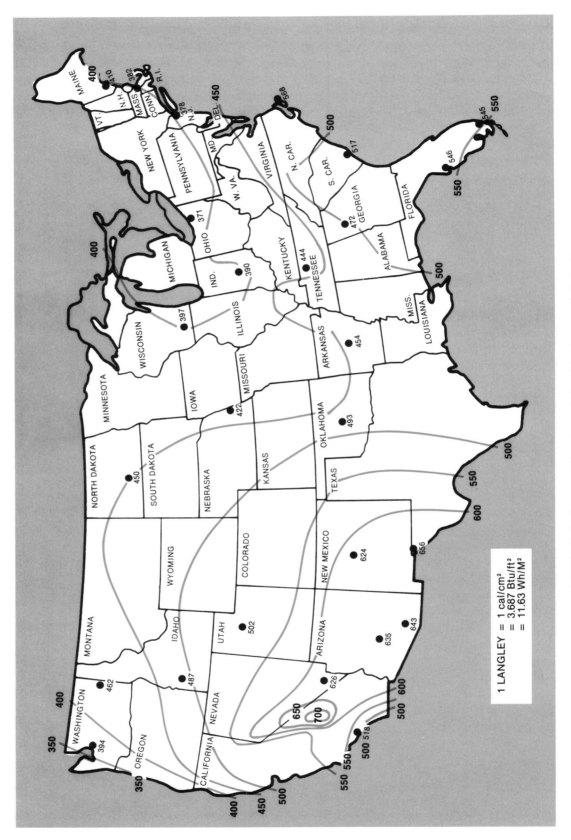

**FIGURE 3–4** Average daily solar radiation (Langleys per day), April.

1 LANGLEY = 1 cal/cm²
= 3.687 Btu/ft²
= 11.63 Wh/M²

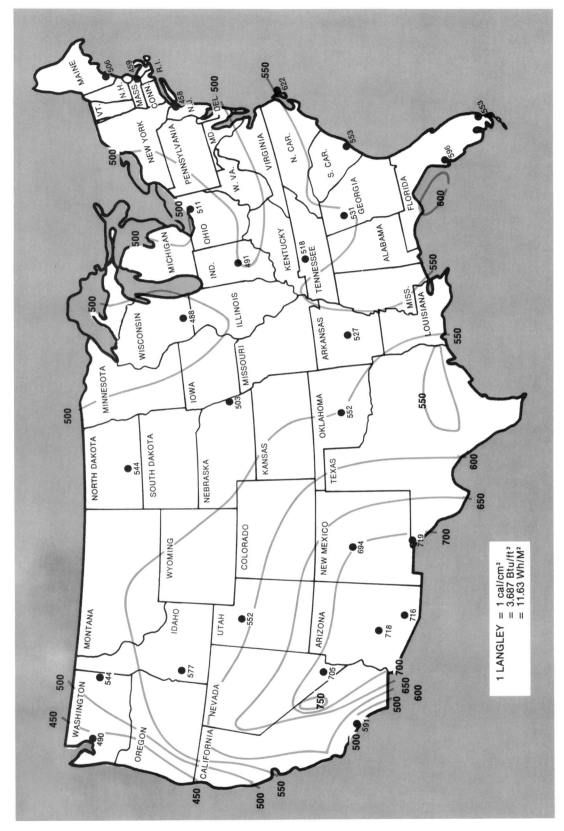

**FIGURE 3–5** Average daily solar radiation (Langleys per day), May.

1 LANGLEY = 1 cal/cm²
= 3.687 Btu/ft²
= 11.63 Wh/M²

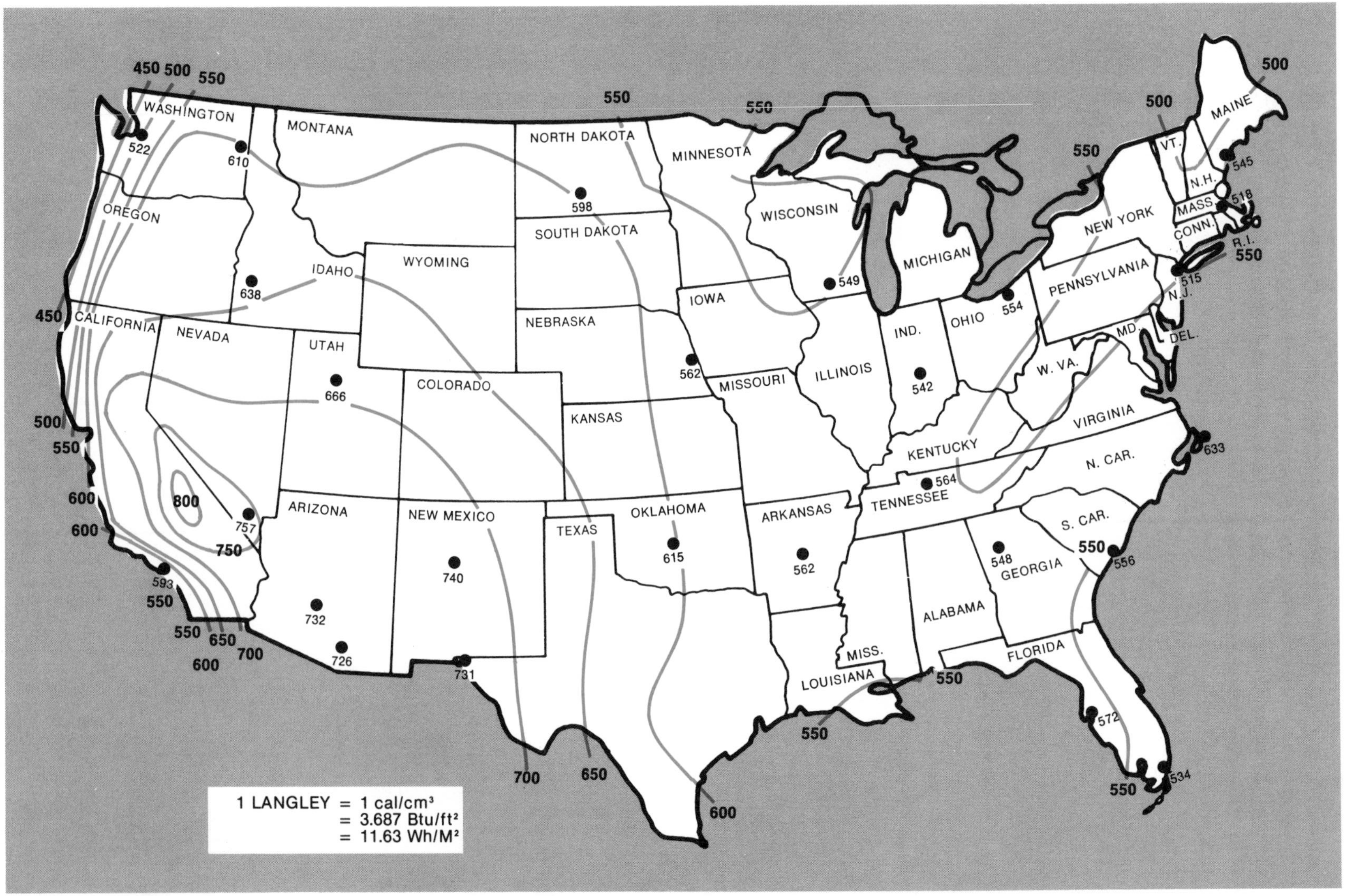

**FIGURE 3–6** Average daily solar radiation (Langleys per day), June.

1 LANGLEY = 1 cal/cm³
= 3.687 Btu/ft²
= 11.63 Wh/M²

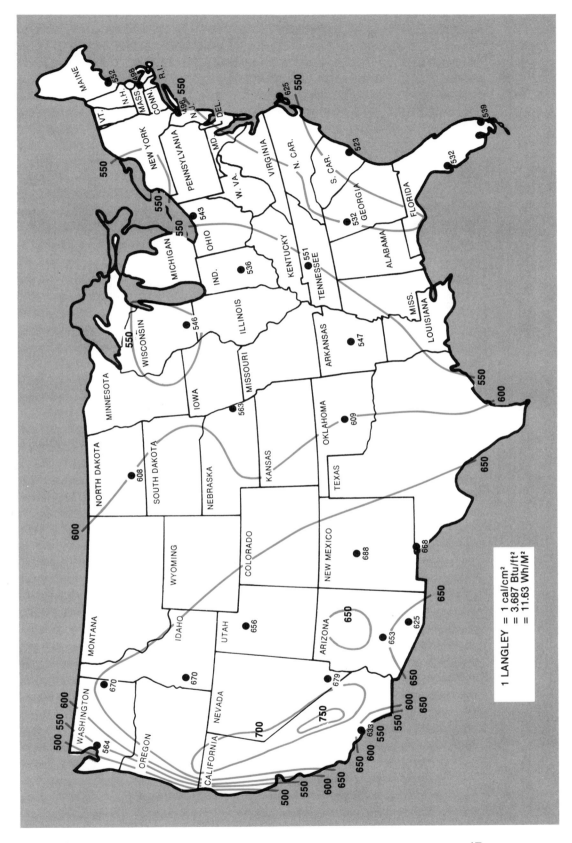

**FIGURE 3–7** Average daily solar radiation (Langleys per day), July.

1 LANGLEY = 1 cal/cm²
= 3.687 Btu/ft²
= 11.63 Wh/M²

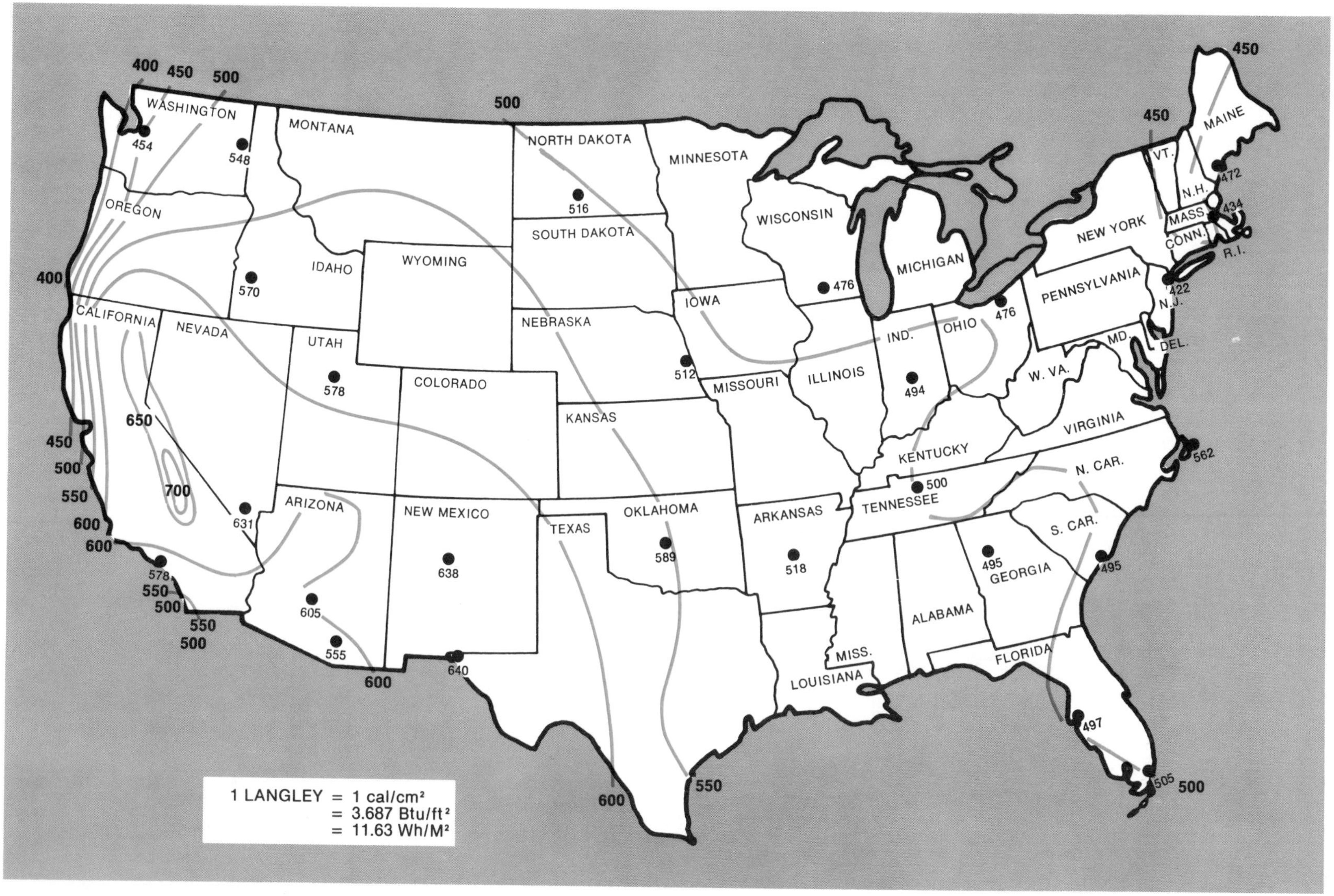

**FIGURE 3–8** Average daily solar radiation (Langleys per day), August.

1 LANGLEY = 1 cal/cm²
= 3.687 Btu/ft²
= 11.63 Wh/M²

48

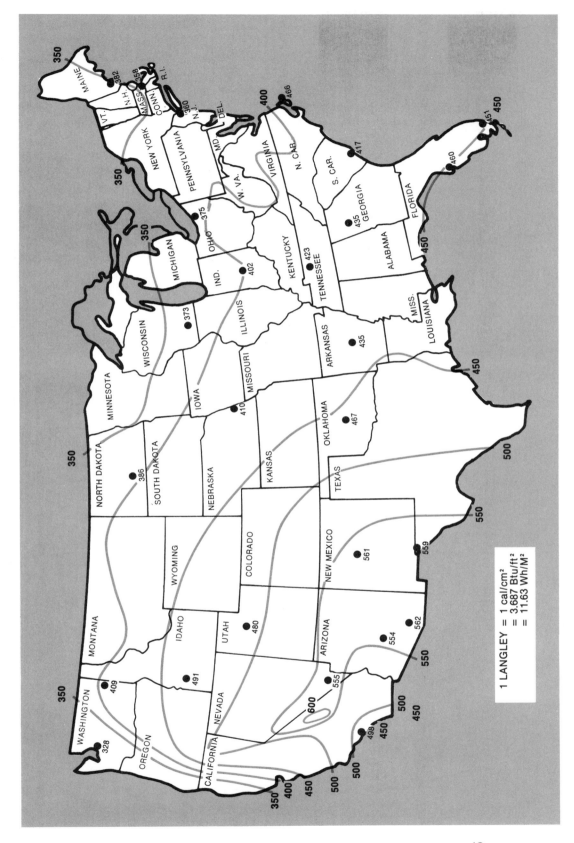

**FIGURE 3–9** Average daily solar radiation (Langleys per day), September.

1 LANGLEY = 1 cal/cm²
= 3.687 Btu/ft²
= 11.63 Wh/M²

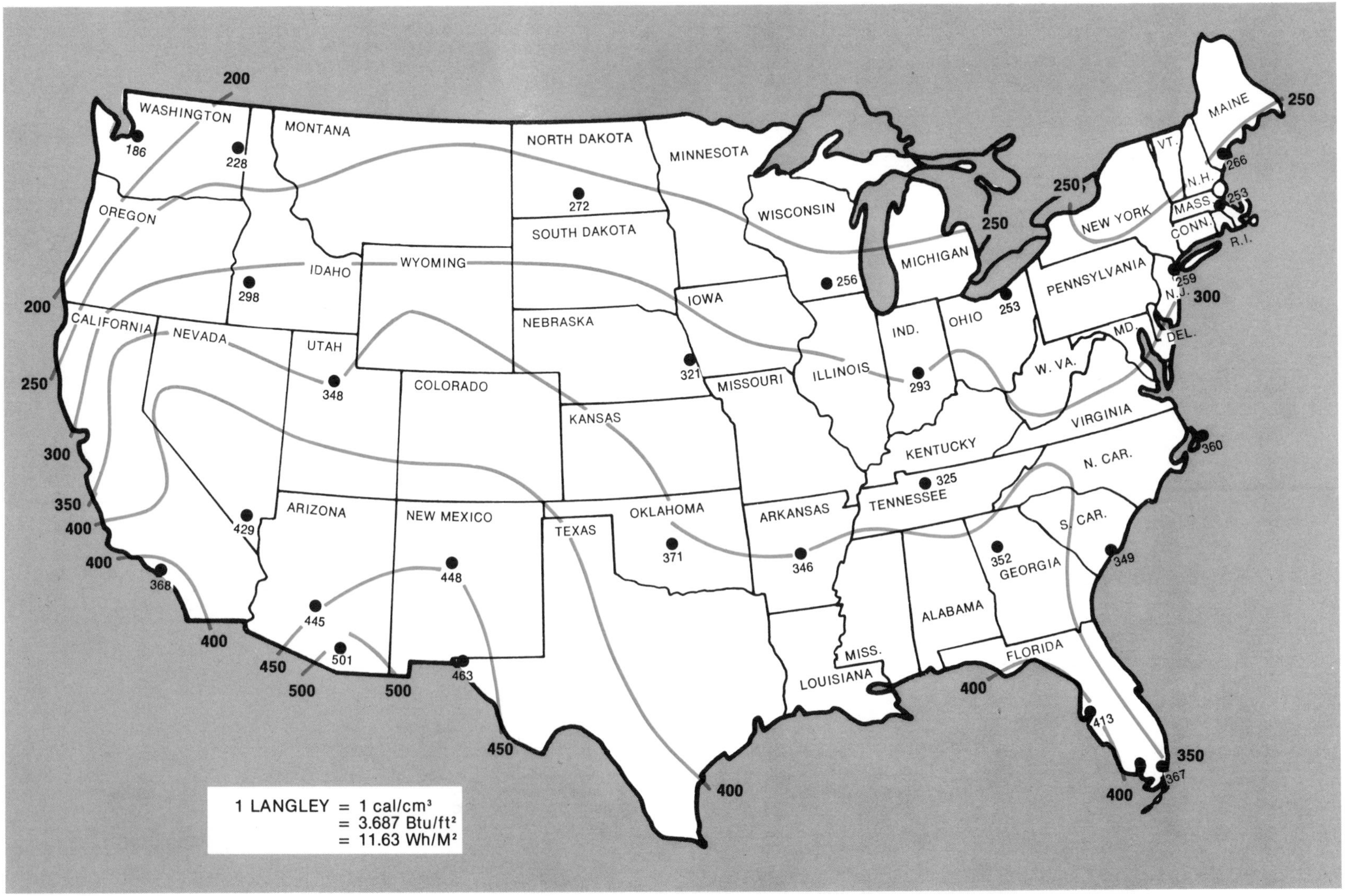

**FIGURE 3–10** Average daily solar radiation (Langleys per day), October.

1 LANGLEY = 1 cal/cm³
= 3.687 Btu/ft²
= 11.63 Wh/M²

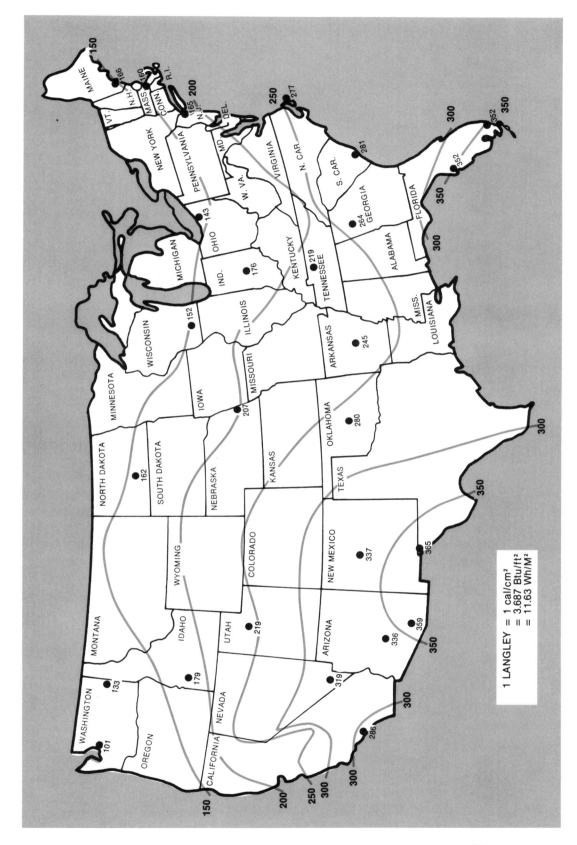

**FIGURE 3–11** Average daily solar radiation (Langleys per day), November.

1 LANGLEY = 1 cal/cm²
= 3.687 Btu/ft²
= 11.63 Wh/M²

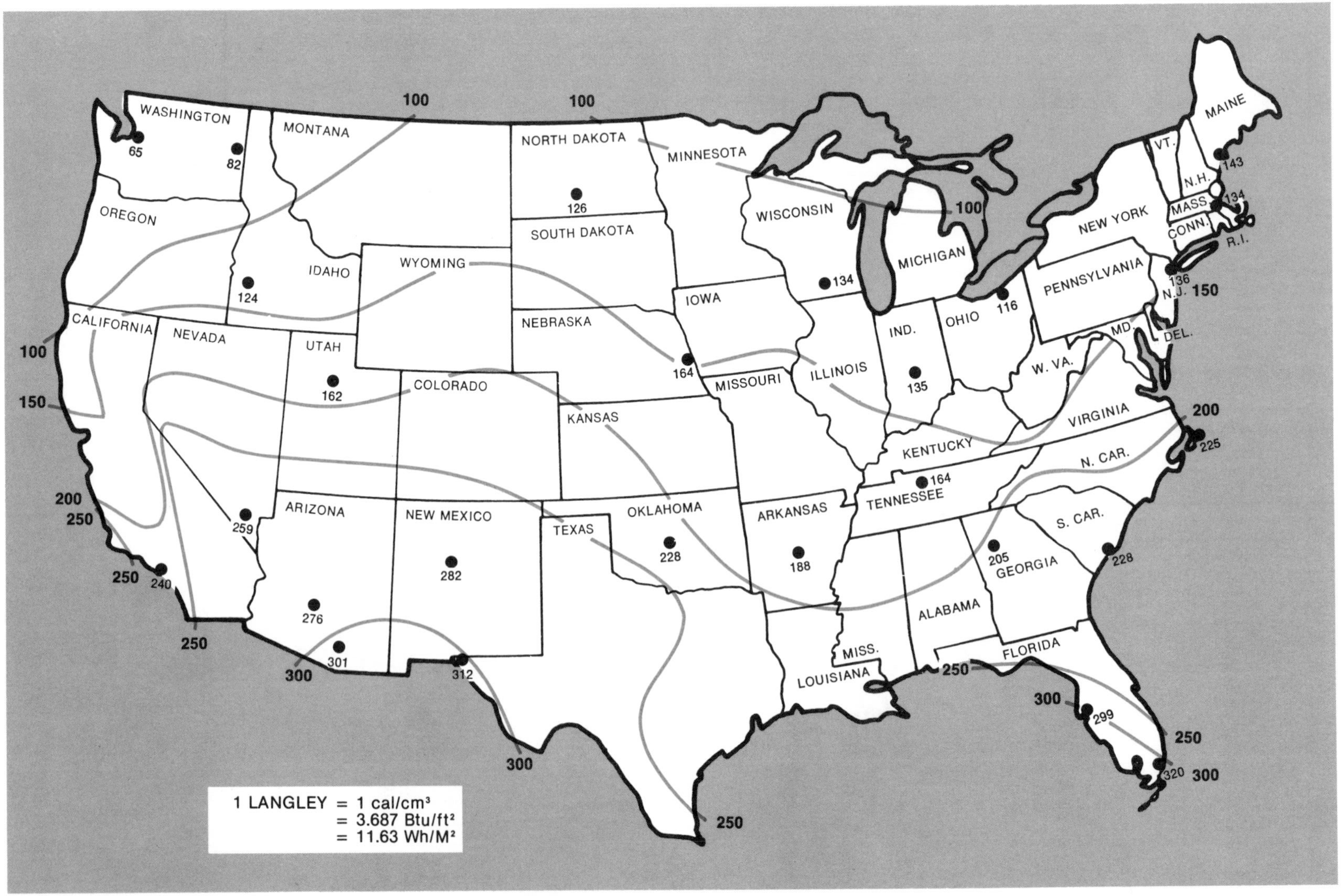

**FIGURE 3–12** Average daily solar radiation (Langleys per day), December.

1 LANGLEY = 1 cal/cm³
= 3.687 Btu/ft²
= 11.63 Wh/M²

energy collection systems deliver zero net energy when the sun is not visible, and diffuse radiation by itself is of little value.

One advantage of a flat plate collector, however, is that it can benefit from the incremental addition of diffuse radiation to direct beam radiation when the sun is visible. This means that if hourly cloud cover information for use in Equation 3–6 is not available, there will be a tendency for the percent sun approach to understate, to some extent, the potential performance of a solar energy system using flat plate collectors. Systems that are based on concentrators which track the sun utilize only the direct normal component of solar irradiance, $I_n$; they cannot collect energy from diffuse radiation.

## AMBIENT TEMPERATURE

Ambient temperature is the environmental temperature in which a solar energy system operates; it is probably the single most important climatic variable to be considered in the design of that solar energy system. The effect of ambient temperatures through the duration of a season is most conveniently expressed in terms of "degree days" of heating or cooling requirements. The degree-day heating requirement is determined by totaling the daily average difference between 65°F and the average ambient temperature; degree-day cooling requirements are determined by totaling the daily average difference between ambient temperature and 74°F. Seasonal degree-day heating and cooling requirements by regional climatic classification and the average daily solar irradiance for heating and cooling seasons are shown in Figures 3–13 and 3–14.

Degree-day information by month for representative cities, data with the precision necessary for use in the worksheets, is given in Table 3–3. Cooling degree-day data can be used in like calculations, but the information is not as accurate in determining air-conditioning energy loads since much of the energy is required to reduce humidity rather than to cool the air. For this reason, the average relative humidity in the same representative cities for each month of the year is provided in Table 3–4.

In the main, a solar system is engineered around the average monthly temperature at the site (shown for selected cities in Table 3–5). Excess thermal energy may be stored during periods when ambient is above this average, utilized for service when ambient falls below the average. It may be preferable, to ensure more complete fulfillment of energy requirements, to design the system around the normal daily *minimum* temperatures as shown in Table 3–6. Systems using the monthly minimum temperatures as the critical parameter will have a greater capacity during those low temperature periods but, as a consequence, they must also be capable, through storage or rejection mechanisms, of coping with the excess energy achieved at the normal daily maximum temperatures shown in Table 3–7. Finally, for purposes of comparison and to provide extreme points for the design of a solar energy system, Tables 3–8 and 3–9 contain the record high and low temperatures for selected cities.

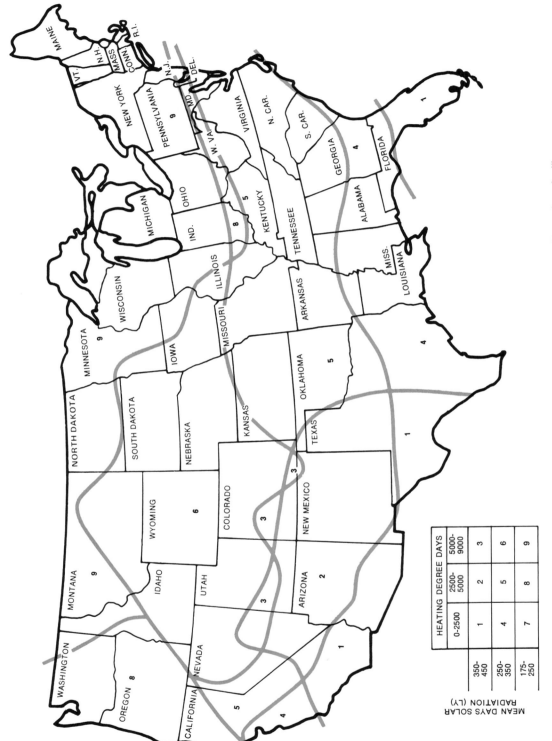

**FIGURE 3–13** Regional climatic classification for the heating season (November–April).

| | HEATING DEGREE DAYS | | |
|---|---|---|---|
| | 0-2500 | 2500-5000 | 5000-9000 |
| 350-450 | 1 | 2 | 3 |
| 250-350 | 4 | 5 | 6 |
| 175-250 | 7 | 8 | 9 |

MEAN DAYS SOLAR RADIATION (LY)

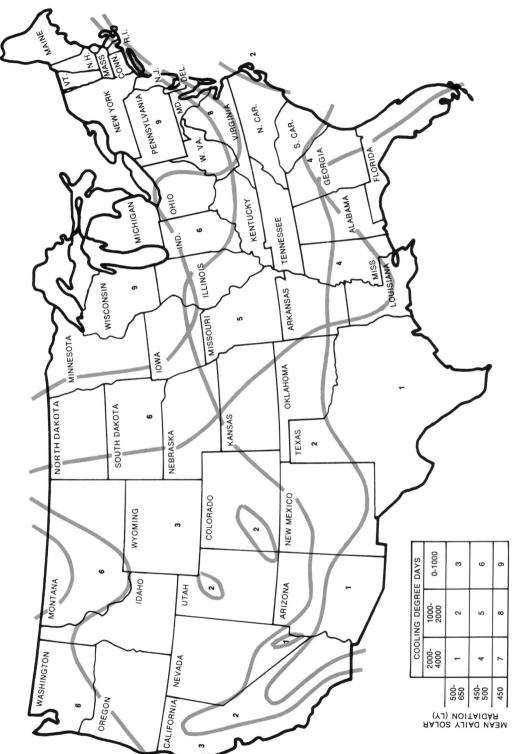

**FIGURE 3–14** Regional classification for the cooling season (May–October).

| MEAN DAILY SOLAR RADIATION (LY) | COOLING DEGREE DAYS | | |
|---|---|---|---|
| | 2000–4000 | 1000–2000 | 0–1000 |
| 500–650 | 1 | 2 | 3 |
| 450–500 | 4 | 5 | 6 |
| 450 | 7 | 8 | 9 |

55

### TABLE 3–3  Normal Monthly and Seasonal Heating Degree Days, 65° Base

(Airport data, except as noted. Based on standard 30-year period, 1941 through 1970.)

| | State and Station | Jan. | Feb. | Mar. | Apr. | May | June | July | Aug. | Sept. | Oct. | Nov. | Dec. | Annual |
|---|---|---|---|---|---|---|---|---|---|---|---|---|---|---|
| AL | Mobile | 451 | 337 | 221 | 40 | – | – | – | – | – | 39 | 211 | 385 | 1,684 |
| AK | Juneau | 1,287 | 1,036 | 1,026 | 783 | 564 | 354 | 288 | 332 | 474 | 719 | 975 | 1,169 | 9,007 |
| AZ | Phoenix | 428 | 292 | 185 | 60 | – | – | – | – | – | 17 | 182 | 388 | 1,552 |
| AR | Little Rock | 791 | 619 | 470 | 139 | 21 | – | – | – | 5 | 143 | 441 | 725 | 3,354 |
| CA | Los Angeles | 331 | 270 | 267 | 195 | 114 | 71 | 19 | 15 | 23 | 77 | 158 | 267 | 1,819 |
| | Sacramento | 617 | 426 | 372 | 227 | 120 | 20 | – | – | 5 | 101 | 360 | 595 | 2,843 |
| | San Francisco | 518 | 386 | 372 | 291 | 210 | 120 | 93 | 84 | 66 | 137 | 291 | 474 | 3,042 |
| CO | Denver | 1,088 | 902 | 868 | 525 | 253 | 80 | – | – | 120 | 408 | 768 | 1,004 | 6,016 |
| CN | Hartford | 1,246 | 1,070 | 911 | 519 | 226 | 24 | – | 12 | 106 | 384 | 711 | 1,141 | 6,350 |
| DE | Wilmington | 1,023 | 879 | 725 | 381 | 128 | – | – | – | 32 | 254 | 579 | 939 | 4,940 |
| DC | Washington | 911 | 776 | 617 | 265 | 72 | – | – | – | 14 | 190 | 510 | 856 | 4,211 |
| FL | Jacksonville | 348 | 282 | 176 | 24 | – | – | – | – | – | 19 | 161 | 317 | 1,327 |
| | Miami | 53 | 67 | 17 | – | – | – | – | – | – | – | 13 | 56 | 206 |
| GA | Atlanta | 701 | 560 | 443 | 144 | 27 | – | – | – | 8 | 137 | 408 | 667 | 3,095 |
| HI | Honolulu | – | – | – | – | – | – | – | – | – | – | – | – | – |
| ID | Boise | 1,116 | 826 | 741 | 480 | 252 | 97 | – | 12 | 127 | 406 | 756 | 1,020 | 5,833 |
| IL | Chicago | 1,262 | 1,053 | 874 | 453 | 208 | 26 | – | 8 | 57 | 316 | 738 | 1,132 | 6,127 |
| | Peoria | 1,277 | 1,044 | 859 | 416 | 180 | 17 | – | 8 | 70 | 327 | 753 | 1,147 | 6,098 |
| IN | Indianapolis | 1,150 | 960 | 784 | 387 | 159 | 11 | – | 5 | 63 | 302 | 699 | 1,057 | 5,577 |
| IO | Des Moines | 1,414 | 1,142 | 964 | 465 | 186 | 26 | – | 13 | 94 | 350 | 816 | 1,240 | 6,710 |
| KS | Wichita | 1,045 | 804 | 671 | 275 | 90 | 7 | – | – | 32 | 211 | 606 | 946 | 4,687 |
| KY | Louisville | 983 | 818 | 661 | 286 | 105 | 5 | – | – | 35 | 241 | 600 | 911 | 4,645 |
| LA | New Orleans | 403 | 299 | 188 | 29 | – | – | – | – | – | 40 | 179 | 327 | 1,465 |
| ME | Portland | 1,349 | 1,179 | 1,029 | 669 | 381 | 106 | 27 | 55 | 200 | 493 | 792 | 1,218 | 7,498 |
| MD | Baltimore | 980 | 846 | 688 | 340 | 110 | – | – | – | 27 | 250 | 567 | 921 | 4,729 |
| MA | Boston | 1,110 | 969 | 834 | 492 | 218 | 27 | – | 8 | 76 | 301 | 594 | 992 | 5,621 |
| MI | Detroit | 1,225 | 1,067 | 918 | 507 | 238 | 26 | – | 11 | 80 | 342 | 717 | 1,097 | 6,228 |
| | Sault Ste. Marie | 1,575 | 1,394 | 1,271 | 804 | 496 | 200 | 96 | 125 | 291 | 583 | 966 | 1,392 | 9,193 |
| MN | Duluth | 1,751 | 1,481 | 1,287 | 792 | 484 | 194 | 67 | 104 | 318 | 611 | 1,098 | 1,569 | 9,756 |
| | Minneapolis-St. Paul | 1,637 | 1,358 | 1,138 | 597 | 271 | 65 | 11 | 21 | 173 | 472 | 978 | 1,438 | 8,159 |
| MS | Jackson | 569 | 442 | 313 | 74 | 6 | – | – | – | – | 91 | 301 | 504 | 2,300 |
| MO | Kansas City | 1,153 | 893 | 745 | 314 | 111 | 12 | – | – | 42 | 235 | 642 | 1,014 | 5,161 |
| | St. Louis | 1,045 | 837 | 682 | 272 | 103 | 10 | – | – | 35 | 224 | 600 | 942 | 4,750 |
| MT | Great Falls | 1,380 | 1,075 | 1,070 | 648 | 367 | 162 | 18 | 42 | 260 | 524 | 912 | 1,194 | 7,652 |
| NE | Omaha | 1,314 | 1,036 | 865 | 391 | 148 | 20 | – | 6 | 71 | 301 | 750 | 1,147 | 6,049 |
| NV | Reno | 1,026 | 781 | 766 | 546 | 328 | 145 | 17 | 50 | 168 | 456 | 747 | 992 | 6,022 |
| NH | Concord | 1,376 | 1,187 | 1,014 | 624 | 315 | 58 | 16 | 45 | 182 | 487 | 810 | 1,246 | 7,360 |
| NJ | Atlantic City | 1,001 | 871 | 741 | 399 | 131 | 9 | – | – | 35 | 262 | 570 | 927 | 4,946 |
| NM | Albuquerque | 924 | 700 | 595 | 282 | 58 | – | – | – | 7 | 218 | 615 | 893 | 4,292 |
| NY | Albany | 1,349 | 1,162 | 980 | 543 | 253 | 39 | 9 | 22 | 135 | 422 | 762 | 1,212 | 6,888 |
| | Buffalo | 1,280 | 1,137 | 1,020 | 603 | 321 | 58 | 12 | 33 | 138 | 419 | 756 | 1,150 | 6,927 |
| | New York* | 1,017 | 885 | 741 | 387 | 137 | – | – | – | 29 | 209 | 528 | 915 | 4,848 |
| NC | Charlotte | 710 | 588 | 461 | 145 | 34 | – | – | – | 10 | 152 | 420 | 698 | 3,218 |
| | Raleigh | 760 | 638 | 502 | 180 | 48 | – | – | – | 12 | 186 | 450 | 738 | 3,514 |
| ND | Bismarck | 1,761 | 1,442 | 1,237 | 660 | 339 | 122 | 18 | 35 | 252 | 564 | 1,083 | 1,531 | 9,044 |
| OH | Cincinnati | 1,020 | 857 | 692 | 307 | 118 | 7 | – | – | 37 | 245 | 612 | 949 | 4,844 |
| | Cleveland | 1,181 | 1,039 | 896 | 501 | 244 | 40 | 9 | 17 | 95 | 354 | 702 | 1,076 | 6,154 |
| | Columbus | 1,135 | 972 | 800 | 418 | 176 | 13 | – | 8 | 76 | 342 | 699 | 1,063 | 5,702 |
| OK | Oklahoma City | 874 | 664 | 532 | 180 | 36 | – | – | – | 12 | 148 | 474 | 775 | 3,695 |
| OR | Portland | 834 | 622 | 598 | 432 | 264 | 128 | 48 | 56 | 119 | 347 | 591 | 753 | 4,792 |
| PA | Philadelphia | 1,014 | 871 | 716 | 367 | 122 | – | – | – | 38 | 249 | 564 | 924 | 4,865 |
| | Pittsburgh | 1,144 | 1,000 | 834 | 444 | 208 | 26 | 7 | 16 | 98 | 372 | 711 | 1,070 | 5,930 |
| RI | Providence | 1,135 | 997 | 871 | 531 | 259 | 36 | – | 10 | 93 | 350 | 651 | 1,039 | 5,972 |
| SC | Columbia | 608 | 493 | 360 | 83 | 12 | – | – | – | – | 112 | 341 | 589 | 2,598 |
| SD | Sioux Falls | 1,575 | 1,277 | 1,085 | 567 | 259 | 65 | 10 | 18 | 165 | 465 | 957 | 1,395 | 7,838 |
| TN | Memphis | 760 | 594 | 457 | 131 | 22 | – | – | – | 7 | 142 | 423 | 691 | 3,227 |
| | Nashville | 828 | 672 | 524 | 176 | 45 | – | – | – | 10 | 180 | 498 | 763 | 3,696 |
| TX | Dallas-Fort Worth | 626 | 456 | 335 | 88 | – | – | – | – | – | 60 | 287 | 530 | 2,382 |
| | El Paso | 663 | 465 | 328 | 89 | – | – | – | – | – | 92 | 402 | 639 | 2,678 |
| | Houston | 416 | 294 | 189 | 23 | – | – | – | 5 | – | 24 | 155 | 333 | 1,434 |
| UT | Salt Lake City | 1,147 | 885 | 787 | 474 | 237 | 88 | – | 5 | 105 | 402 | 777 | 1,076 | 5,983 |
| VT | Burlington | 1,494 | 1,299 | 1,113 | 660 | 331 | 63 | 20 | 49 | 191 | 502 | 840 | 1,314 | 7,876 |
| VA | Norfolk | 760 | 661 | 532 | 226 | 53 | – | – | – | 9 | 141 | 402 | 704 | 3,488 |
| | Richmond | 853 | 717 | 569 | 226 | 64 | – | – | – | 21 | 203 | 480 | 806 | 3,939 |
| WA | Seattle-Tacoma | 831 | 636 | 648 | 489 | 313 | 167 | 80 | 82 | 170 | 397 | 612 | 760 | 5,185 |
| | Spokane | 1,228 | 918 | 853 | 567 | 327 | 144 | 21 | 47 | 196 | 533 | 885 | 1,116 | 6,835 |
| WV | Charleston | 946 | 798 | 642 | 287 | 113 | 10 | – | – | 46 | 267 | 588 | 893 | 4,590 |
| WI | Milwaukee | 1,414 | 1,190 | 1,042 | 609 | 348 | 90 | 15 | 36 | 140 | 440 | 855 | 1,265 | 7,444 |
| WY | Cheyenne | 1,190 | 1,008 | 1,035 | 669 | 394 | 156 | 22 | 31 | 225 | 530 | 885 | 1,110 | 7,255 |
| PR | San Juan | – | – | – | – | – | – | – | – | – | – | – | – | – |

Source: U.S. National Oceanic and Atmospheric Administration, *Comparative Climatic Data.*
Note:– Represents zero.
* City office data.

# TABLE 3–4  Average Relative Humidity

(In percent. Airport data, except as noted. Eastern standard time. For period of record through 1975. Hours selected to give, for most of country, approximation of average highest and average lowest humidity values. Relative humidity observations were made on the half-hour prior to 1957.)

| State and Station | Length of Record (years) | Jan. 7:00 a.m. | Jan. 1:00 p.m. | Feb. 7:00 a.m. | Feb. 1:00 p.m. | Mar. 7:00 a.m. | Mar. 1:00 p.m. | Apr. 7:00 a.m. | Apr. 1:00 p.m. | May 7:00 a.m. | May 1:00 p.m. | June 7:00 a.m. | June 1:00 p.m. |
|---|---|---|---|---|---|---|---|---|---|---|---|---|---|
| AL Mobile | 13 | 82 | 64 | 81 | 56 | 84 | 56 | 87 | 55 | 86 | 53 | 86 | 54 |
| AK Juneau | 32 | 79 | 76 | 81 | 75 | 79 | 69 | 75 | 64 | 74 | 63 | 75 | 64 |
| AZ Phoenix | 15 | 44 | 30 | 37 | 25 | 33 | 23 | 23 | 15 | 17 | 12 | 18 | 12 |
| AR Little Rock | 15 | 82 | 62 | 79 | 57 | 78 | 56 | 82 | 56 | 87 | 57 | 86 | 54 |
| CA Los Angeles | 16 | 54 | 59 | 58 | 62 | 61 | 66 | 60 | 63 | 65 | 66 | 70 | 68 |
| Sacramento | 15 | 86 | 71 | 79 | 61 | 68 | 52 | 58 | 43 | 52 | 37 | 48 | 32 |
| San Francisco | 16 | 79 | 67 | 75 | 65 | 70 | 63 | 65 | 60 | 65 | 61 | 65 | 60 |
| CO Denver | 15 | 45 | 48 | 44 | 43 | 42 | 41 | 38 | 34 | 38 | 37 | 38 | 37 |
| CN Hartford | 16 | 72 | 57 | 73 | 56 | 72 | 53 | 69 | 44 | 74 | 47 | 79 | 53 |
| DE Wilmington | 28 | 75 | 61 | 75 | 58 | 74 | 53 | 73 | 50 | 76 | 53 | 79 | 54 |
| DC Washington | 15 | 69 | 54 | 68 | 53 | 68 | 49 | 68 | 48 | 72 | 51 | 76 | 53 |
| FL Jacksonville | 39 | 88 | 57 | 85 | 53 | 85 | 49 | 85 | 48 | 84 | 49 | 86 | 56 |
| Miami | 11 | 84 | 61 | 82 | 56 | 82 | 56 | 80 | 54 | 83 | 60 | 87 | 67 |
| GA Atlanta | 15 | 80 | 61 | 76 | 55 | 79 | 52 | 80 | 52 | 83 | 55 | 86 | 59 |
| HI Honolulu | 6 | 80 | 64 | 77 | 60 | 75 | 61 | 70 | 59 | 67 | 55 | 66 | 54 |
| ID Boise | 36 | 73 | 71 | 68 | 61 | 55 | 45 | 47 | 36 | 45 | 34 | 42 | 31 |
| IL Chicago | 12 | 73 | 65 | 72 | 62 | 74 | 59 | 74 | 56 | 73 | 53 | 75 | 54 |
| Peoria | 16 | 78 | 69 | 79 | 66 | 82 | 64 | 78 | 57 | 81 | 57 | 82 | 57 |
| IN Indianapolis | 16 | 80 | 69 | 79 | 66 | 79 | 63 | 77 | 55 | 82 | 57 | 82 | 58 |
| IO Des Moines | 14 | 78 | 71 | 80 | 69 | 80 | 64 | 79 | 58 | 79 | 57 | 82 | 58 |
| KS Wichita | 22 | 80 | 64 | 79 | 60 | 76 | 54 | 77 | 52 | 83 | 55 | 83 | 53 |
| KY Louisville | 15 | 76 | 64 | 76 | 62 | 76 | 59 | 75 | 53 | 83 | 56 | 84 | 57 |
| LA New Orleans | 27 | 86 | 67 | 85 | 63 | 85 | 61 | 88 | 61 | 89 | 60 | 89 | 62 |
| ME Portland | 35 | 77 | 62 | 77 | 60 | 75 | 59 | 74 | 55 | 75 | 58 | 79 | 61 |
| MD Baltimore | 22 | 72 | 58 | 72 | 56 | 71 | 51 | 72 | 49 | 77 | 52 | 80 | 53 |
| MA Boston | 11 | 67 | 57 | 68 | 58 | 68 | 58 | 67 | 54 | 71 | 58 | 74 | 60 |
| MI Detroit | 42 | 78 | 69 | 78 | 65 | 77 | 60 | 73 | 53 | 71 | 51 | 74 | 53 |
| Sault Ste. Marie | 34 | 82 | 76 | 82 | 72 | 83 | 68 | 80 | 61 | 79 | 56 | 85 | 62 |
| MN Duluth | 14 | 76 | 69 | 75 | 64 | 78 | 63 | 77 | 58 | 76 | 55 | 82 | 61 |
| Minneapolis-St. Paul | 16 | 74 | 67 | 74 | 65 | 77 | 64 | 76 | 55 | 77 | 54 | 80 | 55 |
| MS Jackson | 12 | 87 | 66 | 86 | 59 | 87 | 57 | 90 | 57 | 91 | 56 | 90 | 55 |
| MO Kansas City | 3 | 75 | 65 | 75 | 65 | 78 | 62 | 74 | 54 | 80 | 56 | 81 | 54 |
| St. Louis | 15 | 82 | 64 | 80 | 61 | 81 | 57 | 78 | 54 | 82 | 56 | 84 | 57 |
| MT Great Falls | 14 | 62 | 61 | 59 | 55 | 53 | 47 | 50 | 42 | 46 | 40 | 45 | 40 |
| NE Omaha | 12 | 76 | 66 | 77 | 62 | 76 | 57 | 76 | 53 | 79 | 54 | 81 | 56 |
| NV Reno | 12 | 66 | 50 | 57 | 38 | 46 | 32 | 38 | 28 | 33 | 25 | 34 | 24 |
| NH Concord | 10 | 74 | 59 | 75 | 59 | 78 | 57 | 77 | 47 | 80 | 49 | 88 | 57 |
| NJ Atlantic City | 11 | 75 | 57 | 77 | 57 | 77 | 55 | 76 | 51 | 79 | 57 | 85 | 60 |
| NM Albuquerque | 15 | 49 | 37 | 43 | 32 | 32 | 23 | 25 | 17 | 22 | 16 | 23 | 17 |
| NY Albany | 10 | 80 | 63 | 77 | 58 | 73 | 54 | 69 | 47 | 74 | 52 | 78 | 56 |
| Buffalo | 15 | 78 | 72 | 79 | 70 | 80 | 68 | 76 | 58 | 76 | 56 | 78 | 57 |
| New York* | 55 | 68 | 60 | 68 | 58 | 67 | 55 | 68 | 51 | 71 | 53 | 74 | 55 |
| NC Charlotte | 15 | 79 | 57 | 76 | 52 | 79 | 50 | 79 | 48 | 84 | 53 | 86 | 57 |
| Raleigh | 11 | 78 | 56 | 74 | 49 | 79 | 49 | 80 | 45 | 86 | 55 | 88 | 57 |
| ND Bismarck | 16 | 73 | 67 | 75 | 68 | 78 | 62 | 79 | 54 | 79 | 50 | 84 | 54 |
| OH Cleveland | 15 | 76 | 69 | 77 | 68 | 77 | 65 | 75 | 57 | 77 | 58 | 79 | 59 |
| Columbus | 16 | 76 | 67 | 76 | 64 | 75 | 59 | 75 | 53 | 79 | 55 | 82 | 55 |
| Dayton | 12 | 75 | 67 | 76 | 63 | 78 | 62 | 75 | 56 | 77 | 54 | 79 | 54 |
| OK Oklahoma City | 10 | 81 | 63 | 79 | 57 | 76 | 53 | 78 | 52 | 82 | 56 | 84 | 56 |
| OR Portland | 35 | 82 | 76 | 79 | 68 | 72 | 60 | 68 | 55 | 66 | 53 | 65 | 49 |
| PA Philadelphia | 16 | 74 | 60 | 71 | 57 | 71 | 53 | 69 | 49 | 75 | 53 | 78 | 55 |
| Pittsburgh | 16 | 76 | 66 | 75 | 64 | 76 | 60 | 73 | 51 | 76 | 52 | 79 | 53 |
| RI Providence | 12 | 71 | 56 | 70 | 55 | 69 | 54 | 69 | 48 | 72 | 52 | 76 | 57 |
| SC Columbia | 9 | 85 | 58 | 82 | 50 | 83 | 48 | 84 | 45 | 88 | 51 | 89 | 54 |
| SD Sioux Falls | 12 | 76 | 67 | 78 | 66 | 81 | 62 | 81 | 56 | 81 | 53 | 83 | 57 |
| TN Memphis | 36 | 80 | 64 | 79 | 60 | 77 | 57 | 79 | 54 | 82 | 55 | 83 | 56 |
| Nashville | 10 | 81 | 66 | 80 | 60 | 79 | 54 | 81 | 53 | 87 | 56 | 88 | 55 |
| TX Dallas-Fort Worth | 12 | 82 | 61 | 80 | 57 | 80 | 57 | 85 | 59 | 88 | 62 | 86 | 57 |
| El Paso | 15 | 43 | 33 | 35 | 25 | 29 | 21 | 21 | 15 | 21 | 14 | 24 | 17 |
| Houston | 6 | 89 | 67 | 87 | 57 | 89 | 61 | 90 | 60 | 93 | 61 | 92 | 59 |
| UT Salt Lake City | 16 | 69 | 67 | 63 | 58 | 51 | 44 | 45 | 39 | 36 | 31 | 32 | 27 |
| VT Burlington | 10 | 69 | 63 | 71 | 61 | 72 | 58 | 75 | 53 | 75 | 52 | 79 | 58 |
| VA Norfolk | 27 | 76 | 60 | 75 | 57 | 73 | 54 | 74 | 51 | 78 | 56 | 80 | 58 |
| Richmond | 41 | 81 | 57 | 79 | 52 | 78 | 48 | 75 | 45 | 79 | 50 | 82 | 53 |
| WA Seattle-Tacoma | 16 | 79 | 75 | 76 | 66 | 74 | 62 | 72 | 59 | 68 | 54 | 67 | 54 |
| Spokane | 16 | 81 | 76 | 78 | 68 | 67 | 54 | 56 | 44 | 51 | 40 | 48 | 35 |
| WV Charleston | 28 | 77 | 62 | 77 | 59 | 75 | 54 | 75 | 48 | 82 | 50 | 86 | 54 |
| WI Milwaukee | 15 | 75 | 68 | 75 | 67 | 80 | 67 | 79 | 63 | 79 | 61 | 82 | 62 |
| WY Cheyenne | 16 | 44 | 48 | 44 | 46 | 45 | 46 | 41 | 39 | 39 | 40 | 40 | 41 |
| PR San Juan | 20 | 81 | 64 | 80 | 62 | 78 | 60 | 75 | 62 | 77 | 65 | 78 | 66 |

Source: U.S. National Oceanic and Atmospheric Administration, *Comparative Climatic Data.*
* City office data.

**TABLE 3–4** Average Relative Humidity (continued)

| July 7:00 a.m. | July 1:00 p.m. | Aug. 7:00 a.m. | Aug. 1:00 p.m. | Sept. 7:00 a.m. | Sept. 1:00 p.m. | Oct. 7:00 a.m. | Oct. 1:00 p.m. | Nov. 7:00 a.m. | Nov. 1:00 p.m. | Dec. 7:00 a.m. | Dec. 1:00 p.m. | Annual 7:00 a.m. | Annual 7:00 p.m. | State | Station |
|---|---|---|---|---|---|---|---|---|---|---|---|---|---|---|---|
| 89 | 61 | 90 | 62 | 88 | 60 | 85 | 52 | 85 | 56 | 84 | 63 | 86 | 58 | AL | Mobile |
| 81 | 70 | 84 | 74 | 87 | 77 | 87 | 80 | 85 | 82 | 82 | 81 | 81 | 73 | AK | Juneau |
| 28 | 20 | 34 | 24 | 31 | 23 | 30 | 22 | 38 | 28 | 48 | 35 | 32 | 22 | AZ | Phoenix |
| 88 | 58 | 88 | 57 | 90 | 60 | 86 | 52 | 83 | 58 | 82 | 63 | 84 | 57 | AR | Little Rock |
| 68 | 68 | 68 | 69 | 65 | 67 | 58 | 64 | 58 | 64 | 55 | 61 | 62 | 65 | CA | Los Angeles |
| 48 | 28 | 49 | 28 | 50 | 31 | 58 | 40 | 77 | 61 | 86 | 73 | 63 | 46 | | Sacramento |
| 66 | 61 | 67 | 62 | 66 | 59 | 68 | 59 | 74 | 65 | 78 | 69 | 70 | 62 | | San Francisco |
| 36 | 36 | 36 | 34 | 40 | 36 | 36 | 36 | 45 | 50 | 45 | 52 | 40 | 40 | CO | Denver |
| 80 | 51 | 84 | 53 | 87 | 55 | 84 | 51 | 80 | 58 | 79 | 63 | 78 | 53 | CN | Hartford |
| 80 | 54 | 84 | 56 | 85 | 55 | 84 | 53 | 80 | 56 | 77 | 60 | 79 | 55 | DE | Wilmington |
| 75 | 52 | 78 | 53 | 79 | 55 | 79 | 51 | 73 | 52 | 71 | 57 | 73 | 52 | DC | Washington |
| 87 | 58 | 90 | 60 | 91 | 62 | 90 | 56 | 89 | 55 | 88 | 58 | 87 | 55 | FL | Jacksonville |
| 86 | 65 | 87 | 66 | 89 | 67 | 87 | 65 | 84 | 60 | 83 | 58 | 84 | 61 | | Miami |
| 90 | 64 | 91 | 62 | 89 | 61 | 84 | 53 | 82 | 54 | 80 | 60 | 83 | 57 | GA | Atlanta |
| 65 | 51 | 67 | 54 | 66 | 52 | 68 | 55 | 74 | 60 | 78 | 61 | 71 | 57 | HI | Honolulu |
| 33 | 21 | 34 | 23 | 39 | 30 | 49 | 41 | 66 | 61 | 74 | 73 | 52 | 44 | ID | Boise |
| 77 | 55 | 80 | 56 | 81 | 56 | 78 | 54 | 79 | 65 | 80 | 72 | 76 | 59 | IL | Chicago |
| 86 | 59 | 88 | 59 | 88 | 60 | 85 | 58 | 83 | 67 | 84 | 73 | 83 | 62 | | Peoria |
| 87 | 60 | 90 | 61 | 90 | 59 | 87 | 57 | 85 | 67 | 83 | 73 | 83 | 62 | ID | Indianapolis |
| 82 | 57 | 85 | 58 | 85 | 61 | 78 | 55 | 80 | 65 | 81 | 72 | 81 | 62 | IO | Des Moines |
| 79 | 49 | 79 | 49 | 82 | 56 | 81 | 53 | 79 | 57 | 80 | 63 | 80 | 55 | KS | Wichita |
| 86 | 58 | 87 | 57 | 89 | 60 | 85 | 55 | 79 | 61 | 77 | 66 | 81 | 59 | KY | Louisville |
| 91 | 66 | 91 | 66 | 89 | 66 | 88 | 59 | 86 | 60 | 86 | 67 | 88 | 63 | LA | New Orleans |
| 81 | 60 | 84 | 59 | 86 | 61 | 85 | 59 | 84 | 64 | 80 | 63 | 80 | 60 | ME | Portland |
| 81 | 53 | 84 | 55 | 85 | 55 | 82 | 53 | 77 | 54 | 75 | 59 | 77 | 54 | MD | Baltimore |
| 73 | 56 | 74 | 56 | 79 | 61 | 75 | 57 | 74 | 61 | 72 | 62 | 72 | 58 | MA | Boston |
| 75 | 51 | 80 | 53 | 83 | 54 | 81 | 55 | 79 | 64 | 79 | 70 | 77 | 58 | MI | Detroit |
| 88 | 62 | 91 | 63 | 92 | 67 | 89 | 67 | 87 | 76 | 85 | 78 | 85 | 67 | | Sault Ste. Marie |
| 84 | 58 | 88 | 63 | 87 | 64 | 82 | 63 | 81 | 71 | 80 | 75 | 80 | 64 | MN | Duluth |
| 82 | 55 | 85 | 56 | 87 | 60 | 83 | 59 | 82 | 67 | 79 | 71 | 80 | 61 | | Minneapolis-St. Paul |
| 93 | 59 | 94 | 60 | 94 | 60 | 93 | 54 | 91 | 58 | 89 | 66 | 90 | 59 | MS | Jackson |
| 77 | 48 | 85 | 58 | 89 | 66 | 81 | 58 | 86 | 65 | 79 | 70 | 80 | 60 | MO | Kansas City |
| 86 | 57 | 89 | 57 | 91 | 61 | 84 | 55 | 84 | 62 | 85 | 70 | 84 | 59 | | St. Louis |
| 37 | 28 | 37 | 29 | 45 | 37 | 46 | 42 | 56 | 54 | 62 | 61 | 50 | 45 | MT | Great Falls |
| 82 | 56 | 86 | 59 | 88 | 61 | 82 | 55 | 81 | 63 | 80 | 68 | 80 | 59 | NE | Omaha |
| 29 | 19 | 31 | 20 | 34 | 21 | 41 | 27 | 57 | 42 | 66 | 53 | 44 | 32 | NV | Reno |
| 89 | 54 | 90 | 54 | 93 | 59 | 89 | 54 | 85 | 63 | 81 | 66 | 83 | 56 | NH | Concord |
| 85 | 59 | 87 | 57 | 88 | 58 | 87 | 56 | 82 | 57 | 77 | 59 | 81 | 57 | NJ | Atlantic City |
| 35 | 28 | 39 | 30 | 40 | 31 | 37 | 29 | 42 | 35 | 51 | 43 | 37 | 28 | NM | Albuquerque |
| 79 | 54 | 83 | 54 | 87 | 58 | 84 | 55 | 82 | 63 | 82 | 68 | 79 | 57 | NY | Albany |
| 79 | 55 | 83 | 59 | 83 | 60 | 82 | 61 | 83 | 71 | 82 | 75 | 80 | 63 | | Buffalo |
| 75 | 55 | 78 | 57 | 79 | 57 | 76 | 55 | 73 | 59 | 70 | 61 | 72 | 56 | | New York* |
| 89 | 60 | 90 | 60 | 90 | 57 | 88 | 53 | 83 | 51 | 80 | 57 | 84 | 55 | NC | Charlotte |
| 91 | 61 | 93 | 62 | 93 | 60 | 90 | 54 | 83 | 49 | 80 | 55 | 84 | 54 | | Raleigh |
| 83 | 47 | 82 | 44 | 82 | 49 | 78 | 50 | 80 | 63 | 76 | 69 | 79 | 56 | ND | Bismarck |
| 82 | 57 | 85 | 60 | 84 | 61 | 80 | 59 | 78 | 67 | 76 | 71 | 79 | 63 | OH | Cleveland |
| 84 | 55 | 88 | 57 | 88 | 58 | 82 | 55 | 82 | 65 | 79 | 70 | 80 | 60 | | Columbus |
| 80 | 54 | 85 | 56 | 86 | 58 | 81 | 56 | 80 | 66 | 79 | 71 | 79 | 60 | | Dayton |
| 81 | 50 | 82 | 51 | 86 | 58 | 81 | 53 | 78 | 55 | 80 | 60 | 81 | 55 | OK | Oklahoma City |
| 61 | 45 | 64 | 46 | 66 | 49 | 79 | 64 | 82 | 74 | 84 | 79 | 72 | 60 | OR | Portland |
| 79 | 54 | 81 | 54 | 83 | 56 | 82 | 53 | 77 | 55 | 74 | 61 | 76 | 55 | PA | Philadelphia |
| 83 | 52 | 86 | 55 | 86 | 57 | 80 | 53 | 79 | 63 | 77 | 68 | 79 | 58 | | Pittsburgh |
| 78 | 57 | 78 | 54 | 82 | 57 | 79 | 53 | 77 | 59 | 76 | 61 | 75 | 55 | RI | Providence |
| 91 | 58 | 93 | 59 | 94 | 58 | 92 | 52 | 88 | 49 | 86 | 56 | 88 | 53 | SC | Columbia |
| 82 | 52 | 84 | 52 | 87 | 58 | 81 | 56 | 84 | 65 | 80 | 71 | 81 | 60 | SD | Sioux Falls |
| 85 | 57 | 86 | 56 | 86 | 56 | 84 | 51 | 80 | 55 | 79 | 62 | 82 | 57 | TN | Memphis |
| 91 | 59 | 92 | 61 | 91 | 61 | 87 | 55 | 81 | 60 | 81 | 66 | 85 | 59 | | Nashville |
| 81 | 50 | 83 | 53 | 88 | 60 | 85 | 56 | 82 | 56 | 81 | 60 | 83 | 57 | TX | Dallas-Fort Worth |
| 39 | 30 | 41 | 32 | 45 | 34 | 35 | 27 | 37 | 31 | 43 | 36 | 34 | 26 | | El Paso |
| 93 | 58 | 95 | 62 | 95 | 66 | 95 | 60 | 90 | 59 | 88 | 62 | 91 | 61 | | Houston |
| 26 | 20 | 28 | 22 | 34 | 27 | 43 | 40 | 58 | 59 | 71 | 72 | 46 | 42 | UT | Salt Lake City |
| 80 | 54 | 84 | 58 | 88 | 64 | 82 | 62 | 82 | 71 | 78 | 72 | 78 | 60 | VT | Burlington |
| 82 | 60 | 85 | 62 | 84 | 62 | 83 | 61 | 78 | 55 | 76 | 59 | 79 | 58 | VA | Norfolk |
| 85 | 57 | 88 | 57 | 89 | 56 | 89 | 52 | 84 | 50 | 81 | 55 | 83 | 53 | | Richmond |
| 66 | 49 | 70 | 52 | 74 | 58 | 80 | 69 | 81 | 75 | 81 | 78 | 74 | 62 | WA | Seattle-Tacoma |
| 38 | 24 | 42 | 27 | 49 | 33 | 66 | 50 | 82 | 75 | 84 | 81 | 62 | 50 | | Spokane |
| 90 | 61 | 92 | 58 | 91 | 55 | 88 | 53 | 80 | 56 | 78 | 62 | 83 | 56 | WV | Charleston |
| 82 | 60 | 87 | 62 | 87 | 63 | 81 | 63 | 80 | 67 | 80 | 73 | 81 | 65 | WS | Milwaukee |
| 34 | 37 | 33 | 34 | 37 | 38 | 37 | 39 | 41 | 47 | 44 | 50 | 40 | 42 | WY | Cheyenne |
| 79 | 66 | 79 | 66 | 80 | 67 | 81 | 66 | 81 | 66 | 80 | 65 | 79 | 64 | PR | San Juan |

## TABLE 3–5 Normal Daily Mean Temperature

(In Fahrenheit degrees. Airport data, except as noted.) Based on standard 30-year period,
1941 through 1970.)

| | State and Station | Jan. | Feb. | Mar. | Apr. | May | June | July | Aug. | Sept. | Oct. | Nov. | Dec. | Annual Average |
|---|---|---|---|---|---|---|---|---|---|---|---|---|---|---|
| AL | Mobile | 51.2 | 54.0 | 59.4 | 67.9 | 74.8 | 80.3 | 81.6 | 81.5 | 77.5 | 68.9 | 58.5 | 52.9 | 67.4 |
| AK | Juneau | 23.5 | 28.0 | 31.9 | 38.9 | 46.8 | 53.2 | 55.7 | 54.3 | 49.2 | 41.8 | 32.5 | 27.3 | 40.3 |
| AZ | Phoenix | 51.2 | 55.1 | 59.7 | 67.7 | 76.3 | 84.6 | 91.2 | 89.1 | 83.8 | 72.2 | 59.8 | 52.5 | 70.3 |
| AR | Little Rock | 39.5 | 42.9 | 50.3 | 61.7 | 69.8 | 78.1 | 81.4 | 80.6 | 73.3 | 62.4 | 50.3 | 41.6 | 61.0 |
| CA | Los Angeles | 54.5 | 55.6 | 56.5 | 58.8 | 61.9 | 64.5 | 68.5 | 69.5 | 68.7 | 65.2 | 60.5 | 56.9 | 61.7 |
| | Sacramento | 45.1 | 49.8 | 53.0 | 58.3 | 64.3 | 70.5 | 75.2 | 74.1 | 71.5 | 63.3 | 53.0 | 45.8 | 60.3 |
| | San Francisco | 48.3 | 51.2 | 53.0 | 55.3 | 58.3 | 61.6 | 62.5 | 63.0 | 64.1 | 61.0 | 55.3 | 49.7 | 56.9 |
| CO | Denver | 29.9 | 32.8 | 37.0 | 47.5 | 57.0 | 66.0 | 73.0 | 71.6 | 62.8 | 52.0 | 39.4 | 32.6 | 50.1 |
| CN | Hartford | 24.8 | 26.8 | 35.6 | 47.7 | 58.3 | 67.8 | 72.7 | 70.4 | 62.8 | 52.6 | 41.3 | 28.2 | 49.1 |
| DE | Wilmington | 32.0 | 33.6 | 41.6 | 52.3 | 62.4 | 71.4 | 75.8 | 74.1 | 67.9 | 57.2 | 45.7 | 34.7 | 54.0 |
| DC | Washington | 35.6 | 37.3 | 45.1 | 56.4 | 66.2 | 74.6 | 78.7 | 77.1 | 70.6 | 59.8 | 48.0 | 37.4 | 57.3 |
| FL | Jacksonville | 54.6 | 56.3 | 61.2 | 68.1 | 74.3 | 79.2 | 81.0 | 81.0 | 78.2 | 70.5 | 61.2 | 55.4 | 68.4 |
| | Miami | 67.2 | 67.8 | 71.3 | 75.0 | 78.0 | 81.0 | 82.3 | 82.9 | 81.7 | 77.8 | 72.2 | 68.3 | 75.5 |
| GA | Atlanta | 42.4 | 45.0 | 51.1 | 61.1 | 69.1 | 75.6 | 78.0 | 77.5 | 72.3 | 62.4 | 51.4 | 43.5 | 60.8 |
| HI | Honolulu | 72.3 | 72.3 | 73.0 | 74.8 | 76.9 | 78.9 | 80.1 | 80.7 | 80.4 | 78.9 | 76.5 | 73.7 | 76.6 |
| ID | Boise | 29.0 | 35.5 | 41.1 | 49.0 | 57.4 | 64.8 | 74.5 | 72.2 | 63.1 | 52.1 | 39.8 | 32.1 | 50.9 |
| IL | Chicago | 24.3 | 27.4 | 36.8 | 49.9 | 60.0 | 70.5 | 74.7 | 73.7 | 65.9 | 55.4 | 40.4 | 28.5 | 50.6 |
| | Peoria | 23.8 | 27.7 | 37.3 | 51.3 | 61.5 | 71.3 | 75.1 | 73.5 | 65.5 | 55.0 | 39.9 | 28.0 | 50.8 |
| IN | Indianapolis | 27.9 | 30.7 | 39.7 | 52.3 | 62.2 | 71.7 | 75.0 | 72.6 | 66.3 | 55.7 | 41.7 | 30.9 | 52.3 |
| IO | Des Moines | 19.4 | 24.2 | 33.9 | 49.5 | 60.9 | 70.5 | 75.1 | 73.3 | 64.3 | 54.3 | 37.8 | 25.0 | 49.0 |
| KS | Wichita | 31.3 | 36.3 | 43.6 | 56.6 | 66.1 | 75.8 | 80.7 | 79.7 | 70.6 | 59.6 | 44.8 | 34.5 | 56.6 |
| KY | Louisville | 33.3 | 35.8 | 44.0 | 55.9 | 64.8 | 73.3 | 76.9 | 75.9 | 69.1 | 58.1 | 45.0 | 35.6 | 55.6 |
| LA | New Orleans | 52.9 | 55.6 | 60.7 | 68.6 | 75.1 | 80.4 | 81.9 | 81.9 | 78.2 | 69.8 | 60.1 | 54.8 | 68.3 |
| ME | Portland | 21.5 | 22.9 | 31.8 | 42.7 | 52.7 | 62.2 | 68.0 | 66.4 | 58.7 | 49.1 | 38.6 | 25.7 | 45.0 |
| MD | Baltimore | 33.4 | 34.8 | 42.8 | 53.8 | 63.7 | 72.4 | 76.6 | 74.9 | 68.5 | 57.4 | 46.1 | 35.3 | 55.0 |
| MA | Boston | 29.2 | 30.4 | 38.1 | 48.6 | 58.6 | 68.0 | 73.3 | 71.3 | 64.5 | 55.4 | 45.2 | 33.0 | 51.3 |
| MI | Detroit | 25.5 | 26.9 | 35.4 | 48.1 | 58.4 | 69.1 | 73.3 | 71.9 | 64.5 | 54.3 | 41.1 | 29.6 | 49.9 |
| | Sault Ste. Marie | 14.2 | 15.2 | 24.0 | 38.2 | 49.0 | 58.7 | 63.8 | 63.2 | 55.3 | 46.2 | 32.8 | 20.1 | 40.0 |
| MN | Duluth | 8.5 | 12.1 | 23.5 | 38.6 | 49.4 | 59.0 | 65.6 | 64.1 | 54.4 | 45.3 | 28.4 | 14.4 | 38.6 |
| | Minneapolis-St. Paul | 12.2 | 16.5 | 28.3 | 45.1 | 57.1 | 66.9 | 71.9 | 70.2 | 60.0 | 50.0 | 32.4 | 18.6 | 44.1 |
| MS | Jackson | 47.1 | 49.8 | 56.1 | 65.7 | 72.7 | 79.4 | 81.7 | 81.2 | 76.0 | 65.8 | 55.3 | 48.9 | 65.0 |
| MO | Kansas City | 27.8 | 33.1 | 41.2 | 55.0 | 65.0 | 73.9 | 78.8 | 77.4 | 68.8 | 58.6 | 43.6 | 32.3 | 54.5 |
| | St. Louis | 31.3 | 35.1 | 43.3 | 56.5 | 65.8 | 74.9 | 78.6 | 77.2 | 69.6 | 59.1 | 45.0 | 34.6 | 55.9 |
| MT | Great Falls | 20.5 | 26.6 | 30.5 | 43.4 | 53.3 | 60.8 | 69.3 | 67.4 | 57.3 | 48.3 | 34.6 | 26.5 | 44.9 |
| NE | Omaha | 22.6 | 28.0 | 37.1 | 52.3 | 63.0 | 72.2 | 77.2 | 75.6 | 66.3 | 55.9 | 40.0 | 28.0 | 51.5 |
| NV | Reno | 31.9 | 37.1 | 40.3 | 46.8 | 54.6 | 61.5 | 69.3 | 66.9 | 60.2 | 50.3 | 40.1 | 33.0 | 49.4 |
| NH | Concord | 20.6 | 22.6 | 32.3 | 44.2 | 55.1 | 64.7 | 69.7 | 67.2 | 59.5 | 49.3 | 38.0 | 24.8 | 45.6 |
| NJ | Atlantic City | 32.7 | 33.9 | 41.1 | 51.7 | 61.6 | 70.3 | 75.1 | 73.4 | 67.1 | 56.7 | 46.0 | 35.1 | 53.7 |
| NM | Albuquerque | 35.2 | 40.0 | 45.8 | 55.8 | 65.3 | 74.6 | 78.7 | 76.6 | 70.1 | 58.2 | 44.5 | 36.2 | 56.8 |
| NY | Albany | 21.5 | 23.5 | 33.4 | 46.9 | 57.7 | 67.5 | 72.0 | 69.6 | 61.9 | 51.4 | 39.6 | 25.9 | 47.6 |
| | Buffalo | 23.7 | 24.4 | 32.1 | 44.9 | 55.1 | 65.7 | 70.1 | 68.4 | 61.6 | 51.5 | 39.8 | 27.9 | 47.1 |
| | New York* | 32.2 | 33.4 | 41.1 | 52.1 | 62.3 | 71.6 | 76.6 | 74.9 | 68.4 | 58.7 | 47.4 | 35.5 | 54.5 |
| NC | Charlotte | 42.1 | 44.0 | 50.6 | 60.8 | 68.8 | 75.9 | 78.5 | 77.7 | 72.0 | 61.7 | 51.0 | 42.5 | 60.5 |
| | Raleigh | 40.5 | 42.2 | 49.2 | 59.5 | 67.4 | 74.4 | 77.5 | 76.5 | 70.6 | 60.2 | 50.0 | 41.2 | 59.1 |
| ND | Bismarck | 8.2 | 13.5 | 25.1 | 43.0 | 54.4 | 63.8 | 70.8 | 69.2 | 57.5 | 46.8 | 28.9 | 15.6 | 41.4 |
| OH | Cincinnati | 32.1 | 34.4 | 42.9 | 55.1 | 64.4 | 73.1 | 76.2 | 75.1 | 68.4 | 57.8 | 44.6 | 34.4 | 54.9 |
| | Cleveland | 26.9 | 27.9 | 36.1 | 48.3 | 58.3 | 67.9 | 71.4 | 70.0 | 63.9 | 53.8 | 41.6 | 30.3 | 49.7 |
| | Columbus | 28.4 | 30.3 | 39.2 | 51.2 | 61.1 | 70.4 | 73.6 | 71.9 | 65.2 | 54.2 | 41.7 | 30.7 | 51.5 |
| OK | Oklahoma City | 36.8 | 41.3 | 48.2 | 60.4 | 68.3 | 76.8 | 81.5 | 81.1 | 73.0 | 62.4 | 49.2 | 40.0 | 59.9 |
| OR | Portland | 38.1 | 42.8 | 45.7 | 50.6 | 56.7 | 62.0 | 67.1 | 66.6 | 62.2 | 53.8 | 45.3 | 40.7 | 52.6 |
| PA | Philadelphia | 32.3 | 33.9 | 41.9 | 52.9 | 63.2 | 72.3 | 76.8 | 74.8 | 68.1 | 57.4 | 46.2 | 35.2 | 54.6 |
| | Pittsburgh | 28.1 | 29.3 | 38.1 | 50.2 | 59.8 | 68.6 | 71.9 | 70.2 | 63.8 | 53.2 | 41.3 | 30.5 | 50.4 |
| RI | Providence | 28.4 | 29.4 | 36.9 | 47.3 | 56.9 | 66.4 | 72.1 | 70.4 | 63.4 | 53.7 | 43.3 | 31.5 | 50.0 |
| SC | Columbia | 45.4 | 47.6 | 54.2 | 64.1 | 72.1 | 78.8 | 81.2 | 80.2 | 74.5 | 64.2 | 53.8 | 46.0 | 63.5 |
| SD | Sioux Falls | 14.2 | 19.4 | 30.0 | 46.1 | 57.7 | 67.6 | 73.3 | 71.8 | 60.9 | 50.2 | 33.1 | 20.0 | 45.4 |
| TN | Memphis | 40.5 | 43.8 | 51.0 | 62.5 | 70.9 | 78.6 | 81.6 | 80.4 | 73.6 | 63.0 | 50.8 | 42.7 | 61.6 |
| | Nashville | 38.3 | 41.0 | 48.7 | 60.1 | 68.5 | 76.6 | 79.6 | 78.5 | 72.0 | 60.9 | 48.4 | 40.4 | 59.4 |
| TX | Dallas-Fort Worth | 44.8 | 48.7 | 55.0 | 65.2 | 72.5 | 80.6 | 84.8 | 84.9 | 77.7 | 67.6 | 55.8 | 47.9 | 65.5 |
| | El Paso | 43.6 | 48.4 | 54.6 | 63.9 | 72.2 | 80.3 | 82.3 | 80.5 | 74.2 | 64.0 | 51.6 | 44.4 | 63.4 |
| | Houston | 52.1 | 55.3 | 60.8 | 69.4 | 75.8 | 81.1 | 83.3 | 83.4 | 79.2 | 70.9 | 61.1 | 54.6 | 68.9 |
| UT | Salt Lake City | 28.0 | 33.4 | 39.6 | 49.2 | 58.3 | 66.2 | 76.7 | 74.5 | 64.8 | 52.4 | 39.1 | 30.3 | 51.0 |
| VT | Burlington | 16.8 | 18.6 | 29.1 | 43.0 | 54.8 | 65.2 | 69.8 | 67.4 | 59.3 | 48.8 | 37.0 | 22.6 | 44.4 |
| VA | Norfolk | 40.5 | 41.4 | 48.1 | 57.8 | 66.7 | 74.5 | 78.3 | 76.9 | 71.8 | 61.7 | 51.6 | 42.3 | 59.3 |
| | Richmond | 37.5 | 39.4 | 46.9 | 57.8 | 66.5 | 74.2 | 77.9 | 76.3 | 70.0 | 59.3 | 49.0 | 39.0 | 57.8 |
| WA | Seattle-Tacoma | 38.2 | 42.3 | 44.1 | 48.7 | 54.9 | 59.8 | 64.5 | 63.8 | 59.6 | 52.2 | 44.6 | 40.5 | 51.1 |
| | Spokane | 25.4 | 32.2 | 37.5 | 46.1 | 54.7 | 61.5 | 69.7 | 68.0 | 59.6 | 47.8 | 35.5 | 29.0 | 47.3 |
| WV | Charleston | 34.5 | 36.5 | 44.5 | 55.9 | 64.5 | 72.0 | 75.0 | 73.6 | 67.5 | 57.0 | 45.4 | 36.2 | 55.2 |
| WI | Milwaukee | 19.4 | 22.5 | 31.4 | 44.7 | 54.2 | 64.5 | 69.9 | 69.2 | 61.1 | 51.0 | 36.5 | 24.2 | 45.7 |
| WY | Cheyenne | 26.6 | 29.0 | 31.6 | 42.7 | 52.4 | 61.3 | 69.1 | 67.6 | 58.2 | 47.9 | 35.5 | 29.2 | 45.9 |
| PR | San Juan | 75.4 | 75.3 | 76.3 | 77.5 | 79.2 | 80.5 | 80.9 | 81.3 | 81.1 | 80.6 | 78.7 | 76.8 | 78.6 |

Source: U.S. National Oceanic and Atmospheric Administration, *Comparative Climatic Data.*
* City office data.

## TABLE 3-6 Normal Daily Minimum Temperature

(In Fahrenheit degrees. Airport data, except as noted. Based on standard 30-year period, 1941 through 1970.)

| | State and Station | Jan. | Feb. | Mar. | Apr. | May | June | July | Aug. | Sept. | Oct. | Nov. | Dec. | Annual Average |
|---|---|---|---|---|---|---|---|---|---|---|---|---|---|---|
| AL | Mobile | 41.3 | 43.9 | 49.2 | 57.7 | 64.5 | 70.7 | 72.6 | 72.3 | 68.4 | 58.0 | 47.5 | 42.8 | 57.4 |
| AK | Juneau | 17.8 | 22.1 | 25.6 | 31.3 | 38.2 | 44.4 | 47.7 | 46.2 | 42.3 | 36.4 | 27.6 | 22.5 | 33.5 |
| AZ | Phoenix | 37.6 | 40.8 | 44.8 | 51.8 | 59.6 | 67.7 | 77.5 | 76.0 | 69.1 | 56.8 | 44.8 | 38.5 | 55.4 |
| AR | Little Rock | 28.9 | 31.9 | 38.7 | 49.9 | 58.1 | 66.8 | 70.1 | 68.6 | 60.8 | 48.7 | 38.1 | 31.1 | 49.3 |
| CA | Los Angeles | 45.4 | 47.0 | 48.6 | 51.7 | 55.3 | 58.6 | 62.1 | 63.2 | 61.6 | 57.5 | 51.3 | 47.3 | 54.1 |
| | Sacramento | 37.1 | 40.4 | 41.9 | 45.3 | 49.8 | 54.6 | 57.5 | 56.9 | 55.3 | 49.5 | 42.4 | 38.3 | 47.4 |
| | San Francisco | 41.2 | 43.8 | 44.9 | 47.0 | 49.9 | 53.0 | 54.0 | 54.3 | 54.5 | 51.6 | 47.2 | 42.9 | 48.7 |
| CO | Denver | 16.2 | 19.4 | 23.8 | 33.9 | 43.6 | 51.9 | 58.6 | 57.4 | 47.8 | 37.2 | 25.4 | 18.9 | 36.2 |
| CN | Hartford | 16.1 | 17.9 | 26.6 | 36.5 | 46.2 | 56.0 | 61.2 | 58.9 | 51.0 | 40.8 | 31.9 | 19.6 | 38.6 |
| DE | Wilmington | 23.8 | 24.9 | 32.0 | 41.5 | 51.6 | 61.1 | 66.1 | 64.3 | 57.6 | 46.5 | 36.2 | 26.3 | 44.3 |
| DC | Washington | 27.7 | 28.6 | 35.2 | 45.7 | 55.7 | 64.6 | 69.1 | 67.6 | 61.0 | 49.7 | 38.8 | 29.5 | 47.8 |
| FL | Jacksonville | 44.5 | 45.7 | 50.1 | 57.1 | 63.9 | 70.0 | 72.0 | 72.3 | 70.4 | 61.7 | 51.0 | 45.1 | 58.7 |
| | Miami | 58.7 | 59.0 | 63.0 | 67.3 | 70.7 | 73.9 | 75.5 | 75.8 | 75.0 | 71.0 | 64.5 | 60.0 | 67.9 |
| GA | Atlanta | 33.4 | 35.5 | 41.1 | 50.7 | 59.2 | 66.6 | 69.4 | 68.6 | 63.4 | 52.3 | 40.8 | 34.3 | 51.3 |
| HI | Honolulu | 65.3 | 65.3 | 66.3 | 68.1 | 70.2 | 72.2 | 73.4 | 74.0 | 73.4 | 72.0 | 69.8 | 67.1 | 69.8 |
| ID | Boise | 21.4 | 27.2 | 30.5 | 36.5 | 44.1 | 51.2 | 58.5 | 56.7 | 48.5 | 39.4 | 30.7 | 25.0 | 39.1 |
| IL | Chicago | 17.0 | 20.2 | 29.0 | 40.4 | 49.7 | 60.3 | 65.0 | 64.1 | 56.0 | 45.6 | 32.6 | 21.6 | 41.8 |
| | Peoria | 15.7 | 19.3 | 28.1 | 40.8 | 50.7 | 60.9 | 64.6 | 62.9 | 54.6 | 44.0 | 31.1 | 20.3 | 41.1 |
| IN | Indianapolis | 19.7 | 22.1 | 30.3 | 41.8 | 51.5 | 61.1 | 64.6 | 62.4 | 54.9 | 44.3 | 32.8 | 23.1 | 42.4 |
| IO | Des Moines | 11.3 | 15.8 | 25.2 | 39.2 | 50.9 | 61.1 | 65.3 | 63.4 | 54.0 | 43.6 | 29.2 | 17.2 | 39.7 |
| KS | Wichita | 21.2 | 25.4 | 32.1 | 45.1 | 55.0 | 65.0 | 69.6 | 68.3 | 59.2 | 47.9 | 33.8 | 24.6 | 45.6 |
| KY | Louisville | 24.5 | 26.5 | 34.0 | 44.8 | 53.9 | 62.9 | 66.4 | 64.9 | 57.7 | 45.9 | 35.1 | 27.1 | 45.3 |
| LA | New Orleans | 43.5 | 46.0 | 50.9 | 58.8 | 65.3 | 71.2 | 73.3 | 73.1 | 69.7 | 59.6 | 49.8 | 45.3 | 58.9 |
| ME | Portland | 11.7 | 12.5 | 22.8 | 32.5 | 41.7 | 51.1 | 56.9 | 55.2 | 47.4 | 38.0 | 29.7 | 16.4 | 34.7 |
| MD | Baltimore | 24.9 | 25.7 | 32.5 | 42.4 | 52.5 | 61.6 | 66.5 | 64.7 | 57.9 | 46.4 | 36.0 | 26.6 | 44.8 |
| MA | Boston | 22.5 | 23.3 | 31.5 | 40.8 | 50.1 | 59.3 | 65.1 | 63.3 | 56.7 | 47.5 | 38.7 | 26.6 | 43.8 |
| MI | Detroit | 19.2 | 20.1 | 27.6 | 38.6 | 48.3 | 59.1 | 63.4 | 62.1 | 54.8 | 45.2 | 34.4 | 23.8 | 41.4 |
| | Sault Ste. Marie | 6.4 | 6.7 | 15.5 | 29.2 | 38.5 | 47.3 | 52.5 | 52.9 | 46.1 | 37.6 | 26.5 | 13.3 | 31.0 |
| MN | Duluth | −.6 | 2.0 | 14.4 | 29.3 | 38.8 | 48.3 | 54.7 | 53.7 | 44.8 | 36.2 | 21.4 | 6.3 | 29.1 |
| | Minneapolis-St. Paul | 3.2 | 7.1 | 19.6 | 34.7 | 46.3 | 56.7 | 61.4 | 59.6 | 49.3 | 39.2 | 24.2 | 10.3 | 34.3 |
| MS | Jackson | 35.8 | 37.8 | 43.4 | 53.1 | 60.4 | 67.7 | 70.6 | 69.8 | 64.0 | 51.5 | 42.0 | 37.3 | 52.8 |
| MO | Kansas City | 19.3 | 24.2 | 31.8 | 45.1 | 55.7 | 65.2 | 69.6 | 68.1 | 58.8 | 48.3 | 34.5 | 24.1 | 45.3 |
| | St. Louis | 22.6 | 26.0 | 33.5 | 46.0 | 55.5 | 64.8 | 68.8 | 67.1 | 59.1 | 48.4 | 35.9 | 26.5 | 46.2 |
| MT | Great Falls | 11.6 | 17.2 | 20.6 | 32.3 | 41.5 | 49.5 | 54.9 | 53.0 | 44.6 | 37.1 | 25.7 | 18.2 | 33.8 |
| NE | Omaha | 12.4 | 17.4 | 26.4 | 40.1 | 51.5 | 61.3 | 65.8 | 64.0 | 54.0 | 42.6 | 29.1 | 18.1 | 40.2 |
| NV | Reno | 18.3 | 23.0 | 24.6 | 29.6 | 37.0 | 42.5 | 47.4 | 44.8 | 38.6 | 30.5 | 23.9 | 19.6 | 31.7 |
| NH | Concord | 9.9 | 11.3 | 22.1 | 31.7 | 41.5 | 51.6 | 56.7 | 54.2 | 46.5 | 36.3 | 28.1 | 14.9 | 33.7 |
| NJ | Atlantic City | 24.0 | 24.9 | 31.5 | 41.0 | 50.7 | 59.7 | 65.4 | 63.8 | 56.8 | 45.9 | 36.1 | 26.0 | 43.8 |
| NM | Albuquerque | 23.5 | 27.4 | 32.3 | 41.4 | 50.7 | 59.7 | 65.2 | 63.4 | 56.7 | 44.7 | 31.8 | 24.9 | 43.5 |
| NY | Albany | 12.5 | 14.3 | 24.2 | 35.7 | 45.7 | 55.6 | 60.1 | 57.8 | 50.1 | 40.0 | 31.1 | 17.7 | 37.1 |
| | Buffalo | 17.6 | 17.7 | 25.2 | 36.4 | 45.9 | 56.3 | 60.7 | 59.1 | 52.3 | 42.7 | 33.5 | 22.2 | 39.1 |
| | New York* | 25.9 | 26.5 | 33.7 | 43.5 | 53.1 | 62.6 | 68.0 | 66.4 | 59.9 | 50.6 | 40.8 | 29.5 | 46.7 |
| NC | Charlotte | 32.1 | 33.1 | 39.0 | 48.9 | 57.4 | 65.3 | 68.7 | 67.9 | 61.9 | 50.3 | 39.6 | 32.4 | 49.7 |
| | Raleigh | 30.0 | 31.1 | 37.4 | 46.7 | 55.4 | 63.1 | 67.2 | 66.2 | 59.7 | 48.0 | 37.8 | 30.5 | 47.8 |
| ND | Bismarck | −2.8 | 2.4 | 14.7 | 31.1 | 41.7 | 51.8 | 57.3 | 54.9 | 43.7 | 33.2 | 18.3 | 5.2 | 29.3 |
| OH | Cincinnati | 24.3 | 25.8 | 33.5 | 44.6 | 53.6 | 62.5 | 65.8 | 64.1 | 57.0 | 46.7 | 36.2 | 27.1 | 45.1 |
| | Cleveland | 20.3 | 20.8 | 28.1 | 38.5 | 48.1 | 57.5 | 61.2 | 59.6 | 53.5 | 43.9 | 34.4 | 24.1 | 40.8 |
| | Columbus | 20.4 | 21.4 | 29.1 | 39.5 | 49.3 | 58.9 | 62.4 | 60.1 | 52.7 | 42.0 | 32.4 | 22.7 | 40.9 |
| OK | Oklahoma City | 26.0 | 30.0 | 36.5 | 49.1 | 57.9 | 66.6 | 70.4 | 69.6 | 61.3 | 50.6 | 37.4 | 29.2 | 48.7 |
| OR | Portland | 32.5 | 35.5 | 37.0 | 40.8 | 46.3 | 51.8 | 55.2 | 55.0 | 50.5 | 44.7 | 38.5 | 35.3 | 43.6 |
| PA | Philadelphia | 24.4 | 25.5 | 32.5 | 42.3 | 52.3 | 61.6 | 66.7 | 64.7 | 57.8 | 46.9 | 36.9 | 27.2 | 44.9 |
| | Pittsburgh | 20.8 | 21.3 | 29.0 | 39.4 | 48.7 | 57.7 | 61.3 | 59.4 | 52.7 | 42.4 | 33.3 | 23.6 | 40.8 |
| RI | Providence | 20.6 | 21.2 | 29.0 | 37.8 | 46.9 | 56.5 | 63.0 | 61.0 | 53.6 | 43.4 | 34.6 | 23.4 | 40.9 |
| SC | Columbia | 33.9 | 35.5 | 41.9 | 51.3 | 59.6 | 67.2 | 70.3 | 69.4 | 63.5 | 51.3 | 40.6 | 34.1 | 51.5 |
| SD | Sioux Falls | 3.7 | 9.0 | 20.2 | 34.4 | 45.7 | 56.3 | 61.5 | 59.8 | 48.7 | 37.6 | 22.7 | 10.4 | 34.2 |
| TN | Memphis | 31.6 | 34.4 | 41.1 | 52.3 | 60.6 | 68.5 | 71.5 | 70.1 | 62.8 | 51.1 | 40.3 | 33.7 | 51.5 |
| | Nashville | 29.0 | 31.0 | 38.1 | 48.8 | 57.3 | 65.7 | 69.0 | 67.7 | 60.5 | 48.6 | 37.7 | 31.1 | 48.7 |
| TX | Dallas-Fort Worth | 33.9 | 37.6 | 43.3 | 54.1 | 62.1 | 70.3 | 74.0 | 73.7 | 66.8 | 56.0 | 44.1 | 37.0 | 54.4 |
| | El Paso | 30.2 | 34.3 | 40.3 | 49.3 | 57.2 | 65.7 | 69.9 | 68.2 | 61.0 | 49.5 | 37.0 | 30.9 | 49.5 |
| | Houston | 41.5 | 44.6 | 49.8 | 59.3 | 65.6 | 70.9 | 72.8 | 72.4 | 68.2 | 58.3 | 49.1 | 43.4 | 58.0 |
| UT | Salt Lake City | 18.5 | 23.3 | 28.3 | 36.6 | 44.2 | 51.1 | 60.5 | 58.7 | 49.3 | 38.4 | 28.1 | 21.5 | 38.2 |
| VT | Burlington | 7.6 | 8.9 | 20.1 | 32.6 | 43.5 | 53.9 | 58.5 | 56.4 | 48.6 | 38.8 | 29.7 | 14.8 | 34.5 |
| VA | Norfolk | 32.2 | 32.7 | 38.9 | 47.9 | 57.2 | 65.5 | 69.9 | 68.9 | 63.9 | 53.3 | 42.6 | 34.0 | 50.6 |
| | Richmond | 27.6 | 28.8 | 35.5 | 45.2 | 54.5 | 62.9 | 67.5 | 65.9 | 59.0 | 47.4 | 37.3 | 28.8 | 46.7 |
| WA | Seattle-Tacoma | 33.0 | 36.0 | 36.6 | 40.3 | 45.6 | 50.6 | 53.8 | 53.7 | 50.4 | 44.9 | 38.8 | 35.5 | 43.3 |
| | Spokane | 19.6 | 25.3 | 28.8 | 35.2 | 42.8 | 49.4 | 55.1 | 54.0 | 46.7 | 37.5 | 29.2 | 24.0 | 37.3 |
| WV | Charleston | 25.3 | 26.8 | 33.8 | 43.8 | 52.3 | 60.6 | 64.3 | 62.8 | 55.9 | 44.8 | 35.0 | 27.2 | 44.4 |
| WI | Milwaukee | 11.4 | 14.6 | 23.4 | 34.7 | 43.3 | 53.6 | 59.3 | 58.7 | 50.7 | 40.6 | 28.5 | 16.8 | 36.3 |
| WY | Cheyenne | 14.9 | 17.3 | 19.6 | 30.0 | 39.7 | 48.1 | 54.5 | 53.2 | 43.5 | 33.9 | 23.5 | 18.1 | 33.0 |
| PR | San Juan | 68.8 | 68.4 | 68.9 | 70.6 | 72.8 | 74.0 | 74.8 | 75.1 | 74.6 | 73.7 | 72.3 | 70.5 | 72.0 |

Source: U.S. National Oceanic and Atmospheric Administration, *Comparative Climatic Data*.
* City office data.

## TABLE 3-7 Normal Daily Maximum Temperature

(In Fahrenheit degrees. Airport data, except as noted. Based on standard 30-year period, 1941 through 1970.)

| | State and Station | Jan. | Feb. | Mar. | Apr. | May | June | July | Aug. | Sept. | Oct. | Nov. | Dec. | Annual Average |
|---|---|---|---|---|---|---|---|---|---|---|---|---|---|---|
| AL | Mobile | 61.1 | 64.1 | 69.5 | 78.0 | 85.0 | 89.8 | 90.5 | 90.6 | 86.5 | 79.7 | 69.5 | 63.0 | 77.3 |
| AK | Juneau | 29.1 | 33.9 | 38.2 | 46.5 | 55.4 | 62.0 | 63.6 | 62.3 | 56.1 | 47.2 | 37.3 | 32.0 | 47.0 |
| AZ | Phoenix | 64.8 | 69.3 | 74.5 | 83.6 | 92.9 | 101.5 | 104.8 | 102.2 | 98.4 | 87.6 | 74.7 | 66.4 | 85.1 |
| AR | Little Rock | 50.1 | 53.8 | 61.8 | 73.5 | 81.4 | 89.3 | 92.6 | 92.6 | 85.8 | 76.0 | 62.4 | 52.1 | 72.6 |
| CA | Los Angeles | 63.5 | 64.1 | 64.3 | 65.9 | 68.4 | 70.3 | 74.8 | 75.8 | 75.7 | 72.9 | 69.6 | 66.5 | 69.2 |
| | Sacramento | 53.0 | 59.1 | 64.1 | 71.3 | 78.8 | 86.4 | 92.9 | 91.3 | 87.7 | 77.1 | 63.6 | 53.3 | 73.2 |
| | San Francisco | 55.3 | 58.6 | 61.0 | 63.5 | 66.6 | 70.2 | 70.9 | 71.6 | 73.6 | 70.3 | 63.3 | 56.5 | 65.1 |
| CO | Denver | 43.5 | 46.2 | 50.1 | 61.0 | 70.3 | 80.1 | 87.4 | 85.8 | 77.7 | 66.8 | 53.3 | 46.2 | 64.0 |
| CN | Hartford | 33.4 | 35.7 | 44.6 | 58.9 | 70.3 | 79.5 | 84.1 | 81.9 | 74.5 | 64.3 | 50.6 | 36.8 | 59.6 |
| DE | Wilmington | 40.2 | 42.2 | 51.1 | 63.0 | 73.1 | 81.6 | 85.5 | 83.9 | 78.2 | 67.8 | 55.2 | 43.0 | 63.7 |
| DC | Washington | 43.5 | 46.0 | 55.0 | 67.1 | 76.6 | 84.6 | 88.2 | 86.6 | 80.2 | 69.8 | 57.2 | 45.2 | 66.7 |
| FL | Jacksonville | 64.6 | 66.9 | 72.2 | 79.0 | 84.6 | 88.3 | 90.0 | 89.7 | 86.0 | 79.2 | 71.4 | 65.6 | 78.1 |
| | Miami | 75.6 | 76.6 | 79.5 | 82.7 | 85.3 | 88.0 | 89.1 | 89.9 | 88.3 | 84.6 | 79.9 | 76.6 | 83.0 |
| GA | Atlanta | 51.4 | 54.5 | 61.1 | 71.4 | 79.0 | 84.6 | 86.5 | 86.4 | 81.2 | 72.5 | 61.9 | 52.7 | 70.3 |
| HI | Honolulu | 79.3 | 79.2 | 79.7 | 81.4 | 83.6 | 85.6 | 86.8 | 87.4 | 87.4 | 85.8 | 83.2 | 80.3 | 83.3 |
| ID | Boise | 36.5 | 43.8 | 51.6 | 61.4 | 70.6 | 78.3 | 90.5 | 87.6 | 77.6 | 64.7 | 48.9 | 39.1 | 62.6 |
| IL | Chicago | 31.5 | 34.6 | 44.6 | 59.3 | 70.3 | 80.6 | 84.4 | 83.3 | 75.8 | 65.1 | 48.1 | 35.3 | 59.4 |
| | Peoria | 31.9 | 36.0 | 46.5 | 61.7 | 72.3 | 81.7 | 85.5 | 84.0 | 76.4 | 65.9 | 48.7 | 35.7 | 60.5 |
| IN | Indianapolis | 36.0 | 39.3 | 49.0 | 62.8 | 72.9 | 82.3 | 85.4 | 84.0 | 77.7 | 67.0 | 50.5 | 38.7 | 62.2 |
| IO | Des Moines | 27.5 | 32.5 | 42.5 | 59.7 | 70.9 | 79.8 | 84.9 | 83.2 | 74.6 | 64.9 | 46.4 | 32.8 | 58.3 |
| KS | Wichita | 41.4 | 47.1 | 55.0 | 68.1 | 77.1 | 86.5 | 91.7 | 91.0 | 81.9 | 71.3 | 55.8 | 44.3 | 67.6 |
| KY | Louisville | 42.0 | 45.0 | 54.0 | 66.9 | 75.6 | 83.7 | 87.3 | 86.8 | 80.5 | 70.3 | 54.9 | 44.1 | 65.9 |
| LA | New Orleans | 62.3 | 65.1 | 70.4 | 78.4 | 84.9 | 89.6 | 90.4 | 90.4 | 86.6 | 79.0 | 70.3 | 64.2 | 77.7 |
| ME | Portland | 31.2 | 33.3 | 40.8 | 52.8 | 63.6 | 73.2 | 79.1 | 77.6 | 69.9 | 60.2 | 47.5 | 34.9 | 55.3 |
| MD | Baltimore | 41.9 | 43.9 | 53.0 | 65.2 | 74.8 | 83.2 | 86.7 | 85.1 | 79.0 | 68.3 | 56.1 | 43.9 | 65.1 |
| MA | Boston | 35.9 | 37.5 | 44.6 | 56.3 | 67.1 | 76.6 | 81.4 | 79.3 | 72.2 | 63.2 | 51.7 | 39.3 | 58.7 |
| MI | Detroit | 31.7 | 33.7 | 43.1 | 57.6 | 68.5 | 79.1 | 83.1 | 81.6 | 74.2 | 63.4 | 47.7 | 35.4 | 58.3 |
| | Sault Ste. Marie | 22.0 | 23.7 | 32.5 | 47.2 | 59.4 | 70.0 | 75.1 | 73.4 | 64.5 | 54.8 | 39.0 | 26.8 | 49.0 |
| MN | Duluth | 17.6 | 22.1 | 32.6 | 47.8 | 60.0 | 69.7 | 76.4 | 74.4 | 64.0 | 54.3 | 35.3 | 22.5 | 48.1 |
| | Minneapolis-St. Paul | 21.2 | 25.9 | 36.9 | 55.5 | 67.9 | 77.1 | 82.4 | 80.8 | 70.7 | 60.7 | 40.6 | 26.6 | 53.8 |
| MS | Jackson | 58.4 | 61.7 | 68.7 | 78.2 | 85.0 | 91.0 | 92.7 | 92.6 | 88.0 | 80.1 | 68.5 | 60.5 | 77.1 |
| MO | Kansas City | 36.2 | 41.9 | 50.5 | 64.8 | 74.3 | 82.6 | 88.0 | 86.7 | 78.8 | 68.9 | 52.7 | 40.4 | 63.7 |
| | St. Louis | 39.9 | 44.2 | 53.0 | 67.0 | 76.0 | 84.9 | 88.4 | 87.2 | 80.1 | 69.8 | 54.1 | 42.7 | 65.6 |
| MT | Great Falls | 29.3 | 35.9 | 40.4 | 54.5 | 65.0 | 72.1 | 83.7 | 81.8 | 70.0 | 59.4 | 43.4 | 34.7 | 55.9 |
| NE | Omaha | 32.7 | 38.5 | 47.7 | 64.4 | 74.4 | 83.1 | 88.6 | 87.2 | 78.6 | 69.1 | 50.9 | 37.8 | 62.8 |
| NV | Reno | 45.4 | 51.1 | 56.0 | 64.0 | 72.2 | 80.4 | 91.1 | 89.0 | 81.8 | 70.0 | 56.3 | 46.4 | 67.0 |
| NH | Concord | 31.3 | 33.8 | 42.4 | 56.7 | 68.6 | 77.7 | 82.6 | 80.1 | 72.4 | 62.3 | 47.9 | 34.6 | 57.5 |
| NJ | Atlantic City | 41.4 | 42.9 | 50.7 | 62.3 | 72.4 | 80.8 | 84.7 | 83.0 | 77.3 | 67.5 | 55.9 | 44.2 | 63.6 |
| NM | Albuquerque | 46.9 | 52.6 | 59.2 | 70.1 | 79.9 | 89.5 | 92.2 | 89.7 | 83.4 | 71.7 | 57.1 | 47.5 | 70.0 |
| NY | Albany | 30.4 | 32.7 | 42.6 | 58.0 | 69.7 | 79.4 | 83.9 | 81.4 | 73.7 | 62.8 | 48.1 | 34.1 | 58.1 |
| | Buffalo | 29.8 | 31.0 | 39.0 | 53.3 | 64.3 | 75.1 | 79.5 | 77.6 | 70.8 | 60.2 | 46.1 | 33.6 | 55.0 |
| | New York* | 38.5 | 40.2 | 48.4 | 60.7 | 71.4 | 80.5 | 85.2 | 83.4 | 76.8 | 66.8 | 54.0 | 41.4 | 62.3 |
| NC | Charlotte | 52.1 | 54.9 | 62.2 | 72.7 | 80.2 | 86.4 | 88.3 | 87.4 | 82.0 | 73.1 | 62.4 | 52.5 | 71.2 |
| | Raleigh | 51.0 | 53.2 | 61.0 | 72.2 | 79.4 | 85.6 | 87.7 | 86.8 | 81.5 | 72.4 | 62.1 | 51.9 | 70.4 |
| ND | Bismarck | 19.1 | 24.5 | 35.4 | 54.8 | 67.1 | 75.8 | 84.3 | 83.5 | 71.3 | 60.3 | 39.4 | 26.0 | 53.5 |
| OH | Cincinnati | 39.8 | 42.9 | 52.2 | 65.5 | 75.2 | 83.6 | 86.6 | 86.0 | 79.8 | 68.8 | 53.0 | 41.8 | 64.6 |
| | Cleveland | 33.4 | 35.0 | 44.1 | 58.0 | 68.4 | 78.2 | 81.6 | 80.4 | 74.2 | 63.6 | 48.8 | 36.4 | 58.5 |
| | Columbus | 36.4 | 39.2 | 49.3 | 62.8 | 72.9 | 81.9 | 84.8 | 83.7 | 77.6 | 66.4 | 50.9 | 38.7 | 62.1 |
| OK | Oklahoma City | 47.6 | 52.6 | 59.8 | 71.6 | 78.7 | 87.0 | 92.6 | 92.5 | 84.7 | 74.2 | 60.9 | 50.7 | 71.1 |
| OR | Portland | 43.6 | 50.1 | 54.3 | 60.3 | 67.0 | 72.1 | 79.0 | 78.1 | 73.9 | 62.9 | 52.1 | 46.0 | 61.6 |
| PA | Philadelphia | 40.1 | 42.2 | 51.2 | 63.5 | 74.1 | 83.0 | 86.8 | 84.8 | 78.4 | 67.9 | 55.5 | 43.2 | 64.2 |
| | Pittsburgh | 35.3 | 37.3 | 47.2 | 60.9 | 70.8 | 79.5 | 82.5 | 80.9 | 74.9 | 63.9 | 49.3 | 37.3 | 60.0 |
| RI | Providence | 36.2 | 37.6 | 44.7 | 56.7 | 66.8 | 76.3 | 81.1 | 79.8 | 73.1 | 63.9 | 52.0 | 39.6 | 59.0 |
| SC | Columbia | 56.9 | 59.7 | 66.5 | 76.9 | 84.5 | 90.3 | 92.0 | 91.0 | 85.4 | 77.1 | 66.9 | 57.9 | 75.4 |
| SD | Sioux Falls | 24.6 | 29.7 | 39.7 | 57.8 | 69.7 | 78.9 | 85.1 | 83.8 | 73.0 | 62.7 | 43.5 | 29.6 | 56.5 |
| TN | Memphis | 49.4 | 53.1 | 60.8 | 72.7 | 81.2 | 88.7 | 91.6 | 90.6 | 84.3 | 74.9 | 61.5 | 51.7 | 71.7 |
| | Nashville | 47.6 | 50.9 | 59.2 | 71.3 | 79.8 | 87.5 | 90.2 | 89.2 | 83.5 | 73.2 | 59.0 | 49.6 | 70.1 |
| TX | Dallas-Fort Worth | 55.7 | 59.8 | 66.6 | 76.3 | 82.8 | 90.8 | 95.5 | 96.1 | 88.5 | 79.2 | 67.5 | 58.7 | 76.5 |
| | El Paso | 57.0 | 62.5 | 68.9 | 78.5 | 87.2 | 94.9 | 94.6 | 92.8 | 87.4 | 78.5 | 66.1 | 57.8 | 77.2 |
| | Houston | 62.6 | 66.0 | 71.8 | 79.4 | 85.9 | 91.3 | 93.8 | 94.3 | 90.1 | 83.5 | 73.0 | 65.8 | 79.8 |
| UT | Salt Lake City | 37.4 | 43.4 | 50.8 | 61.8 | 72.4 | 81.3 | 92.8 | 90.2 | 80.3 | 66.4 | 50.0 | 39.0 | 63.8 |
| VT | Burlington | 25.9 | 28.2 | 38.0 | 53.3 | 66.1 | 76.5 | 81.0 | 78.3 | 70.0 | 58.7 | 44.3 | 30.3 | 54.2 |
| VA | Norfolk | 48.8 | 50.0 | 57.3 | 67.7 | 76.2 | 83.5 | 86.6 | 84.9 | 79.6 | 70.1 | 60.5 | 50.6 | 68.0 |
| | Richmond | 47.4 | 49.9 | 58.2 | 70.3 | 78.4 | 85.4 | 88.2 | 86.6 | 80.9 | 71.2 | 60.6 | 49.1 | 68.8 |
| WA | Seattle-Tacoma | 43.4 | 48.5 | 51.5 | 57.0 | 64.1 | 69.0 | 75.1 | 73.8 | 68.7 | 59.4 | 50.4 | 45.4 | 58.8 |
| | Spokane | 31.1 | 39.0 | 46.2 | 57.0 | 66.5 | 73.6 | 84.3 | 81.9 | 72.5 | 58.1 | 41.8 | 33.9 | 57.2 |
| WV | Charleston | 43.6 | 46.2 | 55.2 | 67.9 | 76.6 | 83.4 | 85.6 | 84.4 | 79.0 | 69.1 | 55.8 | 45.2 | 66.0 |
| WI | Milwaukee | 27.3 | 30.3 | 39.4 | 54.6 | 65.0 | 75.3 | 80.4 | 79.7 | 71.5 | 61.4 | 44.4 | 31.5 | 55.1 |
| WY | Cheyenne | 38.2 | 40.7 | 43.5 | 55.4 | 65.1 | 74.4 | 83.7 | 81.9 | 72.8 | 61.8 | 47.5 | 40.3 | 58.8 |
| PR | San Juan | 81.9 | 82.1 | 83.6 | 84.4 | 85.6 | 87.0 | 87.0 | 87.5 | 87.6 | 87.4 | 85.0 | 83.1 | 85.2 |

Source: U.S. National Oceanic and Atmospheric Administration, *Comparative Climatic Data.*
* City office data.

## TABLE 3–8  Highest Temperature of Record

(In Fahrenheit degrees. Airport data, except as noted. For period of record through 1975.)

| | State and Station | Length of Record (Years) | Jan. | Feb. | Mar. | Apr. | May | June | July | Aug. | Sept. | Oct. | Nov. | Dec. | Annual |
|---|---|---|---|---|---|---|---|---|---|---|---|---|---|---|---|
| AL | Mobile | 14 | 79 | 81 | 89 | 91 | 99 | 101 | 100 | 102 | 98 | 93 | 87 | 81 | 102 |
| AK | Juneau | 32 | 57 | 50 | 55 | 71 | 82 | 86 | 90 | 83 | 72 | 61 | 56 | 54 | 90 |
| AZ | Phoenix | 15 | 88 | 89 | 95 | 101 | 110 | 116 | 115 | 116 | 110 | 103 | 93 | 82 | 116 |
| AR | Little Rock | 16 | 81 | 83 | 91 | 90 | 98 | 102 | 105 | 108 | 102 | 97 | 85 | 79 | 108 |
| CA | Los Angeles | 17 | 87 | 92 | 88 | 95 | 96 | 92 | 92 | 91 | 110 | 106 | 101 | 88 | 110 |
| | Sacramento | 25 | 69 | 76 | 86 | 92 | 102 | 115 | 114 | 108 | 108 | 101 | 87 | 72 | 115 |
| | San Francisco | 16 | 71 | 72 | 79 | 85 | 94 | 106 | 98 | 98 | 103 | 95 | 85 | 72 | 106 |
| CO | Denver | 16 | 69 | 76 | 84 | 84 | 93 | 98 | 103 | 100 | 97 | 87 | 78 | 73 | 103 |
| CN | Hartford | 16 | 65 | 59 | 77 | 94 | 96 | 100 | 102 | 101 | 96 | 91 | 81 | 65 | 102 |
| DE | Wilmington | 28 | 75 | 74 | 86 | 91 | 95 | 99 | 102 | 101 | 100 | 91 | 85 | 72 | 102 |
| DC | Washington | 15 | 76 | 77 | 86 | 91 | 97 | 100 | 101 | 99 | 96 | 91 | 86 | 74 | 101 |
| FL | Jacksonville | 34 | 85 | 88 | 91 | 95 | 100 | 103 | 105 | 102 | 100 | 96 | 88 | 84 | 105 |
| | Miami | 11 | 86 | 88 | 90 | 96 | 93 | 94 | 96 | 96 | 93 | 90 | 87 | 85 | 96 |
| GA | Atlanta | 15 | 77 | 79 | 85 | 88 | 93 | 98 | 98 | 98 | 96 | 88 | 84 | 77 | 98 |
| HI | Honolulu | 6 | 85 | 85 | 87 | 87 | 88 | 90 | 90 | 91 | 92 | 91 | 89 | 85 | 92 |
| ID | Boise | 36 | 63 | 67 | 78 | 92 | 98 | 109 | 111 | 110 | 102 | 91 | 73 | 65 | 111 |
| IL | Chicago | 12 | 65 | 65 | 80 | 88 | 94 | 101 | 99 | 98 | 95 | 94 | 78 | 71 | 101 |
| | Peoria | 16 | 66 | 70 | 81 | 87 | 92 | 100 | 102 | 99 | 94 | 90 | 77 | 71 | 102 |
| IN | Indianapolis | 17 | 70 | 74 | 80 | 89 | 93 | 96 | 99 | 97 | 96 | 88 | 78 | 70 | 99 |
| IO | Des Moines | 15 | 62 | 73 | 83 | 90 | 98 | 99 | 104 | 100 | 95 | 95 | 76 | 67 | 104 |
| KS | Wichita | 23 | 75 | 82 | 89 | 96 | 100 | 106 | 113 | 110 | 103 | 95 | 80 | 83 | 113 |
| KY | Louisville | 15 | 73 | 77 | 83 | 88 | 91 | 97 | 101 | 101 | 96 | 89 | 82 | 73 | 101 |
| LA | New Orleans | 29 | 83 | 85 | 87 | 91 | 96 | 100 | 99 | 100 | 97 | 92 | 86 | 84 | 100 |
| ME | Portland | 35 | 64 | 64 | 86 | 85 | 92 | 97 | 98 | 103 | 95 | 88 | 74 | 62 | 103 |
| MD | Baltimore | 25 | 75 | 76 | 85 | 94 | 98 | 100 | 102 | 102 | 99 | 92 | 83 | 74 | 102 |
| MA | Boston | 11 | 63 | 58 | 70 | 85 | 93 | 97 | 98 | 102 | 95 | 86 | 77 | 70 | 102 |
| MI | Detroit | 41 | 67 | 68 | 82 | 87 | 93 | 104 | 105 | 101 | 100 | 92 | 81 | 66 | 105 |
| | Sault Ste. Marie | 35 | 45 | 45 | 75 | 82 | 89 | 92 | 97 | 98 | 93 | 80 | 66 | 59 | 98 |
| MN | Duluth | 15 | 47 | 44 | 65 | 83 | 90 | 92 | 94 | 95 | 90 | 84 | 69 | 55 | 95 |
| | Minneapolis-St. Paul | 16 | 46 | 54 | 83 | 91 | 95 | 99 | 101 | 98 | 95 | 87 | 74 | 62 | 101 |
| MS | Jackson | 12 | 82 | 82 | 88 | 92 | 99 | 103 | 102 | 99 | 98 | 91 | 88 | 81 | 103 |
| MO | Kansas City | 3 | 61 | 66 | 82 | 85 | 91 | 98 | 107 | 103 | 98 | 89 | 73 | 67 | 107 |
| | St. Louis | 15 | 76 | 85 | 88 | 92 | 92 | 98 | 106 | 105 | 100 | 94 | 81 | 76 | 106 |
| MT | Great Falls | 15 | 61 | 64 | 72 | 83 | 90 | 97 | 105 | 106 | 98 | 85 | 76 | 61 | 106 |
| NE | Omaha | 12 | 64 | 78 | 89 | 93 | 97 | 103 | 110 | 107 | 103 | 95 | 80 | 67 | 110 |
| NV | Reno | 12 | 70 | 74 | 83 | 88 | 95 | 100 | 103 | 103 | 96 | 91 | 76 | 70 | 103 |
| NH | Concord | 10 | 60 | 57 | 68 | 88 | 96 | 97 | 102 | 101 | 93 | 84 | 78 | 61 | 102 |
| NJ | Atlantic City | 11 | 78 | 70 | 81 | 94 | 99 | 106 | 104 | 97 | 93 | 87 | 81 | 72 | 106 |
| NM | Albuquerque | 16 | 69 | 75 | 85 | 89 | 95 | 105 | 104 | 99 | 95 | 87 | 77 | 68 | 105 |
| NY | Albany | 10 | 62 | 57 | 77 | 88 | 92 | 98 | 98 | 97 | 93 | 84 | 77 | 65 | 98 |
| | Buffalo | 15 | 61 | 61 | 78 | 83 | 88 | 94 | 94 | 93 | 90 | 82 | 80 | 66 | 94 |
| | New York* | 107 | 72 | 75 | 86 | 92 | 99 | 101 | 106 | 104 | 102 | 94 | 84 | 70 | 106 |
| NC | Charlotte | 15 | 77 | 78 | 86 | 91 | 95 | 99 | 99 | 100 | 94 | 87 | 85 | 77 | 100 |
| | Raleigh | 11 | 77 | 79 | 89 | 93 | 92 | 95 | 97 | 98 | 93 | 89 | 84 | 77 | 98 |
| ND | Bismarck | 16 | 54 | 61 | 80 | 91 | 95 | 100 | 109 | 107 | 100 | 95 | 75 | 62 | 109 |
| OH | Cincinnati | 60 | 77 | 77 | 88 | 90 | 95 | 102 | 109 | 103 | 101 | 92 | 83 | 71 | 109 |
| | Cleveland | 15 | 68 | 69 | 80 | 85 | 91 | 95 | 98 | 95 | 93 | 86 | 79 | 69 | 98 |
| | Columbus | 16 | 68 | 72 | 80 | 88 | 93 | 96 | 97 | 98 | 96 | 86 | 79 | 72 | 98 |
| OK | Oklahoma City | 10 | 79 | 81 | 93 | 100 | 96 | 102 | 108 | 106 | 101 | 96 | 84 | 80 | 108 |
| OR | Portland | 35 | 62 | 70 | 80 | 87 | 92 | 100 | 107 | 104 | 101 | 90 | 73 | 64 | 107 |
| PA | Philadelphia | 16 | 69 | 69 | 80 | 92 | 96 | 100 | 104 | 99 | 97 | 88 | 81 | 71 | 104 |
| | Pittsburgh | 16 | 68 | 66 | 80 | 87 | 91 | 96 | 98 | 96 | 95 | 87 | 82 | 72 | 98 |
| RI | Providence | 12 | 66 | 59 | 73 | 90 | 94 | 95 | 97 | 104 | 93 | 85 | 78 | 69 | 104 |
| SC | Columbia | 9 | 84 | 81 | 91 | 94 | 96 | 104 | 103 | 106 | 97 | 90 | 89 | 83 | 106 |
| SD | Sioux Falls | 12 | 57 | 59 | 87 | 92 | 100 | 101 | 106 | 108 | 101 | 94 | 76 | 60 | 108 |
| TN | Memphis | 34 | 78 | 81 | 85 | 91 | 97 | 104 | 106 | 105 | 103 | 95 | 85 | 79 | 106 |
| | Nashville | 10 | 78 | 79 | 86 | 88 | 91 | 98 | 103 | 99 | 95 | 90 | 84 | 74 | 103 |
| TX | Dallas-Fort Worth | 12 | 88 | 87 | 96 | 95 | 96 | 105 | 106 | 108 | 102 | 96 | 88 | 84 | 108 |
| | El Paso | 16 | 80 | 83 | 88 | 98 | 101 | 108 | 106 | 105 | 100 | 96 | 84 | 80 | 108 |
| | Houston | 6 | 84 | 82 | 90 | 89 | 93 | 99 | 101 | 101 | 97 | 93 | 88 | 83 | 101 |
| UT | Salt Lake City | 16 | 61 | 69 | 78 | 85 | 92 | 104 | 107 | 103 | 96 | 89 | 75 | 67 | 107 |
| VT | Burlington | 11 | 56 | 51 | 67 | 84 | 91 | 93 | 98 | 99 | 90 | 80 | 71 | 62 | 99 |
| VA | Norfolk | 27 | 78 | 79 | 85 | 97 | 97 | 101 | 103 | 99 | 98 | 95 | 86 | 79 | 103 |
| | Richmond | 46 | 80 | 83 | 93 | 96 | 100 | 104 | 104 | 102 | 103 | 99 | 86 | 80 | 104 |
| WA | Seattle-Tacoma | 16 | 61 | 70 | 71 | 77 | 93 | 94 | 97 | 99 | 93 | 81 | 72 | 60 | 99 |
| | Spokane | 16 | 59 | 60 | 71 | 80 | 92 | 100 | 103 | 108 | 93 | 85 | 67 | 53 | 108 |
| WV | Charleston | 28 | 79 | 77 | 87 | 91 | 93 | 98 | 102 | 100 | 102 | 92 | 85 | 80 | 102 |
| WI | Milwaukee | 16 | 57 | 51 | 77 | 85 | 92 | 95 | 98 | 99 | 94 | 89 | 74 | 63 | 99 |
| WY | Cheyenne | 16 | 62 | 71 | 73 | 82 | 90 | 94 | 98 | 96 | 93 | 83 | 70 | 66 | 98 |
| PR | San Juan | 21 | 90 | 92 | 93 | 93 | 94 | 96 | 93 | 96 | 94 | 95 | 92 | 90 | 96 |

Source: U.S. National Oceanic and Atmospheric Administration, *Comparative Climatic Data.*
* City office data.

# TABLE 3–9 Lowest Temperature of Record

(In Fahrenheit degrees. Airport data, except as noted. For period of record through 1975.)

| | State and Station | Length of Record (Years) | Jan. | Feb. | Mar. | Apr. | May | June | July | Aug. | Sept. | Oct. | Nov. | Dec. | Annual |
|---|---|---|---|---|---|---|---|---|---|---|---|---|---|---|---|
| AL | Mobile | 14 | 8 | 11 | 11 | 36 | 46 | 56 | 62 | 60 | 42 | 38 | 24 | 10 | 8 |
| AK | Juneau | 32 | −22 | −22 | −15 | 6 | 25 | 31 | 36 | 27 | 23 | 12 | −5 | −21 | −22 |
| AZ | Phoenix | 15 | 19 | 26 | 25 | 37 | 40 | 51 | 67 | 61 | 47 | 34 | 31 | 24 | 19 |
| AR | Little Rock | 16 | −4 | 10 | 17 | 28 | 40 | 46 | 54 | 52 | 38 | 31 | 17 | −1 | −4 |
| CA | Los Angeles | 17 | 30 | 37 | 39 | 43 | 45 | 50 | 55 | 58 | 55 | 43 | 38 | 32 | 30 |
| | Sacramento | 25 | 23 | 26 | 26 | 32 | 36 | 41 | 49 | 49 | 43 | 36 | 26 | 20 | 20 |
| | San Francisco | 16 | 29 | 35 | 31 | 38 | 40 | 45 | 48 | 49 | 45 | 39 | 35 | 24 | 24 |
| CO | Denver | 16 | −25 | −18 | −4 | −2 | 26 | 36 | 43 | 41 | 20 | 3 | −2 | −18 | −25 |
| CN | Hartford | 16 | −26 | −21 | −6 | 9 | 30 | 37 | 44 | 36 | 30 | 18 | 12 | −9 | −26 |
| DE | Wilmington | 28 | −4 | −4 | 9 | 22 | 32 | 41 | 50 | 46 | 36 | 24 | 14 | 3 | −4 |
| DC | Washington | 15 | 3 | 4 | 16 | 27 | 36 | 47 | 56 | 51 | 39 | 29 | 20 | 10 | 3 |
| FL | Jacksonville | 34 | 19 | 19 | 25 | 35 | 45 | 56 | 61 | 64 | 50 | 38 | 21 | 12 | 12 |
| | Miami | 11 | 35 | 36 | 37 | 46 | 61 | 67 | 70 | 70· | 70 | 56 | 40 | 34 | 34 |
| GA | Atlanta | 15 | −3 | 8 | 20 | 26 | 37 | 48 | 53 | 56 | 36 | 29 | 14 | 1 | −3 |
| HI | Honolulu | 6 | 53 | 54 | 58 | 59 | 63 | 65 | 67 | 67 | 66 | 64 | 58 | 54 | 53 |
| ID | Boise | 36 | −17 | −10 | 6 | 19 | 26 | 33 | 41 | 37· | 23 | 11 | −3 | −23 | −23 |
| IL | Chicago | 12 | −16 | −9 | 5 | 16 | 29 | 41 | 46 | 43 | 34 | 26 | 7 | −10 | −16 |
| | Peoria | 16 | −20 | −14 | −10 | 17 | 25 | 40 | 47 | 44 | 31 | 19 | 4 | −18 | −20 |
| IN | Indianapolis | 17 | −20 | −10 | −6 | 18 | 28 | 42 | 48 | 41 | 34 | 20 | 4 | −14 | −20 |
| IO | Des Moines | 15 | −24 | −18 | −22 | 9 | 30 | 42 | 47 | 45 | 31 | 14 | −3 | −16 | −24 |
| KS | Wichita | 23 | −12 | −6 | −2 | 15 | 32 | 43 | 51 | 48 | 35 | 23 | 1 | −5 | −12 |
| KY | Louisville | 15 | −20 | −4 | 15 | 24 | 31 | 42 | 50 | 49 | 37 | 25 | 10 | −3 | −20 |
| LA | New Orleans | 29 | 14 | 19 | 26 | 32 | 41 | 55 | 60 | 60 | 42 | 35 | 24 | 17 | 14 |
| ME | Portland | 35 | −26 | −39 | −21 | 8 | 23 | 33 | 40 | 33 | 23 | 18 | 5 | −21 | −39 |
| MD | Baltimore | 25 | −7 | −1 | 6 | 20 | 32 | 40 | 52 | 48 | 35 | 25 | 13 | 0 | −7 |
| MA | Boston | 11 | −4 | −3 | 6 | 22 | 37 | 46 | 54 | 47 | 38 | 30 | 17 | −3 | −4 |
| MI | Detroit | 41 | −13 | −16 | −1 | 14 | 30 | 38 | 42 | 43 | 32 | 24 | 5 | −5 | −16 |
| | Sault Ste. Marie | 35 | −28 | −28 | −24 | 1 | 18 | 28 | 36 | 32 | 25 | 16 | −5 | −20 | −28 |
| MN | Duluth | 15 | −39 | −32 | −28 | −5 | 17 | 27 | 36 | 33 | 23 | 12 | −23 | −29 | −39 |
| | Minneapolis-St. Paul | 16 | −34 | −28 | −32 | 2 | 18 | 37 | 43 | 39 | 26 | 15 | −17 | −24 | −34 |
| MS | Jackson | 12 | 7 | 11 | 18 | 30 | 38 | 49 | 51 | 55 | 35 | 30 | 17 | 14 | 7 |
| MO | Kansas City | 3 | −13 | −4 | −1 | 12 | 34 | 47 | 52 | 50 | 39 | 28 | 7 | −7 | −13 |
| | St. Louis | 15 | −11 | −2 | 3 | 22 | 31 | 43 | 51 | 47 | 36 | 25 | 1 | −6 | −11 |
| MT | Great Falls | 15 | −37 | −24 | −19 | −6 | 19 | 32 | 40 | 37 | 23 | −3 | −15 | −43 | −43 |
| NE | Omaha | 12 | −22 | −19 | −1 | 5 | 31 | 40 | 44 | 43 | 31 | 13 | −9 | −13 | −22 |
| NV | Reno | 12 | −11 | 0 | 0 | 15 | 18 | 29 | 33 | 29 | 20 | 8 | 5 | −16 | −16 |
| NH | Concord | 10 | −29 | −27 | −16 | 8 | 21 | 30 | 35 | 29 | 22 | 10 | −1 | −18 | −29 |
| NJ | Atlantic City | 11 | −8 | −7 | 7 | 12 | 25 | 37 | 46 | 40 | 32 | 23 | 11 | 0 | −8 |
| NM | Albuquerque | 16 | −17 | 1 | 9 | 22 | 28 | 42 | 54 | 52 | 37 | 25 | 10 | 3 | −17 |
| NY | Albany | 10 | −28 | −21 | −10 | 10 | 26 | 36 | 43 | 37 | 28 | 16 | 5 | −22 | −28 |
| | Buffalo | 15 | −11 | −20 | −4 | 13 | 29 | 36 | 46 | 38 | 32 | 20 | 9 | −4 | −20 |
| | New York* | 107 | −6 | −15 | 3 | 12 | 32 | 44 | 52 | 50 | 39 | 28 | 5 | −13 | −15 |
| NC | Charlotte | 15 | 4 | 7 | 18 | 25 | 32 | 45 | 53 | 53 | 39 | 24 | 13 | 2 | 2 |
| | Raleigh | 11 | 0 | 5 | 17 | 23 | 33 | 43 | 48 | 46 | 39 | 24 | 11 | 9 | 0 |
| ND | Bismarck | 16 | −42 | −37 | −28 | −12 | 15 | 30 | 35 | 33 | 11 | 5 | −29 | −43 | −43 |
| OH | Cincinnati | 60 | −17 | −9 | 3 | 18 | 28 | 40 | 48 | 43 | 32 | 20 | 1 | −13 | −17 |
| | Cleveland | 15 | −19 | −15 | 4 | 10 | 25 | 31 | 41 | 41 | 34 | 22 | 13 | −4 | −19 |
| | Columbus | 16 | −15 | −11 | −2 | 18 | 25 | 35 | 43 | 39 | 31 | 20 | 11 | −10 | −15 |
| OK | Oklahoma City | 10 | −1 | 3 | 9 | 21 | 39 | 51 | 53 | 54 | 37 | 31 | 13 | 1 | −1 |
| OR | Portland | 35 | −2 | −3 | 19 | 29 | 29 | 39 | 43 | 44 | 34 | 26 | 13 | 6 | −3 |
| PA | Philadelphia | 16 | −5 | −4 | 9 | 24 | 28 | 44 | 51 | 45 | 35 | 25 | 17 | 3 | −5 |
| | Pittsburgh | 16 | −18 | −9 | −1 | 15 | 26 | 34 | 42 | 40 | 31 | 16 | 8 | −5 | −18 |
| RI | Providence | 12 | −5 | −5 | 1 | 19 | 32 | 41 | 49 | 40 | 34 | 21 | 14 | −4 | −5 |
| SC | Columbia | 9 | 5 | 5 | 18 | 29 | 36 | 46 | 59 | 53 | 40 | 25 | 12 | 15 | 5 |
| SD | Sioux Falls | 12 | −36 | −30 | −14 | 5 | 17 | 33 | 38 | 37 | 22 | 9 | −17 | −26 | −36 |
| TN | Memphis | 34 | −4 | −11 | 12 | 29 | 38 | 48 | 52 | 48 | 36 | 25 | 9 | −13 | −13 |
| | Nashville | 10 | −6 | −5 | 14 | 24 | 35 | 42 | 54 | 51 | 37 | 27 | 12 | 6 | −6 |
| TX | Dallas-Ft. Worth | 12 | 4 | 12 | 19 | 30 | 42 | 51 | 59 | 56 | 46 | 37 | 22 | 10 | 4 |
| | El Paso | 16 | −8 | 11 | 14 | 24 | 31 | 51 | 59 | 56 | 42 | 25 | 18 | 10 | −8 |
| | Houston | 6 | 19 | 22 | 25 | 31 | 46 | 52 | 62 | 62 | 48 | 39 | 24 | 21 | 19 |
| UT | Salt Lake City | 16 | −18 | −4 | 2 | 22 | 25 | 35 | 40 | 37 | 27 | 16 | 11 | −15 | −18 |
| VT | Burlington | 11 | −27 | −25 | −13 | 2 | 24 | 33 | 40 | 35 | 29 | 15 | 2 | −23 | −27 |
| VA | Norfolk | 27 | 8 | 8 | 20 | 28 | 36 | 45 | 56 | 52 | 45 | 29 | 20 | 14 | 8 |
| | Richmond | 46 | −12 | −10 | 11 | 26 | 31 | 40 | 51 | 46 | 35 | 21 | 10 | −1 | −12 |
| WA | Seattle-Tacoma | 16 | 12 | 18 | 23 | 29 | 33 | 41 | 45 | 45 | 35 | 30 | 22 | 6 | 6 |
| | Spokane | 16 | −19 | −12 | 1 | 17 | 26 | 34 | 38 | 35 | 25 | 13 | −2 | −25 | −25 |
| WV | Charleston | 28 | −12 | −6 | 7 | 19 | 26 | 33 | 46 | 41 | 34 | 17 | 6 | −2 | −12 |
| WI | Milwaukee | 16 | −24 | −15 | −10 | 13 | 21 | 36 | 40 | 44 | 28 | 21 | 6 | −15 | −24 |
| WY | Cheyenne | 16 | −27 | −24 | −11 | −8 | 21 | 34 | 39 | 36 | 22 | 7 | −8 | −24 | −27 |
| PR | San Juan | 21 | 61 | 62 | 60 | 64 | 66 | 69 | 69 | 70 | 69 | 67 | 66 | 63 | 60 |

Source: U.S. National Oceanic and Atmospheric Administration, *Comparative Climatic Data.*
* City office data.

63

## TABLE 3-10  Average Wind Speed

(In miles per hour. Airport data, except as noted. For period of record through 1975.)

| State and Station | | Length of Record (Years) | Jan. | Feb. | Mar. | Apr. | May | June | July | Aug. | Sept. | Oct. | Nov. | Dec. | Annual Average |
|---|---|---|---|---|---|---|---|---|---|---|---|---|---|---|---|
| AL | Mobile | 27 | 10.8 | 11.0 | 11.3 | 10.7 | 9.1 | 7.9 | 7.1 | 6.9 | 8.2 | 8.4 | 9.6 | 10.3 | 9.3 |
| AK | Juneau | 32 | 8.5 | 8.8 | 8.8 | 8.9 | 8.4 | 7.9 | 7.6 | 7.6 | 8.0 | 9.7 | 8.7 | 9.4 | 8.5 |
| AZ | Phoenix | 30 | 5.1 | 5.7 | 6.4 | 6.8 | 6.8 | 6.8 | 7.1 | 6.5 | 6.2 | 5.7 | 5.2 | 5.0 | 6.1 |
| AR | Little Rock | 33 | 8.9 | 9.4 | 10.1 | 9.6 | 8.1 | 7.6 | 7.0 | 6.6 | 7.0 | 7.0 | 8.2 | 8.5 | 8.2 |
| CA | Los Angeles | 27 | 6.7 | 7.3 | 8.0 | 8.4 | 8.2 | 7.8 | 7.6 | 7.5 | 7.1 | 6.8 | 6.6 | 6.6 | 7.4 |
| | Sacramento | 27 | 8.0 | 8.0 | 9.0 | 9.1 | 9.4 | 10.0 | 9.2 | 8.7 | 7.9 | 6.9 | 6.5 | 7.2 | 8.3 |
| | San Francisco | 48 | 7.1 | 8.5 | 10.3 | 12.1 | 13.1 | 13.9 | 13.6 | 12.8 | 11.0 | 9.2 | 7.2 | 6.8 | 10.5 |
| CO | Denver | 27 | 9.3 | 9.3 | 10.0 | 10.4 | 9.5 | 9.1 | 8.6 | 8.2 | 8.2 | 8.2 | 8.7 | 9.0 | 9.0 |
| CN | Hartford | 21 | 9.6 | 9.9 | 10.5 | 10.7 | 9.5 | 8.5 | 7.9 | 7.7 | 7.8 | 8.2 | 8.8 | 9.0 | 9.0 |
| DE | Wilmington | 27 | 9.7 | 10.5 | 11.2 | 10.5 | 9.0 | 8.4 | 7.7 | 7.4 | 7.8 | 8.1 | 9.1 | 9.3 | 9.1 |
| DC | Washington | 27 | 9.9 | 10.4 | 10.9 | 10.5 | 9.2 | 8.7 | 8.1 | 8.0 | 8.2 | 8.5 | 9.2 | 9.4 | 9.2 |
| FL | Jacksonville | 26 | 8.4 | 9.6 | 9.6 | 9.3 | 8.8 | 8.5 | 7.6 | 7.4 | 8.4 | 8.8 | 8.3 | 8.2 | 8.6 |
| | Miami | 26 | 9.3 | 10.0 | 10.3 | 10.6 | 9.4 | 8.2 | 7.8 | 7.7 | 8.2 | 9.3 | 9.4 | 9.0 | 9.1 |
| GA | Atlanta | 37 | 10.5 | 11.0 | 11.0 | 10.1 | 8.6 | 7.9 | 7.4 | 7.1 | 8.0 | 8.3 | 9.1 | 9.8 | 9.1 |
| HI | Honolulu | 26 | 10.0 | 10.8 | 11.4 | 12.1 | 12.2 | 12.9 | 13.6 | 13.6 | 11.7 | 10.9 | 11.2 | 11.1 | 11.8 |
| ID | Boise | 36 | 8.6 | 9.4 | 10.4 | 10.3 | 9.6 | 9.2 | 8.5 | 8.3 | 8.3 | 8.6 | 8.7 | 8.5 | 9.0 |
| IL | Chicago | 33 | 11.4 | 11.6 | 11.9 | 11.8 | 10.4 | 9.4 | 8.3 | 8.1 | 9.0 | 9.8 | 11.3 | 11.1 | 10.3 |
| | Peoria | 32 | 11.2 | 11.6 | 12.3 | 12.3 | 10.4 | 9.1 | 8.0 | 7.8 | 8.8 | 9.5 | 11.2 | 10.9 | 10.3 |
| IN | Indianapolis | 27 | 11.1 | 11.2 | 11.9 | 11.5 | 9.7 | 8.5 | 7.4 | 7.2 | 8.1 | 8.9 | 10.7 | 10.5 | 8.7 |
| IO | Des Moines | 26 | 11.7 | 11.8 | 13.1 | 13.4 | 11.6 | 10.5 | 9.0 | 8.8 | 9.6 | 10.6 | 11.7 | 11.5 | 11.1 |
| KS | Wichita | 22 | 12.5 | 13.1 | 14.5 | 14.7 | 13.1 | 12.6 | 11.3 | 11.3 | 11.7 | 12.3 | 12.4 | 12.2 | 12.6 |
| KY | Louisville | 28 | 9.6 | 9.8 | 10.4 | 10.1 | 8.1 | 7.4 | 6.7 | 6.4 | 6.8 | 7.1 | 9.0 | 9.3 | 8.4 |
| LA | New Orleans | 27 | 9.5 | 10.1 | 10.3 | 9.8 | 8.3 | 7.0 | 6.3 | 6.1 | 7.5 | 7.6 | 8.9 | 9.2 | 8.4 |
| ME | Portland | 35 | 9.2 | 9.6 | 10.1 | 10.0 | 9.2 | 8.2 | 7.7 | 7.6 | 7.8 | 8.5 | 8.8 | 9.0 | 8.8 |
| MD | Baltimore | 25 | 9.9 | 10.7 | 11.2 | 11.0 | 9.6 | 8.7 | 8.2 | 8.2 | 8.4 | 8.9 | 9.5 | 9.4 | 9.5 |
| MA | Boston | 18 | 14.2 | 14.2 | 14.0 | 13.4 | 12.2 | 11.3 | 10.8 | 10.8 | 11.4 | 12.2 | 13.1 | 13.8 | 12.6 |
| MI | Detroit | 41 | 11.6 | 11.5 | 11.5 | 11.1 | 9.9 | 9.1 | 8.3 | 8.1 | 8.9 | 9.5 | 11.3 | 11.3 | 10.2 |
| | Sault Ste. Marie | 34 | 10.1 | 10.0 | 10.5 | 10.8 | 10.2 | 8.9 | 8.3 | 8.1 | 9.0 | 9.6 | 10.0 | 10.0 | 9.6 |
| MN | Duluth | 26 | 12.1 | 11.9 | 12.1 | 13.3 | 12.3 | 10.8 | 9.9 | 9.8 | 10.7 | 11.5 | 12.2 | 11.4 | 11.5 |
| | Minneapolis-St. Paul | 37 | 10.4 | 10.6 | 11.3 | 12.4 | 11.4 | 10.5 | 9.3 | 9.1 | 9.9 | 10.5 | 11.0 | 10.3 | 10.6 |
| MS | Jackson | 12 | 9.1 | 9.0 | 9.6 | 9.1 | 7.3 | 6.4 | 6.1 | 5.8 | 6.7 | 6.5 | 7.9 | 8.7 | 7.7 |
| MO | Kansas City | 3 | 10.7 | 11.7 | 11.7 | 11.8 | 9.6 | 9.4 | 7.9 | 8.6 | 8.3 | 10.0 | 11.6 | 10.9 | 10.2 |
| | St. Louis | 26 | 10.3 | 10.8 | 11.8 | 11.4 | 9.4 | 8.6 | 7.3 | 6.9 | 7.4 | 7.9 | 8.5 | 9.2 | 9.5 |
| MT | Great Falls | 34 | 15.9 | 14.8 | 13.5 | 13.2 | 11.5 | 11.4 | 10.3 | 10.5 | 11.6 | 13.8 | 15.0 | 16.1 | 13.1 |
| NE | Omaha | 40 | 11.1 | 11.5 | 12.7 | 13.2 | 11.4 | 10.5 | 9.1 | 9.2 | 9.7 | 10.1 | 11.2 | 10.8 | 10.9 |
| NV | Reno | 33 | 6.0 | 6.1 | 7.6 | 8.0 | 7.6 | 7.2 | 6.6 | 6.2 | 5.4 | 5.3 | 5.3 | 5.1 | 6.4 |
| NH | Concord | 34 | 7.3 | 7.9 | 8.2 | 7.9 | 7.0 | 6.3 | 5.6 | 5.3 | 5.4 | 5.9 | 6.5 | 7.0 | 6.7 |
| NJ | Atlantic City | 17 | 11.8 | 12.2 | 12.5 | 12.3 | 10.8 | 9.7 | 9.1 | 8.7 | 9.3 | 9.7 | 11.1 | 11.3 | 10.7 |
| NM | Albuquerque | 36 | 8.0 | 8.8 | 10.0 | 10.9 | 10.5 | 10.0 | 9.1 | 8.1 | 8.5 | 8.3 | 7.8 | 7.7 | 9.0 |
| NY | Albany | 37 | 9.8 | 10.3 | 10.6 | 10.5 | 9.1 | 8.1 | 7.3 | 6.9 | 7.3 | 7.9 | 8.9 | 9.1 | 8.8 |
| | Buffalo | 36 | 14.5 | 14.1 | 13.8 | 13.0 | 11.8 | 11.3 | 10.6 | 10.1 | 10.6 | 11.4 | 13.0 | 13.4 | 12.3 |
| | New York* | 56 | 10.7 | 10.9 | 11.1 | 10.5 | 8.8 | 8.1 | 7.7 | 7.7 | 8.1 | 9.0 | 9.9 | 10.4 | 9.4 |
| NC | Charlotte | 26 | 8.0 | 8.5 | 9.0 | 9.1 | 7.6 | 6.9 | 6.6 | 6.5 | 6.8 | 7.1 | 7.3 | 7.4 | 7.6 |
| | Raleigh | 26 | 8.7 | 9.2 | 9.6 | 9.4 | 7.9 | 7.1 | 6.8 | 6.5 | 7.0 | 7.3 | 7.9 | 8.2 | 8.0 |
| ND | Bismarck | 36 | 10.1 | 10.1 | 11.3 | 12.6 | 12.2 | 11.0 | 9.6 | 9.9 | 10.4 | 10.2 | 10.4 | 9.6 | 10.6 |
| OH | Cincinnati | 43 | 8.3 | 8.4 | 9.0 | 8.4 | 6.7 | 6.4 | 5.2 | 5.1 | 5.4 | 6.1 | 7.7 | 7.9 | 7.1 |
| | Cleveland | 34 | 12.5 | 12.3 | 12.5 | 11.9 | 10.4 | 9.5 | 8.7 | 8.4 | 9.1 | 10.0 | 12.1 | 12.3 | 10.8 |
| | Columbus | 26 | 10.3 | 10.5 | 10.8 | 10.2 | 8.6 | 7.5 | 6.7 | 6.4 | 6.8 | 7.6 | 9.5 | 9.8 | 8.7 |
| OK | Oklahoma City | 27 | 13.3 | 13.7 | 15.1 | 15.1 | 13.3 | 12.7 | 11.3 | 10.8 | 11.5 | 12.2 | 12.7 | 12.8 | 12.9 |
| OR | Portland | 27 | 10.1 | 8.8 | 8.3 | 7.2 | 6.9 | 6.9 | 7.4 | 7.0 | 6.4 | 6.4 | 8.4 | 9.6 | 7.8 |
| PA | Philadelphia | 35 | 10.3 | 11.1 | 11.5 | 11.1 | 9.7 | 8.8 | 8.1 | 7.9 | 8.3 | 8.9 | 9.7 | 10.1 | 9.6 |
| | Pittsburgh | 23 | 10.7 | 11.0 | 11.1 | 10.8 | 9.3 | 8.2 | 7.5 | 7.3 | 7.7 | 8.4 | 10.1 | 10.5 | 9.4 |
| RI | Providence | 22 | 11.5 | 11.9 | 12.4 | 12.5 | 11.1 | 10.1 | 9.5 | 9.5 | 9.6 | 9.7 | 10.6 | 11.0 | 10.8 |
| SC | Columbia | 27 | 7.1 | 7.7 | 8.4 | 8.5 | 7.0 | 6.7 | 6.5 | 6.0 | 6.2 | 6.1 | 6.5 | 6.6 | 6.9 |
| SD | Sioux Falls | 27 | 11.0 | 11.2 | 12.6 | 13.6 | 12.1 | 10.7 | 9.7 | 9.8 | 10.3 | 10.8 | 11.6 | 10.7 | 11.2 |
| TN | Memphis | 27 | 10.6 | 10.6 | 11.3 | 10.8 | 9.0 | 8.1 | 7.6 | 7.1 | 7.6 | 7.8 | 9.4 | 10.1 | 9.2 |
| | Nashville | 34 | 9.2 | 9.4 | 10.0 | 9.6 | 7.6 | 7.0 | 6.4 | 6.1 | 6.4 | 6.5 | 8.4 | 8.8 | 7.9 |
| TX | Dallas | 22 | 11.5 | 12.3 | 13.3 | 13.1 | 11.4 | 11.0 | 9.7 | 9.3 | 9.7 | 9.9 | 10.9 | 11.2 | 11.1 |
| | El Paso | 33 | 9.2 | 10.0 | 11.9 | 12.0 | 11.0 | 10.1 | 8.9 | 8.4 | 8.4 | 8.2 | 8.6 | 8.6 | 9.6 |
| | Houston | 6 | 8.1 | 8.6 | 9.4 | 9.5 | 7.8 | 7.3 | 6.3 | 5.1 | 6.7 | 6.3 | 7.9 | 7.7 | 7.6 |
| UT | Salt Lake City | 46 | 7.7 | 8.2 | 9.2 | 9.5 | 9.3 | 9.3 | 9.4 | 9.5 | 9.0 | 8.5 | 7.8 | 7.5 | 8.7 |
| VT | Burlington | 32 | 9.7 | 9.4 | 9.3 | 9.3 | 8.8 | 8.2 | 7.8 | 7.4 | 8.0 | 8.9 | 9.5 | 9.7 | 8.8 |
| VA | Norfolk | 27 | 11.7 | 12.1 | 12.5 | 11.9 | 10.3 | 9.6 | 8.8 | 8.7 | 9.6 | 10.4 | 10.7 | 11.1 | 10.6 |
| | Richmond | 27 | 7.9 | 8.6 | 9.0 | 8.9 | 7.8 | 7.2 | 6.7 | 6.3 | 6.6 | 6.8 | 7.4 | 7.5 | 7.6 |
| WA | Seattle-Tacoma | 27 | 10.4 | 9.9 | 10.2 | 9.8 | 9.2 | 9.0 | 8.4 | 8.1 | 8.3 | 8.9 | 9.4 | 10.0 | 9.3 |
| | Spokane | 28 | 9.0 | 9.1 | 9.5 | 9.7 | 8.8 | 8.8 | 8.2 | 8.0 | 8.1 | 8.2 | 8.3 | 8.8 | 8.7 |
| WV | Charleston | 28 | 7.6 | 7.9 | 8.6 | 7.9 | 6.3 | 5.6 | 5.0 | 4.5 | 4.8 | 5.3 | 6.9 | 7.2 | 6.5 |
| WI | Milwaukee | 35 | 12.9 | 12.8 | 13.2 | 13.2 | 12.0 | 10.5 | 9.7 | 9.6 | 10.7 | 11.5 | 12.8 | 12.5 | 11.8 |
| WY | Cheyenne | 18 | 15.8 | 15.3 | 14.9 | 15.0 | 13.1 | 11.8 | 10.5 | 10.8 | 11.5 | 12.5 | 13.8 | 15.0 | 13.3 |
| PR | San Juan | 20 | 9.3 | 9.3 | 9.7 | 9.4 | 8.7 | 9.0 | 9.9 | 9.2 | 7.6 | 6.9 | 7.7 | 8.9 | 8.8 |

Source: U.S. National Oceanic and Atmospheric Administration, *Comparative Climatic Data.*
* City office data.

64

## WIND EFFECTS

Wind must be considered during the design of solar energy systems since it directly affects the collection equipment and the building which it serves, and thereby indirectly affects the design of the system equipment which connects them. The *direct effect* is one of heat loss from building and collector due to convection and air leakage. A location with a relatively high average wind speed can be expected to suffer greater losses than one of relative calm. Average wind speeds for selected cities are provided in Table 3–10.

## OTHER FACTORS

Atmospheric dust and moisture, air pollution and pollens, fog or mist in valleys and coves, can all affect the performance of solar energy systems. But their effects, like other environmental problems, such as bird or insect interference, are so transitory in nature or so easily remedied that they need not be considered here.

## SUMMARY AND APPROXIMATIONS

Because climate is the prime external influence in many solar energy applications, it is vitally important that the planning of solar systems include precise calculations of the specific effects of the climate at the site. Temperature, cloud cover, humidity, and wind speed will all affect performance and may dictate the use of design refinements which might otherwise be unneeded. The use of a selective absorber can change from optional to obligatory if high system efficiency must be achieved under mediocre climatological conditions. These refinements are more costly, but they can also be the difference between long-term cost effectiveness and mediocrity.

It is possible, however, to utilize averaged climatological data and generalized cost figures to arrive at a total system expenditure for general areas of the United States. It is worth repeating that these approximations cannot substitute for precise definitions of on-site realities, since within these general geographical zones significant climatological variations occur. But it can be helpful to identify a reasonable first approximation of the required collection area, energy savings, and system costs.

The DOE has made relevant approximations in publication SE-101, "Solar Energy for Space Heating and Hot Water," from which the following maps, charts and tables are taken. As a first approximation the contiguous United States has been divided into 12 solar climate zones which are shown graphically in Figure 3–15. For each of these zones, the approximate collector and storage tank requirements have been computed. The dimensions necessary to provide

**FIGURE 3–15** Solar climatic zones as established by the Department of Energy.

**TABLE 3–11** Approximate Collector and Storage Tank Sizes Required to Provide the Heating and Hot Water Needs of a 1500-Square-Foot Home

| Climatic Zone | Percent of Energy Supplied by Solar | Collector Area, Square Feet | Representative Collector Dimensions | | Storage Tank Capacity, Gallons | Representative Cylindrical Storage Tank Dimensions | |
|---|---|---|---|---|---|---|---|
| | | | No. of 8-Ft.-High Rows | Length of Each Row, Ft. | | Diameter, Inches | Length, Inches |
| 1 | 71 | 800 | 3 | 33 | 1,500 | 48 | 200 |
| 2 | 72 | 500 | 2 | 31 | 750 | 42 | 138 |
| 3 | 66 | 800 | 3 | 33 | 1,500 | 48 | 200 |
| 4 | 73 | 300 | 1 | 37.5 | 500 | 48 | 78 |
| 5 | 75 | 200 | 1 | 25 | 280 | 42 | 60 |
| 6 | 70 | 750 | 3 | 31 | 1,500 | 48 | 200 |
| 7 | 70 | 500 | 2 | 31 | 750 | 42 | 138 |
| 8 | 71 | 200 | 1 | 25 | 280 | 42 | 60 |
| 9 | 72 | 600 | 2 | 37.5 | 1,000 | 48 | 132 |
| 10 | 53 | 500 | 2 | 31 | 750 | 42 | 138 |
| 11 | 85 | 200 | 1 | 25 | 280 | 42 | 60 |
| 12* | 85 | 45 | 1 | 5.5 | 80 | 20 | 63 |

* Includes only hot water needs.

the indicated percentage of yearly heating and hot water energy demand for a typical 1500-square-foot home are shown in Table 3–11, those for a 10,000-square-foot building in Table 3–12. These figures are, as noted, only approximations and the dimensions of actual installation equipment can vary above or below the sizes shown. Storage tanks are especially variable, depending upon the defined objective of the system's operation. A good rule of thumb is that in a system using water for heat transfer and storage, about 1½ gallons of storage capacity should be available for each square foot of collector area. Heated air systems, which usually store energy in rocks, require about three times the volume storage capacity of a water system.

The performance of a generalized solar energy system in providing for heating and hot water requirements in a home of 1500 square feet is depicted by month for each climatological zone in Figures 3–16 through 3–18. Estimated total requirements are indicated in the right-hand bar for comparison with the left-hand bar which combines the incident solar energy on the indicated square footage of collectors tilted to an appropriate angle (approximately the local latitude angle) and the amount of energy which is actually collected. Note that for the months of November through March energy requirements are greater than solar energy collected, while during the remainder of the year collection usually exceeds demand. A greater proportion of winter heating needs could be met by installing more collection area but this adds significantly to costs. Instead, it is usually more economical to reduce heating requirements by installing more building insulation, storm windows, and other energy conserving measures. As is evident from the chart examples, solar energy supplies be-

67

**TABLE 3–12**  Approximate Collector and Storage Tank Sizes Required to Provide the Heating Needs of a 10,000-Square-Foot Building

| Climatic Zone | Percent of Energy Supplied by Solar | Collector Area, Square Feet | Representative Collector Dimensions | | Storage Tank Capacity, Gallons | Representative Cylindrical Storage Tank Dimensions | |
|---|---|---|---|---|---|---|---|
| | | | No. of 8-Ft.-High Rows | Length of Each Row, Ft. | | Diameter, Inches | Length, Inches |
| 1 | 74 | 5,330 | 7 | 95 | 10,000 | 96 | 326 |
| 2 | 71 | 3,330 | 5 | 83 | 5,000 | 72 | 300 |
| 3 | 68 | 5,330 | 7 | 95 | 10,000 | 96 | 326 |
| 4 | 73 | 2,000 | 4 | 62.5 | 4,000 | 72 | 246 |
| 5 | 75 | 1,210 | 3 | 50.5 | 2,000 | 54 | 215 |
| 6 | 72 | 5,000 | 7 | 80 | 7,500 | 84 | 322 |
| 7 | 71 | 3,330 | 5 | 83 | 5,000 | 72 | 300 |
| 8 | 74 | 1,330 | 3 | 55.5 | 2,000 | 54 | 215 |
| 9 | 75 | 4,000 | 6 | 83 | 6,000 | 72 | 354 |
| 10 | 60 | 3,330 | 5 | 83 | 5,000 | 72 | 300 |
| 11 | 77 | 1,000 | 3 | 41.5 | 1,500 | 54 | 163 |
| 12* | | | | | | | |

* There is essentially no heating requirement in this zone.

tween 55 and 85 percent of the yearly heating and hot water requirements. Most solar systems are designed toward that range of supplying 50%–80% of energy demand, since this most often represents an acceptable compromise between costs and capacity.

The actual costs of solar energy systems, like the costs of any acquisition, are of prime importance. Too often, though, many purchases are made on the basis of initial cost with scant attention to the secondary but no less important costs of maintenance and operation. Unfortunately, those purchases where continuing costs can most easily become burdensome—homes, automobiles, air conditioners, and heating systems, among others—are most often those in which initial cost assumes an unwise ascendancy. Recognition of the importance of continuing costs is especially important with energy consuming items. A conventional heating or cooling system may require a lower initial investment than does a solar system but that advantage, extended over the continual use and long life of the equipment, is soon lost.

An illustration of the long-term superiority of solar systems is easily shown. Based upon the collector areas, heating and hot water requirements of Tables 3–11 and 3–12 for buildings of 1500 and 10,000 square feet, the energy savings for solar systems, which consume little external fuels, are given in Tables 3–13 and 3–14. The range of prices which a home or building owner can expect to pay for such systems—total costs which include design, materials and installation—are given in Table 3–15. The costs are not modest when viewed through the historic frame of energy expense, but after the economy of long-term operation costs, the inevitability of rising traditional fuel prices, and the expected decrease in solar costs are factored into the equation, solar systems

**FIGURE 3–16** Performance of a generalized solar energy system: energy available, collected, required for a 1500-square-foot home.

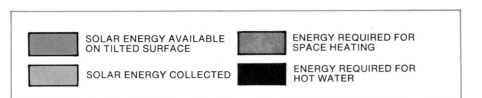

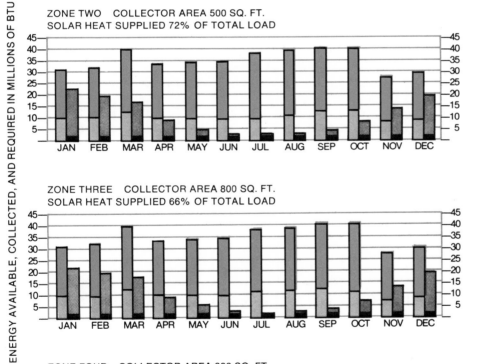

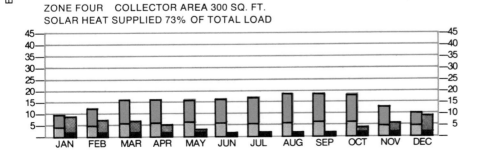

**FIGURE 3–17**  Performance of a generalized solar energy system.

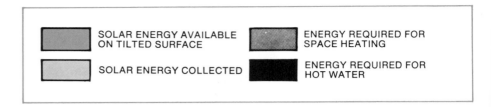

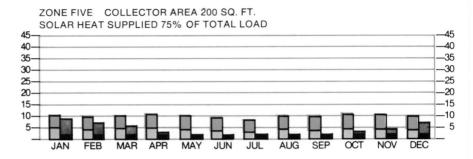

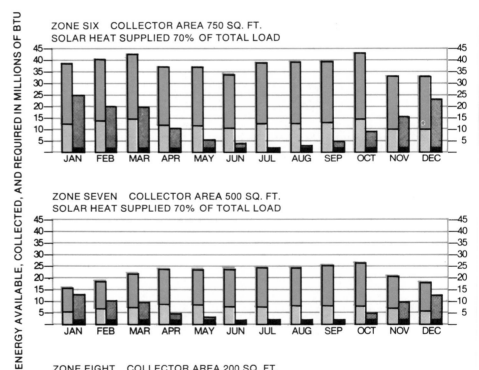

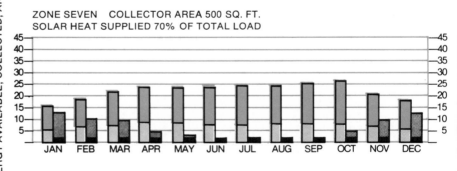

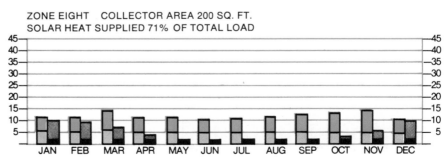

**FIGURE 3–18** Performance of a generalized solar energy system.

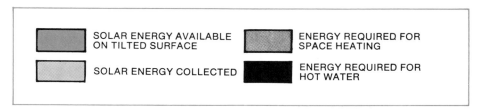

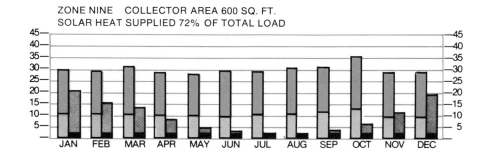

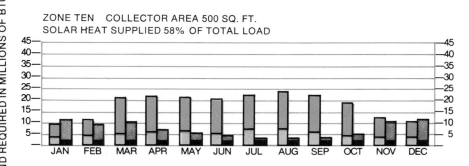

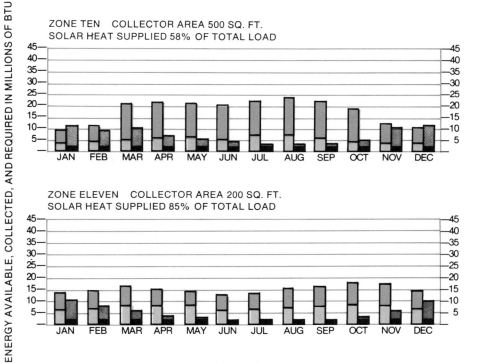

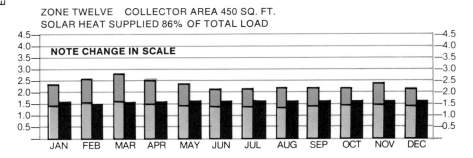

**TABLE 3–13**  Energy Supplied and Annual Dollar Savings
(1500-Square-Foot Building)

| Climatic Zone | Energy Savings, Millions of BTUs | Comparison with Oil | | Comparison with Electricity | | |
|---|---|---|---|---|---|---|
| | | Equivalent Gallons | Dollar Savings* | Equivalent Kilowatt Hours | Dollar Savings at Indicated Cost | Cost per kWh (Cents) |
| 1 | 67.9 | 757 | 303 | 19,900 | 995 | 5 |
| 2 | 54.9 | 612 | 245 | 16,000 | 720 | 4.5 |
| 3 | 82.0 | 914 | 366 | 24,000 | 960 | 4.5 |
| 4 | 41.8 | 466 | 186 | 12,200 | 488 | 4 |
| 5 | 33.8 | 377 | 151 | 9,900 | 347 | 3.5 |
| 6 | 98.9 | 1,103 | 441 | 29,000 | 1,015 | 3.5 |
| 7 | 50.6 | 564 | 226 | 14,800 | 518 | 3.5 |
| 8 | 89.0 | 435 | 174 | 11,400 | 399 | 3.5 |
| 9 | 74.6 | 832 | 333 | 21,900 | 767 | 3.5 |
| 10 | 46.5 | 518 | 207 | 13,600 | 272 | 2 |
| 11 | 43.7 | 487 | 195 | 12,800 | 512 | 4 |
| 12 | 16.7 | 186 | 74 | 4,900 | 196 | 4 |

* 65% furnace efficiency at 40¢/gallon.

**TABLE 3–14**  Solar Energy Supplied and Annual Dollar Savings
(10,000 Square Foot Building)

| Climatic Zone | Energy Savings, Millions of BTUs | Comparison with Oil | | Comparison with Electricity | | |
|---|---|---|---|---|---|---|
| | | Equivalent Gallons | Dollar Savings† | Equivalent Kilowatt Hours | Dollar Savings at Indicated Cost | Cost per kWh (Cents) |
| 1 | 372 | 4,147 | 1,659 | 109,000 | 5,450 | 5 |
| 2 | 268 | 2,988 | 1,195 | 78,500 | 3,530 | 4.5 |
| 3 | 472 | 5,262 | 2,015 | 138,300 | 6,220 | 4.5 |
| 4 | 186 | 2,074 | 830 | 54,500 | 2,180 | 4 |
| 5 | 128 | 1,427 | 571 | 37,500 | 1,310 | 3.5 |
| 6 | 587 | 6,544 | 2,618 | 172,000 | 6,020 | 3.5 |
| 7 | 253 | 2,821 | 1,128 | 74,100 | 2,594 | 3.5 |
| 8 | 174 | 1,940 | 776 | 51,000 | 1,784 | 3.5 |
| 9 | 418 | 4,660 | 1,864 | 122,500 | 4,290 | 3.5 |
| 10 | 242 | 2,698 | 1,079 | 70,900 | 1,420 | 2 |
| 11 | 165 | 1,839 | 736 | 48,300 | 1,930 | 4 |
| 12 | * | | | | | |

* There is essentially no heating requirement in this zone.
† 65% furnace efficiency at 40¢/gallon.

**TABLE 3–15**  Range of Solar Heating and Hot Water System Costs for 1,500 and 10,000-Square-Foot Buildings

| Climatic Zone | Range of Solar System Costs for a 1,500-Square-Foot Home | Range of Solar System Costs for a 10,000-Square-Foot Building |
|---|---|---|
| 1 | $8,000–$24,000 | $53,000–$159,000 |
| 2 | 5,000– 15,000 | 33,000– 90,000 |
| 3 | 8,000– 24,000 | 53,000– 159,000 |
| 4 | 3,000– 9,000 | 20,000– 60,000 |
| 5 | 2,000– 6,000 | 13,000– 30,000 |
| 6 | 7,500– 22,500 | 50,000– 150,000 |
| 7 | 5,000– 15,000 | 33,000– 90,000 |
| 8 | 2,000– 6,000 | 13,000– 30,000 |
| 9 | 6,000– 18,000 | 40,000– 120,000 |
| 10 | 5,000– 15,000 | 33,000– 99,000 |
| 11 | 2,000– 6,000 | 13,000– 39,000 |
| 12 | 450– 1,350 | — |

can be considered immediately competitive and, in the near future, superior. In addition, these figures represent the "arm chair" expenditures, i.e., the costs to be expected if the entire system installation is performed by a commercial contractor with no contribution by the home or building owner. There are any number of areas where substantial cost reductions may be effected through do-it-yourself efforts. The addition of more building insulation to reduce demand (and therefore equipment requirements) is perhaps the easiest method of reducing costs. Other contributions, depending upon the owner's skills and inclinations, such as the construction of collector plate support structures, the plumbing of pipe connections, and the installation of in-home delivery equipment, can produce larger economies.

Solar energy utilization, no less than the business of accomplishing a day's work, is a matter of making intelligent choices. The diversity of potential applications, each with differing demands and each operating under unique climatological conditions, can be met with an equal diversity of equipment, operational options to counter the most difficult problems, and a wide range of costs. Whether to use more costly materials to reduce collection area or to improve performance at the expense of greater weight are general questions addressed to specific problems. Making those decisions intelligently is not difficult once expertise is applied to the problem. The methods of precise delineation of system needs, exact determinations of environmental and operational parameters, and the implementation necessary for an economical and efficient system—in short, application expertise—are the subjects of the remainder of this handbook.

# PART II

# Solar Thermal Technology

# 4  Collection Techniques

A summary list of solar energy collection techniques contains a diversity of entries, some of which may be a little surprising. One method has been in use for thousands of years while another, though perhaps not always recognized as a collection process, is practiced in site and construction planning. Each method has its own special advantages and disadvantages which must be considered before choosing the method most appropriate for a particular application.

Collection by passive solar techniques consists of incorporating the sun into the architecture of a building. An appreciation of the possibilities has been a consideration in some construction planning for many years, often as a subsidiary concern, but the pressures of the contemporary fuel markets have forced a greater awareness. The orientation of buildings, the design of eaves on the glazed walls facing south, the choice of deciduous trees on the south and evergreens on the north, roof angles, interior coloration, and accommodation for thermal insulation are a few examples of architectural renderings which collect or reject solar energy in a purposeful way. Each has value in reducing a building's energy demand when a conventional system is used (it is obviously self-defeating to cool a building while simultaneously allowing the sun to heat it through large areas of window glass), but each has greater importance in a solar system application.

The combination of "passive solar" based on architectural design and building siting, with active or semi-active components or systems such as movable-insulated curtains, controlled air movement over masonry storage elements, or the thermal and electrical solar devices described below, can result in a highly effective energy conservation system.

## PASSIVE SOLAR COLLECTION

The relationship between the sun and the building is important when designing a solar energy system to meet the most efficient energy requirements. Two basic seasonal requirements must be considered when building the passive solar system: 1) Winter or cool weather, when the goal is to maximize the capture of solar energy and retain as much of this energy as possible in heat storing elements that are an integral part of the building. 2) Summer or hot weather, when the goal is to minimize the entry of solar energy into the building and then dissipate as much of any retained heat as possible by natural or forced ventilation. Both of these objectives require proper utilization of external and internal components of the building structure, and proper positioning of the building on the site. Utilization of various technologies which minimize heat loss from or heat gain to the building should be constructed along with the solar system.

These fundamental objectives are best achieved by beginning them in the preliminary stages of the building design. This introduces aesthetic as well as economic and engineering considerations that are beyond the scope of this book.

Of course, some of the elements of effective passive solar design can be incorporated into already existing plans or buildings. Obviously building siting cannot be revised (at least not without exceptional effort and cost), nor can major structural changes be economically justified. But certain passive techniques can be used in almost any building, even though their effectiveness will not approach that of a building designed from inception with passive solar collection in mind.

## ACTIVE SOLAR COLLECTION

The *flat plate solar collector* is the most economical *active* method of collecting solar energy. It consists of an assembly of transparent covers over an absorber plate which is backed with thermal insulation. Typically, a flat plate collector is used to heat air or water that is then used in hot water space heating and other services, or stored for later use. Because of the importance of the flat plate collector in solar energy applications, it is considered in greater detail in the next two chapters.

*Concentrating collectors* deflect sunlight from a large area into a smaller region where the concentration of light can be used to yield temperatures higher than those obtainable from flat plates. Focusing mirrors or lenses are used to effect the concentration but they must be aligned more precisely toward the sun for greatest efficiency. This often requires a tracking system—electrical or mechanical techniques, though more costly, yield greater energy collection than a stationary plate—to move the focusing mechanism in concert with the sun's motion.

Some concentration techniques can be used without tracking systems (the Winston design is one example), but there are practical and theoretical limits to the degree of concentration and efficiency that can be achieved without tracking.

*Photovoltaic collectors* convert sunlight directly into electricity through the action of the sun on a semiconductor junction, producing an electrical voltage that supplies current through conducting wires. These collectors, often called PV or solar cells, can be incorporated into a flat plate structure, or used as a conversion element in concentrating collectors to simultaneously produce both thermal and electrical energy for immediate use or for storage in thermal tanks and batteries.

In another application, a photovoltaic cell is used to protect bridges and other outdoor structures from electrochemical corrosion resulting from the contact potential produced between the metallic structure and the moist earth. The voltage provided by a photovoltaic solar cell mounted on the structure is used to counter the contact potential. As they become more affordable, solar cells can be used to replace batteries or to charge batteries which are used only intermittently: in boats, golf carts, or lawn care equipment which may stand idle for long periods.

In summary, photovoltaic collectors provide the most versatile form of energy—electricity—but at present are cost-effective in a very limited range of applications, such as powering space vehicles and in remote locations.

*Bioconversion* is a general category of solar energy utilization techniques wherein the sun powers the vegetative photosynthesis which converts $CO_2$ and water to combustible carbohydrates. Wood fuels supply energy stored by the living tree; an ancient source of energy but one which is founded on the sun. The method is inefficient since the energy yield of the combustion process is much less than the energy input to the plant, but the process is being improved.

Several fuel alcohols (methanol, ethanol) and gases (methane) can be distilled by synthetic or natural processes to be used as fuels, a possibility in underdeveloped nations where the availability of land otherwise unsuitable for plant food production and cheap labor resources can contribute to cost effectiveness. It is probably not an economical solar energy technique for industrialized nations.

Power generation through *natural geographic features* by using windmills, ocean thermal gradients, tides, and waterfalls are all solar sources, since the sun is the basis of provision for each of these opportunities. Though they are not direct utilizations of solar energy, they have the advantage of being as effective at night as they are during the day; but they depend upon natural formations for power generation at the site with the necessity for a distribution mechanism for supply elsewhere. Southern Florida has no waterfalls, Kansas has no tides, Los Angeles experiences long periods with little wind: all are obvious and severe limitations, like those attendant to geothermal power.

*Extraterrestrial production* of solar energy refers to the use of artificial earth satellites that would collect solar energy in space, convert it to microwave radio beams, and transmit it to earth. This technique has never been demonstrated but has been proposed by the Arthur D. Little Company, which holds a patent on the process.

Other extraterrestrial techniques could function by orbiting large automated process plants in highly elliptical orbits around the sun similar to the path of the minor planet Icarus (named for the legendary wingclad aviator whose wax-constructed wings melted when he flew too close to the sun). Such a process plant could convert low-grade meteoric ores into highly concentrated, scarce metals during its close path to the sun for delivery upon return to the vicinity of the earth. These concepts may seem too fantastic, but no more so than the concept of television if introduced to those living a century ago.

## FLAT PLATE COLLECTION

The primary active technique is the flat plate collector, an assembly of transparent covers, absorber plate, insulation, and an enclosure, as illustrated in the sectional drawing in Figure 4–1. The covers serve three functions: to prevent convection losses from the collector to the air while transmitting the sun's rays, to reduce thermal radiation losses from the collector, and to protect the collector's absorber plate against environmental hazards. The specifics

**FIGURE 4–1** Flat plate solar collector.

of a given site and the collector's application will determine the precise cover configuration; some collectors will operate most efficiently with one, two or three covers, while others may need none. The absorber is a coated plate upon which the sun's energy is converted to heat. The choice of coating material is, again, a function of the demand which the system must serve and the on-site conditions under which it must operate; but all absorbers conduct the heat generated on their surfaces to a site for transfer of the heat to either a liquid or to air. Since heat absorption and transfer is the *raison d'etre* of the collector assembly, heat losses must be kept to a minimum. Good commercially produced collectors provide insulation behind the collector plate with additional insulation between the edges of the plate and its frame to further reduce losses there. The frame performs the obvious function of holding the assembly together and, depending upon the construction, of providing support to the absorber plate. In some designs, the pan in which the insulation is mounted is attached to, or is a part of, the frame, while other collectors will use two frames: an inner frame to support the absorber plate and an outer frame to contain insulation, hold the covers, and provide further support. The objective of the entire assembly is to achieve energy gathering with a minimum of the potential loss factors illustrated in Figure 4–2.

## ABSORBER PLATE CONSTRUCTION

The principal methods used in the construction of absorber plates for *liquid* heating are shown in Figure 4–3. Integral construction, shown in drawing (*a*), is the best but most expensive choice. It provides the most efficient heat transfer, flexibility in fluid flow path design capability, can be mass produced, and uses the minimum weight of material per unit of total weight. Solder bond construction (Figure 4–3*b*) is the most easily constructed and can be made from readily available copper which will conduct heat and is easily soldered, points which have not been lost on the individual hobbyist. The wired construction has rather poor thermal contact between plate and conduit. (Figure 4–

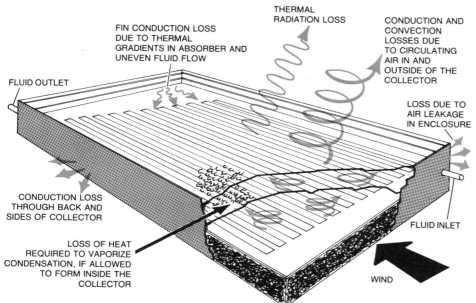

FIN CONDUCTION LOSS
DUE TO THERMAL
GRADIENTS IN ABSORBER AND
UNEVEN FLUID FLOW

THERMAL
RADIATION LOSS

CONDUCTION AND
CONVECTION
LOSSES DUE
TO CIRCULATING
AIR IN AND
OUTSIDE OF THE
COLLECTOR

FLUID OUTLET

LOSS DUE TO
AIR LEAKAGE
IN ENCLOSURE

CONDUCTION LOSS
THROUGH BACK AND
SIDES OF COLLECTOR

FLUID INLET

LOSS OF HEAT
REQUIRED TO VAPORIZE
CONDENSATION, IF ALLOWED
TO FORM INSIDE THE
COLLECTOR

WIND

**FIGURE 4–2**  Factors that contribute to losses in a flat plate collector.

**FIGURE 4–3**  Collector absorber plate construction.

**(a)** *Integral Construction*

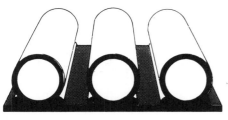

**(b)** *Solder Bond*

**(c)** *Wired      Pressed      Clamped*

**(d)** *Riveted Sheets*

3c). Pressed and clamped constructions increase thermal contact but they use much more metal, suffer corrosion under moist conditions when dissimilar metals are used, and are usually limited to linear conduit configurations. The spot-welded construction in Figure 4–3d has good thermal contact characteristics and is economical since it uses a minimum of material, but all edges must be sealed in a separate operation and its longevity is limited by the susceptibility of spot-welds to fail as tension and temperature cycling take their effects.

Absorber plates offer an equally wide choice of construction materials. Rubber and plastics impregnated with carbon black will perform satisfactorily in simple installations such as swimming pool heaters, but those systems under greater demand require the greater efficiency obtained by construction with metals of high thermal conductance. Copper, aluminum, and steel qualify and are the most common absorber plate materials, but none of the three is without disadvantage. Copper has the highest thermal conductance but is quite expensive. Steel is usually more economical but is heavy and may rust. Aluminum is low in weight but can corrode when in contact with other metals. Any problem inherent in a particular material can be overcome or minimized so the ultimate choice of collector material must be made in response to the special conditions of each application.

## TRICKLE COLLECTORS

Flat plate collectors may be designed so that the water trickles down the face of the collector rather than flowing through enclosed tubes—a much more simple construction, but one with limited use. Corrugated metal, as shown in Figure 4–3d, is often used in this construction since the corrugations provide ready-made flow paths. Or the fluid may be directed, as pictured in Figure 4–4, in a slightly more complex manner. Both flow configurations are referred to as "open-channel flow," but there are serious limitations to the effectiveness of all trickle collectors. The action of water can damage the absorbing surface and, conversely, the surface can contaminate the water with a consequent reduction in the available options for absorptive coatings. Free moisture between the plate and transparent covers is unavoidable with open-channel flow, since the fluid alternately evaporates from the hot absorber surface and condenses on the underside of the cover; this results in increased convection losses and partial blockage of sunlight transmitted through the cover. Also, the flow in trickle collectors must be directed downward, requiring greater pumping effort rather than taking advantage of the siphoning effect and the natural tendency of a heated fluid to rise in a closed loop. Despite their limitations, these variously named "trickle collectors" are low in initial cost and, when the additional area and mechanical support for them is available, can be used to advantage.

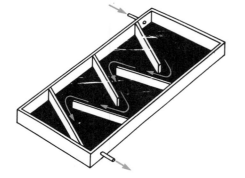

**FIGURE 4–4** A low-cost but inefficient trickle collector.

## FLOW CONFIGURATIONS IN CLOSED CONDUITS

The flow *through* an absorber plate can be either in series or in parallel configurations, as compared in Figure 4–5. Differing characteristic temperature and pressure gradients provide several points of comparison which are worthy of note.

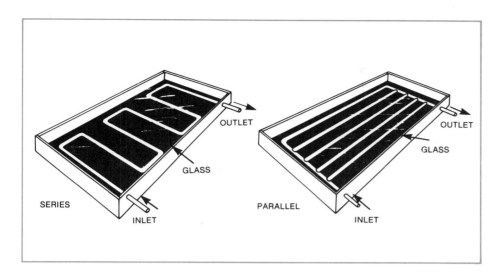

**FIGURE 4–5**  Series and parallel flow configurations.

If all the tubes are soldered to the same plate, the series configuration yields a non-uniform heating pattern since the fluid is heated as it progresses along the first path, then is cooled as it passes back toward the inlet region of the plate. In the parallel configuration, by contrast, fluid temperature rises continuously as the fluid proceeds along parallel paths. In addition, the parallel configuration is more advantageous because the flow rates in the paths are self-adjusting by the thermosiphon effect (i.e., the tendency of a fluid to rise as heat reduces its density). This upward motive effect is of sufficient magnitude in a thermosiphon system (detailed in a later section) that no additional pumping is required to move the fluid. An upward directed flow is desirable to either configuration but, with all else equal, the higher average absorber-plate temperature-gradient of the series configuration makes it less efficient.

The series configuration also requires a longer flow path than does the parallel, with a number of related consequences. Given the same flow rate at the inlet, and equivalent tubing diameters in both configurations, the series path has a greater fluid velocity, with more resistance to flow, and therefore a greater pressure loss across the collector. Minimizing this problem is best accomplished in the planning stage of the solar system by calculating and comparing the values for pressure loss for each of the collectors under consideration for the application. The pressure difference between the inlet and outlet of a tube length, $L$, and inside diameter, $d$, is expressed in terms of the equivalent head loss, $h$, in units of length. The head loss, $h$, is identical to the pressure produced by a column of water of height, $h$.

With an open outlet the head loss, $h$, is approximated by:

$$h = \frac{32VL\mu}{g\rho d^2} + (Z_o - Z_i) \tag{4–1}$$

where:  $h$  is the head loss across the collector (ft)

$Z_o$  is the height of the outlet (ft)

$Z_i$  is the height of the inlet (ft)

$V$  is the velocity of fluid in the tube (ft/sec)

$L$   is the length of the tube (ft)

$\mu$   is the dynamic viscosity of the fluid (lb-sec$^2$/ft)

$d$   is the inside diameter of the tube (ft)

$\rho$   is the mass density of the fluid (lb/ft$^3$)

$g$   is the acceleration of gravity = 32.2 ft/sec$^2$.

The units may be either in the metric or English system, but must be consistent. In a closed loop system, the $(Z_o - Z_i)$ term is zero and drops out of the equation.

Air-heating flat-plate collectors offer some theoretical advantages over collectors which use a liquid as the heat transfer medium. Among those advantages are potential cost savings since less plumbing is required, the complications of leakage of a liquid transfer medium (though not of thermal losses) are ameliorated, the possibility of plumbing corrosion is reduced, and the collector itself may be of a more simple construction. Further, the demands of choosing and constructing the heart of the collector—its absorber plate—are less stringent, since there is no need for integral conduits or the necessity to conduct heat from one area of the absorber to another. Because any blackened absorber surface will transfer heat to air blown over it, though with varying degrees of effectiveness, there are any number of possibilities for use as an absorber. Several of those possibilities, black gauze, glass plates, and metal plates with fin backing, as well as various construction configurations for air-heating collectors, are illustrated in Figure 4–6.

Air flow characteristics through the collector are of importance. Normally, air will flow in a *laminar* fashion, i.e., a layering of the air moving through the collector, with the layer nearest the absorber plate absorbing most of the heat and relatively less heat transfer (and cooler temperatures) in the overlying layers. A better choice is to cause *turbulent* air flow where these layers are forced to mix as they proceed through the collector. Corrugations in the plate, fins, louvers, or the partitions of a maze-box collector all serve to achieve turbulence. At the same time, however, over-restriction of the air flow in an effort to achieve turbulence can have negative effects, notably a greater pressure drop across the collector. In many cases, this situation will necessitate the use of a larger fan. Another general point worth notice on the matter of air flow is that better designs of air-heating collectors direct the air flow away from the transparent covers after heating by the absorber plate, or pass the air behind the absorber plate.

Damage from leakage of liquid transfer medium from system plumbing is not a concern with air-heating collectors, but leakage remains a serious concern. Because the air flowing through the system is under pressure, any fault in the system's integrity, however small, will reduce collector efficiency. Loss of solar-heated air through cracks in the collector or ductwork and, conversely, infiltration of cooler air anywhere in the system, can be crucial. In addition, though not strictly a leakage problem, the necessary turbulence of the air flow in the collector causes a relatively larger thermal loss through convection than is the case with a similar liquid transfer unit, unless the air is passed behind the absorber plate.

Slight leakage can also be beneficial. The following points are addressed by

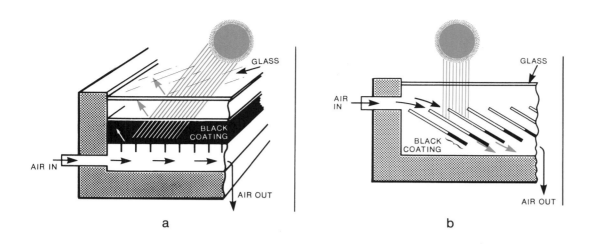

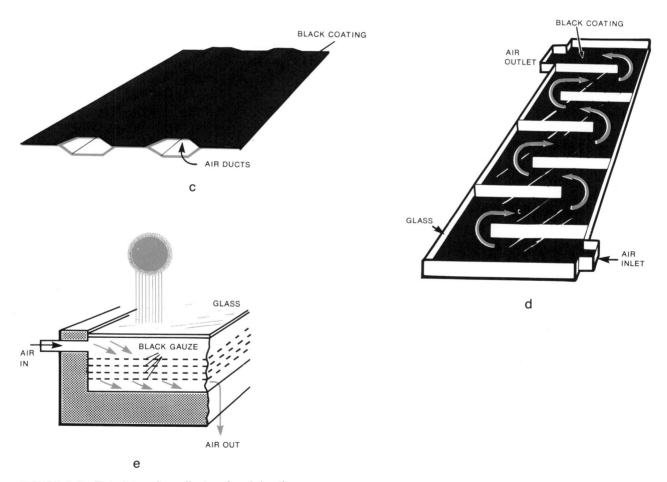

**FIGURE 4-6** Flat plate solar collectors for air heating.

George O. G. Löf, Solar Energy Applications Laboratory, Engineering Research Center at the Colorado State University.

"On the leakage question, your points are sound if the leakage into the air system resulted in added cold air supply to the building. However, unless excessive, this *ventilation* replaces (a) fresh air if 100 percent fresh air is used in the system, or (b) some of the make-up air if a partially recirculating system is used, or (c) some of the infiltration air if fresh air is supplied by that means. The first two cases are obvious, but the replacement of infiltration is not so clear. [MARKS' Standard Handbook for Mechanical Engineers] page 12-86 says, 'Infiltration losses are minimized with warm-air heating systems or air-conditioning systems that pressurize the heated areas.' The effect of air drawn into the collector is to slightly pressurize the building, thereby reducing infiltration, as indicated in the quotation. In ASHRAE *Fundamentals* book, page 21-31, the following statements are made; 'When inside the outside pressures on the windward side are balanced, the outdoor air required will be roughly equal to the infiltration that would otherwise occur---.' The other quotation is; 'Where mechanical ventilating systems are designed to produce positive or negative pressures in an enclosure, if the specified rate at which air is supplied or removed exceeds the calculated infiltration rate, it is common practice to use the greater value in calculating heating or cooling load.' The evidence, then, is that added air supply via the collector *replaces,* on an approximately equal volume basis, some of the infiltration air. For example, if the air leakage in an air collector is 10 percent of a 1000 cfm flow, equivalent to 100 cfm, about one-third of the usually estimated infiltration into a 2000 square foot residence (one air change per hour) would be replaced.

There is the added benefit that the air thus supplied is at least partially preheated in the collector, thereby reducing cold drafts in the building. In our work with CSU Solar House II, we have found that these effects indeed occur. We can state with no hesitation that 10 percent air leakage into the collector is an insignificant factor in the building load and is a modest benefit to the collector efficiency. Air leakage is not really the bugaboo that is often asserted, provided that professionally designed and installed systems are utilized.''

Although air collectors find their greatest applications where home heating is the primary objective, since the heat transfer medium can be used directly and the collector can operate at a lower overall temperature, they have some disadvantages. In the specific case of domestic water heating, the use of air collectors results in an increased collector area requirement of 1.3 to 1.7 times that of a liquid cooled collector. Part of this is caused by the poor heat transfer from air to liquid and part is due to the nature of air collector construction. Air collectors have been sold for domestic water heating mainly on the basis that air does not freeze, increasing the collector's reliability and durability. However, it is possible that a failure in the fan control switch could keep air circulating when it should not and consequent freezing of the heat exchange coil could occur.

Presently, both liquid- and air-cooled collectors applied to home heating can provide all hot water needs, especially in the summer months when they have no other use.

Air has a significantly lower capacity to contain heat. The specific heat of air is 0.24, while the specific heat of water is 1.0, more than four times greater. Further, the lower density of air (approximately 0.075 pounds per cubic foot) as compared to water (more than 62 pounds per cubic foot) carries with it other consequences. Much greater volumes of air must be moved through the collector to transfer an equivalent amount of thermal energy. Both of these factors, in turn, necessitate the use of increased spaces through which to move the air, not only in the collector but in the service ductwork, and greater surface area in ductwork can mean greater thermal losses from that surface. Larger ductwork space within the collector will decrease the pressure drop across the collector but, at the same time, reduce heat transfer between plate and air. As is the case so often in solar system design, and as will be shown throughout succeeding chapters, final system design is usually a product of trade-offs in many areas. It is reasonable to project, however, that unless a second energy-generating system (such as a photovoltaic cell) is combined with it, the air-heating collector will find utility mainly in applications such as crop drying, home heating, or preheat for plant air make-up where space heating is the long objective.

## SELECTIVE COATINGS

As incident solar energy is absorbed by the collector, the absorber plate temperature will rise to an equilibrium at that point where the energy delivered plus the energy lost is equal to energy absorbed. High collector plate temperatures are desirable but, since all three types of heat losses (conduction, radiation, and convection) increase as plate temperature rises—with a consequent decrease in energy delivered and system efficiency—strict control of losses is necessary. As has been shown, conduction and convection losses are countered with multiple covers and back insulation, leaving losses by radiation to assume an increasing importance. Radiation losses increase as the fourth power of the absolute temperatures (see Equation A–2) with the maximum losses occurring in the wavelength region given by the Wein Displacement Law (Equation A–4). Wavelengths in the thermal region are significantly longer than the average of wavelengths of the solar radiation region.

If convection and conduction losses were completely eliminated by surrounding the collector with a perfect vacuum, and if the solar constant was absorbed perfectly by the absorber plate (as in a blackbody in space), the highest equilibrium temperature is a function of the thermal emissivity. The relationship is expressed:

$$T_{eq} = \sqrt[4]{\frac{I_o}{\sigma \varepsilon}} \qquad\qquad (4\text{--}2)$$

where:  $T_{eq}$  is the highest equilibrium temperature (°K)

$I_o$  is the solar constant (see Table 2–1)

$\sigma$  is the Stefan-Boltzmann constant, Equation A–2

$\varepsilon$  is the emissivity of the absorber at wavelengths near those given by Equation A–4.

For $\varepsilon = 1$, such as is defined for a blackbody, $T_{eq} = 124°C$. For most real objects the thermal emissivity ranges from 0.6 to 0.9 and in that range $T_{eq}$ can range from approximately 170°C to 130°C, respectively. Table 4–1 shows the highest equilibrium temperature, $T_{eq}$ in °F for various values of emissivity, $\varepsilon$, under the assumptions that: no other losses are present, there is perfect absorptance, and irradiance occurs at the solar constant.

It is apparent from the data that the optimum highest possible equilibrium temperature results when collectors are constructed of materials with the lowest thermal emissivity. Unfortunately, most materials with low thermal emissivity are also low in solar absorptance with the effect that equilibrium temperatures above 500°F have not been achieved with flat plate collectors.

A selective surface is a coating or surface preparation, black in appearance

**TABLE 4–1** Effect of Emissivity on Equilibrium Temperature under Idealized Conditions (in Vacuum with No Covers)

| Emissivity, $\varepsilon$ | Equilibrium Temperature, $T_{eq}$ (°C) | Equilibrium Temperature, $T_{eq}$ (°F) |
|---|---|---|
| 0.05 | 552.7 | 1,027 |
| 0.10 | 421.6 | 791 |
| 0.20 | 311.1 | 592 |
| 0.30 | 255 | 491 |
| 0.40 | 218.3 | 425 |
| 0.50 | 191.6 | 377 |
| 0.60 | 170.5 | 339 |
| 0.70 | 153.8 | 309 |
| 0.80 | 140 | 284 |
| 0.90 | 127.7 | 262 |
| 1.00 | 117.7 | 244 |

since the sensitivity of the human eye is limited to that part of the solar spectrum where the surface is an absorber, which combines low thermal emissivity with high absorptance. Many coatings provide high absorptance, but to simultaneously gain low thermal emissivity, it may be necessary for the coating to have high thermal and electrical conductance as well. The emissivity of a material is related to its electrical resistivity by Drude's relation:

$$\varepsilon = 0.365 \sqrt{\frac{r}{\lambda}} \qquad\qquad (4–3)$$

where:  $\varepsilon$  is the thermal emissivity

$r$  is the electrical resistivity at wavelength, $\lambda$

$\lambda$  is the wavelength.

Drude's relation is valid where the surface of the material is smooth and maintains its structural integrity over distances significantly larger than the wavelength. Drude's relation serves to show why most conventional black paints (which are non-conducting and, thus, have high resistivity) are not useful as selective coatings.

Many materials and surface preparation techniques have been explored for possible application as selective surfaces. The synthesis of a selective surface technique that is economical and long-lasting is still the subject of basic research, but practical selectively-absorbing surfaces are now being produced.

Polished metals typically absorb very little of the solar radiation that is incident upon them. As conductors of electricity, metals have an abundance of highly-mobile surface electrons that quickly respond to incident electromagnetic fields and reflect them away from the metal. At the same time, however, these mobile electrons are unresponsive to the thermal mechanical agitation of their respective metallic atoms and do not emit radiation in response to heat as well as other substances. Thus, most metals have both low absorptivity and low emissivity. The mobility of the electrons on a polished metal surface varies with the frequency (or wavelength) of the incident electromagnetic wave or thermal mechanical agitation, however, so that although absorptivity and emissivity must be equal at any particular wavelength, they can vary considerably between the short wavelengths of solar energy and the longer wavelengths of thermal mechanical agitation. Table D–5 shows the solar absorptivity and thermal emissivity of commonly used metal surfaces.

Iron and zinc have both been considered as naturally selective materials. Since galvanized steel (which consists of a zinc coating on steel) is inexpensive, it seems an attractive possibility. Its solar absorptivity, however, is only 0.5, which while higher than other metals, is only slightly more than half as absorptive as black paint. Consequently, some experimentation is being conducted on galvanized sheet steel that has been folded accordion-fashion into V grooves; sunlight is forced to reflect several times with successive reductions in net reflectance. The result is some improvement in absorptivity at the expense of an increase in emissivity, resulting from a larger emission surface area for thermal radiation.

Another approach to the synthesis of a selective surface is to coat a metal surface with tiny conducting spheres or holes that absorb light but are smaller than the wavelengths that are characteristic of thermal radiation. Because of this, the tiny black spheres are transparent to thermal radiation and retain the low emissivity of the underlying metal surface. This is the concept used in applying very thin layers of paint containing carbon black, miniscule agglomerated metal droplets, or other optically absorbing ingredients to a metal backing. Difficulties arise in finding a suitable matrix or binder for the spheres, in obtaining a uniform particle size and spacing, and in providing a uniform thickness.

One of the most successful methods to date of producing selective surfaces involves the use of semiconducting materials. Some semiconductors are similar to metals in thermal emission and are similar to non-metals in their ability to absorb solar energy. The semiconducting oxides of iron, cobalt, vanadium, nickel, chromium, copper, and lead all have selective properties and have been shown to perform as selective surfaces. Table 4–2 lists the properties of some of the selective coatings that are practical for use and commercially available.

**TABLE 4–2**  Practical Selective Surface Materials

| Material | Developer | Solar Absorptance $\alpha$ | Thermal Emittance $\varepsilon$ |
|---|---|---|---|
| Semiconductor Lead Dioxide (on copper)[1] | AMETEK | 0.99 | 0.25 |
| Black Chromium on Bright Nickel (on aluminum or steel)[2] | Olympic Honeywell NASA-Lewis | 0.95 0.977 0.93 | 0.10 0.19 0.12 |
| Black Nickel[2] (on nickel or copper)[3] (on galvanized iron)[3] | Honeywell Tabor Tabor | 0.90 0.90 0.89 | 0.08 0.05 0.12 |
| Copper Oxide (on copper)[3] | | 0.89 | 0.17 |

[1] U.S. Patent #3,958,554, Dr. Ferenc Schmidt; assigned to AMETEK, Inc.
[2] R. L. Lincoln, D. K. Deardorff and R. Blickensderfer. "Development of ZrOxNy Films for Solar Absorbers," Society of Photooptical Instrumentation Engineers. Vol. 68:1975, page 161.
[3] H. Tabor. "Selective Surfaces for Solar Collectors," Chapter IV of *Low Temperature Engineering Application of Solar Energy*, published by ASHRAE.

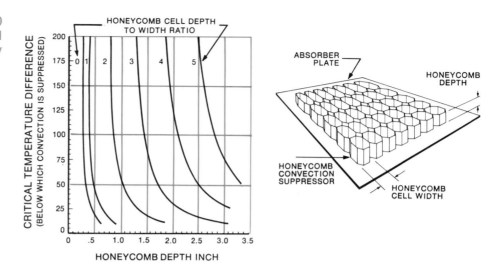

**FIGURE 4–7** The critical temperature difference below which convection is suppressed by a transparent honeycomb structure. (After K. G. T. Hollands)

The selective surface materials listed in Table 4–2 are limited to surfaces that can be prepared by electrodeposition with relatively cost-effective raw materials. There are hundreds of techniques of producing selective surfaces by vacuum deposition but the process is expensive when large, planar areas must be covered. Meinel and Meinel[1] have discussed many selective surfaces, including those requiring vacuum deposition, and those with multiple layer interference stacks which operate on principles similar to those of optical antireflection coatings. These more advanced types of selective coatings are not cost effective for flat-plate collectors which require large planar areas, but they may have significant application in concentrating collectors where the active absorber surface has considerably less area than the solar interception aperture area.

## CONVECTION SUPPRESSION WITH HONEYCOMBS

Honeycomb devices—planar arrays of closely packed hexagonal or square cells—may be used to suppress convection currents which contribute to the upward losses in flat plate collectors. K. G. T. Hollands reported on the effectiveness of honeycomb structures of this application.[2]

Honeycomb structures will suppress convection currents up to a critical temperature difference that depends upon the honeycomb depth and the depth-to-width ratio. Figure 4–7 shows this relationship as determined by Hollands for a square honeycomb array.

The heat transfer coefficients across a honeycomb panel include terms for convection, conduction, and radiation. If the honeycomb walls are very thin

1. Meinel and Meinel. *Applied Solar Energy,* Ch. 9. Addison Wesley, 1976.

2. Hollands, K. G. T. "Natural convection in horizontal thin-walled honeycomb panels. *J. Heat Transfer* 95:1973, pages 439–444.

and highly transparent to solar radiation, conduction and radiation losses are minimized. A reduction in convection loss by nearly 80 percent is theoretically possible with an idealized honeycomb structure, but this is not achieved because of the added conduction and radiation losses contributed by the honeycomb structure itself. Typically, the reduction in net upward loss is from 30 to 70 percent, depending upon the numerous factors which must be considered and the operating conditions in which the improvement is calculated or measured.

Honeycomb structures are currently being investigated for their effect on the performance of solar collectors. A firm conclusion upon their efficacy must await the resolution of questions which remain, not the least of which are those related to the high-temperature stability of the low-cost plastic materials which are the most easily fabricated into honeycomb structures.

## VACUUM CONVECTION SUPPRESSION

Convection loss can be very effectively suppressed by removing the air altogether from the space between the absorber plate and the transparent covers. Evacuated collectors have been conceived and built using posts to support the cover against collapse due to the pressure of the outside atmosphere. In addition, long evacuated glass tubes have operated efficiently as transparent covers when used to enclose black absorber surfaces. Arrays of evacuated glass tubes offer advantages of good performance but high costs, and the difficulties of the numerous vacuum seals (required in large planar areas of collectors) leave some room for improvement.

Evacuated tube collectors, because of the circular cross-section of the tubes and the spacing between them, do not capture as much of the incident sun as a flat plate collector. The performance is less affected by small deviations from normal in the incident angle, however, so that overall performance is quite good. Since trough-type reflectors are often used behind the evacuated tubes to improve the proportion of incident solar energy captured, evacuated tube collectors are making some use of concentration techniques. A complete discussion of evacuated tube collectors requires methodology that goes beyond the techniques developed for the analysis of flat plates, so it is difficult to generalize about them. However, several observations can be made. Convection and conduction losses are reduced to near zero in an evacuated tube collector so the importance of a selective coating to reduce radiation loss is more heavily emphasized. Also, the vacuum environment may provide important constraints, both restrictive and beneficial, to the selective coating options to be considered. Needless to say, the evacuated tube collector is an area of active research and development interest in the solar industry.

## ENCLOSURE DESIGN

The enclosure design must reflect consideration for thermal performance, structural integrity, long term durability, and cost-effective manufacturing.

The cover assembly and absorber plate require support; ideally, the absorber plate should be supported independently of any thermally-conducting outer frame to limit heat losses by conduction and to allow for thermal expansion. If multiple glazings are used, the inner glazing should also be supported independently since its temperature will rise, but the outer cover can be supported by the external enclosure or frame since it is already in contact with the outer air.

Wind must be considered, since it directly affects the structural requirements of the collector, the structural support, and the building on which the collector is installed. Both the design of structures and component parts of structures are presented in ANSI A58.1, Building Code Requirements for Minimum Design Loads in Buildings and Other Structures. The following procedure is outlined for developing design specifications.

1. Select the mean recurrence interval. This is generally 25, 50, or 100 years, with the latter interval providing the most severe wind speeds and most conservative design (see Figure 4–8).

2. Select the wind speed appropriate for the general geographical location and any special local conditions.

3. Determine the effective velocity pressure, $q_p$, for the structure at the appropriate exposure, height, and selected wind speed.

4. Determine the maximum positive and negative pressure coefficients, $C_F$, appropriate for the tilt angle, critical wind direction, and type of supporting enclosure. The range of $C_F$ is approximately $-1$ to 2.0 for common collector mounting designs. Negative coefficients refer to a suction-type load.

5. Calculate the $q_p C_F$ product to determine the structural rating required at the specific load per unit area.

6. The load per unit area multiplied by the gross collector area will result in the total design load to be used for mounting hardware and for the balance of the structural design.

Table 4–3 summarizes rating requirements for a variety of exposures, wind speeds, heights, and pressure coefficients. The numbers in orange indicate those applications with a wind load which requires a collector rated for 50 pounds per square foot.

The unit must be able to breathe in a controlled fashion to prevent condensation from building up in the enclosure. It is impractical to build a sealed unit, because under no-flow conditions the temperature increases the internal pressure to a level exceeding that of hurricane force winds. Controlled breathing that expels the moisture condensing on the inside of the cover glass before it can drop down into the enclosure has proven to be a successful solution.

Figure 4–9 diagrammatically compares two enclosure designs. Figure 4–9a for a double-glazed collector used a rigid enclosure, independent support for absorber plate, and inner glass cover. The design in Figure 4–9b for single-glazing offers advantages in cost as well as a refined transparent cover-seal design.

Designing for durability requires proper material selection for long-term thermal effects and environmental exposure. With long payback periods, most solar systems are expected to remain operable for twenty to thirty years with minimal maintenance and minimum reduction in system performance. Most solar collector designs allow atmospheric air and moisture inside the collector enclosure. When coupled with the 300–400°F stagnation temperatures of the

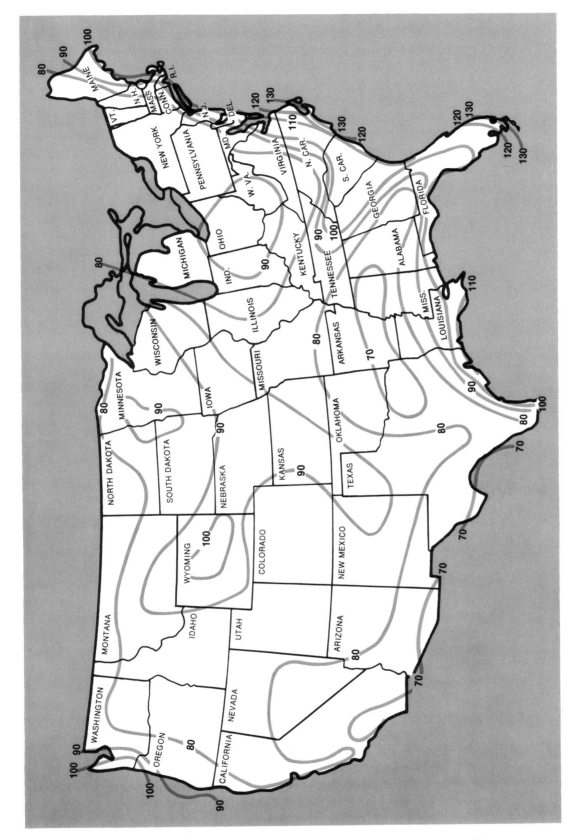

**FIGURE 4–8** Annual extreme wind speeds in miles per hour (100-year cycle).

93

**TABLE 4–3**  Collector Structural Design Rating Requirements
(pounds per square foot)

Exposure A:  Basic wind speed (mph) for centers of large cities and very rough, hilly terrain

| Height (ft) | 60 | | 100 | | 130 | |
|---|---|---|---|---|---|---|
| | $C_F = -1$ | $C_F = 2$ | $C_F = -1$ | $C_F = 2$ | $C_F = -1$ | $C_F = 2$ |
| 30 | −5* | 10 | −13 | 26 | −21 | 42 |
| 200 | −10 | 10 | −29 | 58 | −49 | 98 |
| 800 | −20 | 40 | −57 | 114 | −96 | 192 |

Exposure B:  Basic wind speed (mph) for suburban areas, towns, city outskirts, wooded areas and rolling terrain

| Height (ft) | 60 | | 100 | | 130 | |
|---|---|---|---|---|---|---|
| | $C_F = -1$ | $C_F = 2$ | $C_F = -1$ | $C_F = 2$ | $C_F = -1$ | $C_F = 2$ |
| 30 | −8 | 16 | −28 | 56 | −40 | 80 |
| 200 | −15 | 30 | −42 | 84 | −72 | 144 |
| 800 | −24 | 48 | −67 | 134 | −113 | 226 |

Exposure C:  Basic wind speed (mph) for flat open country, open flat coastal belts and grasslands

| Height (ft) | 60 | | 100 | | 130 | |
|---|---|---|---|---|---|---|
| | $C_F = -1$ | $C_F = 2$ | $C_F = -1$ | $C_F = 2$ | $C_F = -1$ | $C_F = 2$ |
| 30 | −14 | 28 | −38 | 76 | −64 | 128 |
| 200 | −20 | 40 | −57 | 114 | −96 | 192 |
| 800 | −28 | 56 | −76 | 152 | −129 | 258 |

* Negative design ratings refer to suction-type loads.

absorber plate, solar collectors must be designed to withstand low and high temperature, high humidity, and environmental pollutants both inside and outside the enclosure.

Thermal expansion poses an interesting challenge. As absorber plate temperatures rise, heat loss to the transparent cover (normally glass) increases. Cover temperatures are therefore different than enclosure temperatures and in most cases dissimilar materials are used. The gasket-seal design should therefore account for differential thermal exposure. An improved gasket-seal design is shown in Figure 4–9b.

## CONCENTRATING COLLECTORS

Concentrating collectors use mirrors or lenses to focus light from a large area onto a small area, where the more intense light can achieve higher temperatures and a greater conversion into usable energy with less receiving area.

Concentration with mirrors is possible in several ways. To focus light onto a small area, conical, spherical, or paraboloidal mirrors are used. Linear, circular, or parabolic reflectors will focus light onto a narrow line. A large number of

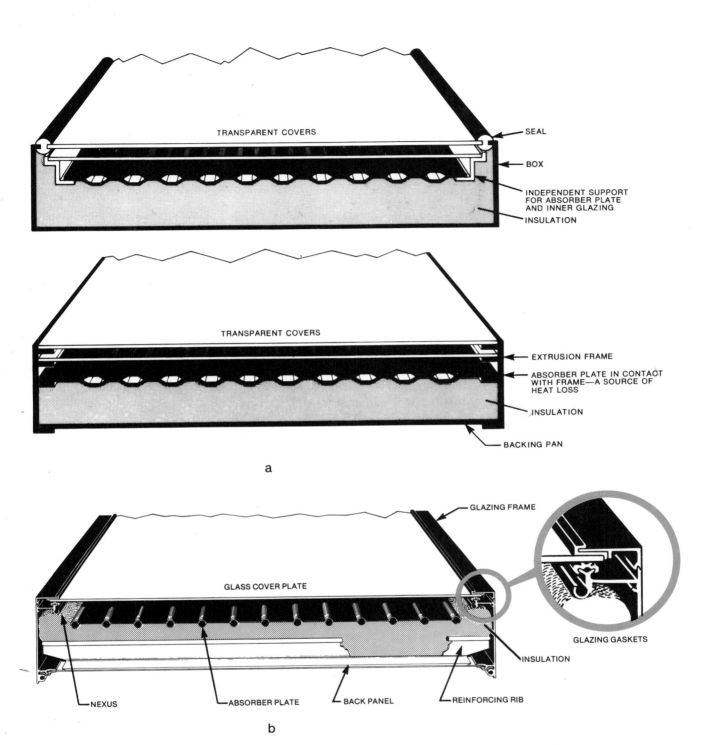

**FIGURE 4-9** Enclosure designs: (a) double glazed; (b) single glazed. *Designs patented by AMETEK, Inc.*

flat mirrors (or a Fresnel reflector) can be used to concentrate light onto a single place or a narrow line.

Lenses can also be used for light concentration. A plano-convex lens, similar to the familiar magnifying glass, will focus light onto a small area. Convex, cylindrical lenses will yield a narrow line. A Fresnel lens uses adjacent

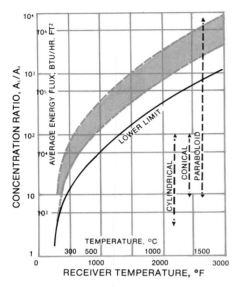

**FIGURE 4–10** The effect of concentration ratio on receiver temperature.

flat refractors to accomplish the same effect but can be made in a flat sheet rather than a convex shape.

No matter what the means employed, concentration is a matter of degree which is specified by the concentration ratio (CR); the concentration ratio of a concentrating collector is the ratio of area of sunlight intercepted ($A_c$) to the area onto which it is concentrated ($A_x$). Figure 4–10 depicts the effect of concentration ratio, CR, on the equilibrium temperature of an object placed at the focus of a concentrating collector. The lower limit shown represents the concentration ratio required to offset thermal losses. Any useful gain would require the higher CR shown in the shaded area.

Concentrating schemes can only make use of the direct beam contribution of total solar irradiance, so the total energy collected will be less than that collected by a flat plate intercepting the same area, since the flat plate collects both direct and diffuse radiation. Thus, the principal reason for using concentration is to increase the operating temperature. It has been argued that, given a stated interception area of solar irradiance, a reflecting concentrating collector may be more cost effective than a flat plate absorbing area, but the cost advantage is usually more than equalled by higher initial, maintenance, and operating costs. Moreover, dirt and scratches are more degrading for concentrating collectors. While all these considerations result in fewer applications, concentrating collectors still comprise the second largest group of solar energy systems, with a number of variations worth noting.

## MULTIPLE HELIOSTAT CONCENTRATORS

The first concentrating technique, of use primarily in large, central utility installations, is the concentration of sunlight by reflecting the solar energy from a number of flat mirrors onto a central receiver. Each of the flat mirrors must be individually tilted on two axes with a central computer providing the necessary tracking information. These rotating mirrors are called "heliostats" and a multiple heliostat concentrator is shown diagrammatically in Figure 4–11. The degree of concentration attained in a multiple heliostat concentrator is large enough to provide temperatures necessary for the generation of electricity from super-heated steam. Thus, although it is suitable for solar energy utilization in utility power plants, it is not considered appropriate for supplying solar energy directly to individual buildings and houses.

The technologies employed in the central receiver can be more advanced because its cost is based on the large area of sunlight intercepted by the heliostats rather than upon the small area of the receiver itself.

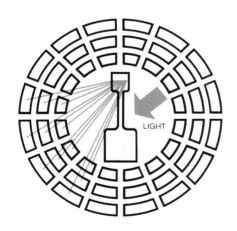

**FIGURE 4–11** Central receiver.

## FOCUSING REFLECTORS

Reflecting bowl-shaped surfaces may be used to concentrate solar energy onto an area smaller than the area of sunlight intercepted by the bowl. The highest concentration ratio is obtained with a paraboloidal surface that focuses sunlight onto a small disc representing the optical image of the sun, in a manner similar to that employed in reflecting telescopes. In fact, any portion of such a surface, on or off its principal axis, can serve as a focusing device. Like each of the mirrors in a multiple heliostat concentrator, a paraboloid must be rotated on two axes to follow the sun. With one axis properly aligned in parallel to the earth's axis of rotation, the tracking during any particular day can be achieved on one axis, while seasonal tracking is obtained on the other. A paraboloidal reflector is shown diagrammatically in Figure 4–12.

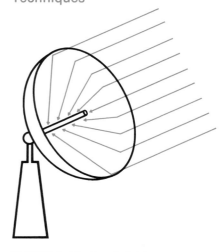

**FIGURE 4–12** Paraboloid.

A reflecting spherical bowl can also be used to concentrate sunlight. Since the requirement for a well-defined solar image is not fully essential for most concentrating applications, it is possible to sacrifice a small amount of concentration ratio to gain other benefits. Because a spherical surface is symmetrical to incident angles of light, tracking of the sun does not require rotation of the entire surface, only the receiver. This principle may be applied where very large stationary bowls are fabricated by mounting reflectors into extinct volcano calderas, meteor craters, or other depressions in the ground with a suspended, mobile, central receiver to track the moving focal region of the sun's energy. The same technique is used in some radar antennae and large radio telescopes.

A reflecting conical bowl will concentrate light also, but to a lesser degree than a spherical or paraboloidal reflector. Its primary advantage is that it can be fabricated from a flat sheet of reflecting material because it is a simple curved surface, in contrast to the others which are compound curve surfaces.

Focusing collectors can also be made from troughed surfaces that focus sunlight onto a long pipe or linear region, as shown diagrammatically in Figure 4–13. The concentration ratio is fundamentally smaller than that for bowl collectors but this trough type does have many practical advantages. The reflecting surfaces, whether of parabolic, circular, or V cross section, are simple curved surfaces capable of fabrication from flat sheets. The focal region of a trough is well suited to the transfer of heat to conventional fluid pipes, and the tracking requirements along at least one of the two axes are less stringent if the orientation of the collector is properly chosen.

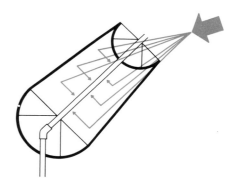

**FIGURE 4–13** Trough concentrator.

## LENSES AND REFRACTION CONCENTRATORS

The refracting property of transparent materials can be utilized by constructing lenses of glass or any of a wide variety of transparent plastics to concentrate light onto small focal regions. Three of the most commonly seen lens geometrics are diagrammatically shown in Figures 4–14, 4–15, and 4–16. Just as with reflectors, refracting devices can be designed to concentrate light onto a region surrounding a single point or onto a line conforming to the geometry of fluid pipes. The constraints for tracking are also quite similar to their geometric counterparts among reflectors.

Refracting elements offer some advantages over reflecting techniques since the refracting element can serve the dual purpose of acting as a convection suppressing cover plate, as well as a concentrating element. Under the most practical conditions of using the most economical materials, the transmittance of most refracting elements is greater than the reflectance of reflecting surfaces. Thus, refracting concentrators can usually collect more energy per area of sunlight intercepted than would reflecting techniques.

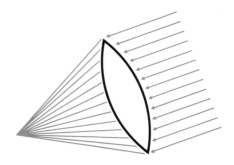

**FIGURE 4–14** Convex lens (edge view).

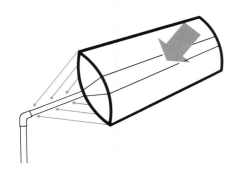

**FIGURE 4–15**   Linear convex lens.

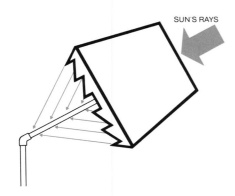

**FIGURE 4–16**   Linear Fresnel lens.

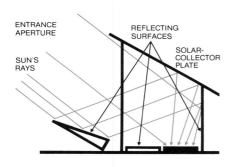

**FIGURE 4–17**   Pyramidal reflector.

## FIXED CONCENTRATORS

Because it is widely acknowledged that one of the principal constraints in using concentrating collectors is the necessity for tracking the sun, a great deal of attention and effort have been directed toward the development of a fixed concentrator. Fixed concentrators are usually designed to operate within a prescribed acceptance angle over which the concentrator can collect sunlight.

The acceptance angle of a fixed concentrator is dependent upon the concentration ratio in a manner which is a function of the geometry of the collector. Invariably, the acceptance angle tends to increase as concentration ratio is decreased. Ideal reflector geometries which optimize the combination of acceptance angle and concentration ratio have been described, and recently patented by Winston, so collectors of this type are often called Winston Concentrators or compound parabolic concentrators (CPCs).

Another class of fixed concentrators is known as pyramidal concentrators. These make use of multiple reflections into more confined areas where the solar collection element is located. A typical example of a pyramidal reflector for a fixed concentrator is shown in Figure 4–17. Among the limitations of these reflectors (which use flat reflecting surfaces in the main) is the necessity for multiple reflections which successively reduce the incident energy intensity; a second drawback is the large reflector area compared to collection aperture area. In addition, the external reflector is vulnerable to the effects of wind and snow.

CPCs and other ideal geometries require curved surfaces that may be difficult to fabricate and support. Many of the CPC geometries, including the more exotic "seashell" curves, toroidal and involuted shapes, despite their undeniable elegance, require large surface areas of reflector as compared to the aperture area of sunlight collected.

## TRACKING TECHNIQUES

Concentrating collectors, for the most part, must follow the sun as it moves across the sky to gain the efficiencies that justify their added costs. The accuracy with which a concentrating collector must follow the sun increases with the concentration ratio, CR. As has been shown, some collector geometrics, particularly those with spherical or cylindrical reflectors, allow the reflector surface to remain fixed while the receiver is tracked. It is also possible to rotate a flat reflecting surface around the earth's polar axis and reflect the sun's rays in a fixed direction along the polar axis so that concentration can be achieved without tracking any of the collector components. Figure 4–18 shows a few of the techniques used for tracking in order of their increasing accuracy.

The action of sunlight on a temperature sensitive bimetallic helix will provide sufficient tracking force for the lightweight thermal heliotrope shown in Figure 4–18a. As the collector approaches its correct position, the shade automatically regulates the amount of sunlight heating the coil, a continual process which allows the collector to progressively advance in concert with the sun.

Paired photocells (Figure 4–18b) are used to provide an electrical signal, which is a function of the degree of tilt of the collector surface from the normal

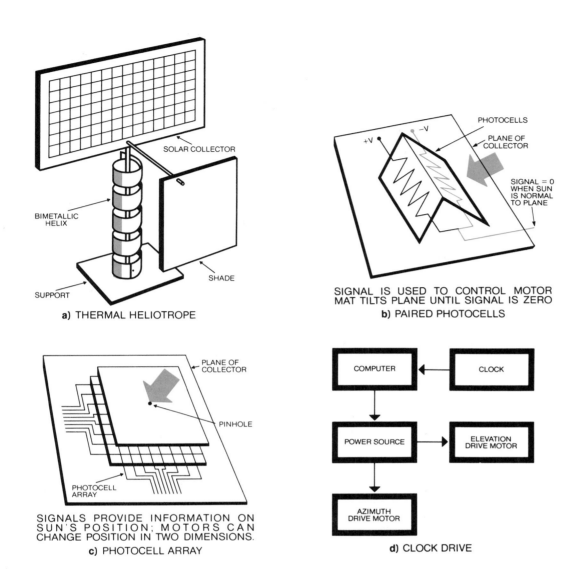

**a)** THERMAL HELIOTROPE

SIGNAL IS USED TO CONTROL MOTOR
MAT TILTS PLANE UNTIL SIGNAL IS ZERO
**b)** PAIRED PHOTOCELLS

SIGNALS PROVIDE INFORMATION ON
SUN'S POSITION; MOTORS CAN
CHANGE POSITION IN TWO DIMENSIONS.
**c)** PHOTOCELL ARRAY

**d)** CLOCK DRIVE

**FIGURE 4–18** Tracking techniques.

to the sun's direction. Motor drives are set to respond to the signal by tilting the surface in the same direction that reduces the signal, thus following the sun.

Tracking on two axes can be achieved by the use of a pinhole mounted over a photocell array, as shown in Figure 4–18c. Two motors are operated by signals from the opposing pairs of photodetectors. In the same manner, an array of photocells under a pinhole can be used to maintain an adjustable tilt angle.

Clock drives (Figure 4–18d), operating from the precise determinations of the sun's position made possible with computers, are used in tracking large or heavy collectors in the same manner as their use in telescope tracking. Unlike other tracking techniques, a clock drive can operate when the sun is obscured and, indeed, will track any object in the sky, visible or invisible, which follows a known or predictable path. It is therefore the most accurate technique.

# 5 Flat Plate Collector Efficiency

The efficiency of a flat plate solar collector is an important parameter for determining the collector area required to provide a specified amount of solar thermal energy. When purchasing the collectors for a system, efficiency is an important measure to be used in conjunction with other properties (shape, weight, strength, and durability, to name a few) as a means of comparison between various collector designs.

While efficiency is a vital concern, it is also one of the most difficult properties of a collector to measure accurately; it is potentially subject to misinterpretation and abuse since it is neither an obvious property nor a single quantity. Furthermore, the value of efficiency that pertains to conditions at the solar noon hour is quite different from the efficiency that applies for the course of a day, month, or year. This is because the instantaneous efficiency changes with different levels of solar irradiance, temperature, and wind speed in a manner prescribed by the curves shown in Figures 5–6, 5–7, 5–9, and 5–10. Before progressing to the mathematical relationship which defines efficiency, it is of value to first review the sources of potential collector gains and losses with a brief note of the counter-measure specified in Chapter 4. Also relevant is the consideration of the accuracy of the means of measurement for all collector properties (Appendix C).

## COLLECTOR GAIN AND LOSS SUMMARY

The goal of a solar collector designer is to maximize the energy gains while minimizing the energy losses. The most efficient collector would be an unglazed blackened plate operating near or below ambient temperature. Swimming pool heating meets this requirement at times, but for most other practical applications, the collector plate will be at some elevated temperature. For reasons outlined in Chapter 4, selective coatings offer compromises in gains while reducing losses at higher temperatures. The higher the operating temperature is above the ambient, the more important these factors become. The wind plays an important part in collector design, necessitating glazing in applications where it normally would not be required. For example, a brisk breeze below 70°F will cool unglazed swimming pool collectors faster than the sun can warm them.

Fin conduction loss due to thermal gradients on the absorber plate occurs when heat generated by the absorptance of sunlight on the absorber plate is not transferred into the fluid. This loss results when portions of the absorber plate are not in close proximity to the fluid conduits, when the thermal conductivity of the absorber plate is poor, or when the fluid flow is poorly distributed. Fin conduction loss can be minimized by the proper choice of plate materials, construction method, and fluid flow pattern in the absorber plate.

Thermal radiation loss results from the net electromagnetic radiation that is emitted from any part of the collector that is heated. This loss tends to increase as the fourth power of the absolute temperature rises in accordance with the Stefan-Boltzmann equation. It can be reduced by using selective coatings of low emissivity and high absorptance for solar radiation on the absorber plate. Another technique is to use selectively-reflecting "heat mirrors" that reflect the emitted energy back onto the absorber plate without disturbing the incoming solar energy.

Convection losses are due to the effects of air circulation between surfaces of different temperatures. As the absorber plate temperature rises, convective movements of air carry its heat onto adjacent cooler surfaces. Suppression can be accomplished by using multiple covers or honeycomb suppressors that reduce the freedom of air to move without disturbing the incoming solar radiation.

Conduction losses occur through materials that are in direct contact with the absorber plate. Such losses can be reduced by insulating the back and edges of the absorber plate with materials which have low thermal conductivity. Fastening the absorber plate rigidly to prevent movement during shipment in a manner which prevents thermal losses is similarly important.[1]

Losses due to condensation and air leakage in the collector enclosure are minimized by using tight construction for the enclosure and by properly directing the air that daily enters and leaves the collector as its temperature rises and falls, allowing condensation to run directly out of the collector and preventing condensation accumulation.

## INSTANTANEOUS EFFICIENCY

The instantaneous efficiency of a flat plate collector is defined as:

$$\eta = \frac{\dot{m} C p \Delta T}{A_c I_t} \tag{5-1}$$

where:  $\eta$  is the efficiency of the collector expressed as a number between 0 and 1

$\dot{m}$  is the rate of fluid mass flow through the collector (lb/hr)

$C_p$  is the heat capacity of the fluid (Btu/lb-°F; water = 1.0, air = .24)

$A_c$  is the gross area of the collector (ft$^2$)

$I_t$  is the total solar irradiance (Btu/hr-ft$^2$)

$\Delta T$  is the temperature difference between the outlet and inlet fluid
$\Delta T = T_0 - T_i$ (°F).

1. Patented by AMETEK, Inc.

The efficiency of a flat plate collector may also be expressed in terms of the solar irradiance, $I_t$, intercepted and the fundamental quantities related to the plate's construction. Following Duffie and Beckmann:[2]

$$\eta = F_R A_x \left[ K_{\tau\alpha}(\tau\alpha)_{e,n} - U_L \frac{(T_c - T_a)}{I_t} \right] \quad (5\text{--}2)$$

where:

$$F_R = \frac{\dot{m}C_p}{U_L} \left[ 1 - e^{\frac{-U_L F'}{\dot{m}C_p}} \right] \quad (5\text{--}3)$$

$$F' = \frac{1}{w\left[ \dfrac{1}{F(w-u)+u} + \dfrac{U_L}{h_o\pi(u-2t)} + \dfrac{U_L}{C_b} \right]} \quad (5\text{--}4)$$

and where:

$\eta$ = thermal efficiency of the collector

$F_R$ = overall heat transfer factor

$F'$ = collector geometry efficiency factor

$F$ = fin efficiency factor (Equation 5–8)

$A_x$ = $A_p/A_g$

$A_p$ = area of the absorber plate

$A_g$ = gross collector area

$K_{\tau\alpha}$ = incident angle modifier (see Equation 5–5)

$(\tau\alpha)_{e,n}$ = effective transmission absorptance factor at normal incidence

$\bar{T}_c$ = average collector temperature = $\dfrac{\tau\alpha I}{U_L} + \bar{T}_a$

$\bar{T}_a$ = average ambient temperature

$U_L$ = collector heat loss coefficient

$\dot{m}$ = mass flow rate

$C_p$ = specific heat of fluid

$w$ = distance between tubes

$u$ = outside diameter of circular risers in the plate

$t$ = tube wall thickness

$h_o$ = convection coefficient between the tube wall and circulating fluid

$C_b$ = bond conductance between fin and riser where

$$C_b = \frac{k_b b}{\gamma}$$

$k_b$ = thermal conductivity of the material bonding the plate and fin

$b$ = bond length

$\gamma$ = bond thickness

2. Duffie, J. A., and Beckman, W. A. *Solar Engineering of Thermal Processes.* John Wiley & Sons, 1980.

The equation for efficiency, 5–2, is divided into two terms within the brackets. The first term describes the solar irradiance which is absorbed, a quantity which is proportional to the intensity and angle of incidence of solar radiation hitting the plate. Irradiance is reduced at angles from the perpendicular direct beam irradiance by the cosine law (Equation 2–12), increased reflection from the outermost glazing (Equation 5–6), and shadows cast upon the plate by the edges of enclosure (Equation 5–8). These off-angle reductions act only on the first term in the efficiency equation (see Appendix B). The first term is also influenced by the fin efficiency factor F and the collector overall heat loss coefficient $U_L$. The fin efficiency is determined by how far the heat must travel to a heat removal conduit. A fully wetted plate such as that shown in Figure 4–3d and approximated by Figure 4–3a have fin efficiencies approaching 1.0.

## FIN EFFICIENCY FACTOR

Tube spacing in a flat-plate collector depends upon the tubing diameter, thermal conductivity and thickness of the absorber plate, and the heat transfer coefficients between both fluid and plate and plate and surroundings. An approximate guide to the tube spacing sufficient to provide a 95 percent efficiency in the transfer of heat from the absorber plate to the fluid is shown in Table 5–1.

A useful expression for the fin efficiency of an absorber plate is derived by Duffie and Beckman:

$$F = \frac{\tanh\left[\left(\sqrt{\frac{U_L}{k\delta}}\right)\left(\frac{w-u}{2}\right)\right]}{\left(\sqrt{\frac{U_L}{k\delta}}\right)\left(\frac{(w-u)}{2}\right)} \qquad (5-5)$$

where:  $F$  is the fin efficiency expressed as a number between 0 and 1

tanh  is the hyperbolic tangent, where $\tanh(x) = \dfrac{e^x - e^{-x}}{e^x + e^{-x}}$

$U_L$  is the overall loss coefficient of the absorber plate from Equation (5–10) in Btu/hr-ft²-°F

**TABLE 5–1**   Tube Spacing for 95% Plate Efficiency

| Fin Thickness | | Collector Plate Material Tube Spacing | | | | | |
|---|---|---|---|---|---|---|---|
| | | Copper (w) | | Aluminium (w) | | Steel (w) | |
| (mm) | (in) | (cm) | (in) | (cm) | (in) | (cm) | (in) |
| 0.1 | .0039 | 9.0 | 3.5 | 7.5 | 2.95 | 4.0 | 1.57 |
| 0.2 | .0079 | 11.5 | 4.53 | 9.5 | 3.74 | 6.0 | 2.36 |
| 0.3 | .0118 | 12.5 | 4.92 | 10.5 | 4.13 | 6.5 | 2.56 |
| 0.5 | .0197 | 14.5 | 5.71 | 13.0 | 5.12 | 7.5 | 2.95 |
| 1.0 | .0393 | 16.0 | 6.3 | 15.0 | 5.90 | 9.5 | 3.74 |
| 1.5 | .0590 | | | 16.5 | 6.5 | 10.5 | 4.13 |
| 2.0 | .0787 | | | 18.0 | 7.08 | 11.5 | 4.53 |
| 3.0 | .1181 | | | | | 13.0 | 5.12 |

$k$ is the thermal conductivity of the absorber plate in Btu-in/hr-ft$^2$-°F

$\delta$ is the thickness of the fin in inches

$w$ is the distance between centers of fluid conduits in inches

$u$ is the outside diameter of the fluid conduit in inches.

## INCIDENT ANGLE MODIFIER

Since the effective transmittance absorptance product $(\tau\alpha)_e$ varies with the incident angle, $\theta$, and depends on the properties and number of transparent covers, a short cut has been devised to describe the variation of $(\tau\alpha)_e$ with incident angle.

The short cut involves defining an incident angle modifier as the ratio of $(\tau\alpha)_e$ at incident angle, $\theta$, to its value at normal incidence, $\theta = 0$. The incident angle modifier is:

$$K_{\tau\alpha} = \frac{(\tau\alpha)_e}{(\tau\alpha)_{e,n}} \qquad\qquad \textbf{(5–6)}$$

It has been shown that the incident angle modifier can be approximated by the expression:

$$K_{\tau\alpha} = 1 - b_0 \left[ \frac{1}{\cos\theta} - 1 \right] \qquad\qquad \textbf{(5–7)}$$

for most flat plate collector geometries and for values of $\theta$ that are within the usual operating region of practical interest (i.e., not too close to 90°).

The determination of the relationship when $\Delta T/I = 0$ is one of the requirements of the ASHRAE 93-77 test procedure. Although there are several problems with the incident angle data obtained from ASHRAE 93-77, this procedure simultaneously measures the reductions due to reflections from the surface of both the cover glazing and the absorber plate, and the additional effect of the edge shadow modifier (which is discussed separately below).

The data problems presently being considered by the 93-77R committee are as follows:

1. It is allowable to test at solar noon with the collector tilted away from the sun at the prescribed angles. This results in a higher irradiance level and less percentage diffuse light than normally occurs when the collector is in a fixed south facing mount. The effect of higher irradiance levels is greater on poor absorber plates since their ability to transfer high flux rates is limited. Therefore at the 60° incident angle test point, low values should theoretically be obtained when the irradiance is higher than normal.

2. In order to obtain the value of $b_o$ in Equation 5–6, a linear fit through the four data points (0°, 30°, 45°, 60°) is executed. This is usually done by computer, although many scientific hand-held calculators can easily perform the regression calculations. When computers are used, the best "fit" normally occurs when the equation does not go through 1.000. The Solar Rating and Certification Corporation's First Edition shows $b_o$ values as high as 1.007 and as low as 0.986. This affects the near noon values when calculating all-day performance for rating purposes. Figure 5–1 shows data taken with a fixed-mount, single-glazed, black-chrome collector. Both morning and afternoon values were tabulated and aver-

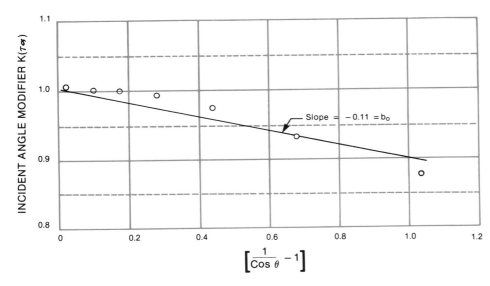

**FIGURE 5-1** Incident angle modifier for D-189 SunJammer solar collector.

aged. Data taken approximately every half hour showed resultant average angles 2.3°, 5.3°, 23.2°, 30.7°, 38.1°, 45.6°, 53.2°, and 60.5°. If the data are forced through 1.0 at the intercept, the use of a linear plot of $\left[\dfrac{1}{\cos\theta} - 1\right]$ in a prediction of all-day data will underestimate (by a few percentage points) the performance of the collector within the 0 to 45° incident angle range. This is the most important part of the day. For use in rating programs, data linearized by approximations should not be used; instead more data near noon should be taken and the *real* data used.

3. A method must be found to weight the modifiers so that diffuse radiation is recognized. Even clear days which meet the extreme steadiness of radiation required to do ASHRAE 93-77 testing will have diffuse radiation of 8–10 percent at noon time and 20–30 percent at a 60° incident angle. Most days have much more diffuse radiation at off-angle times of the day.

The brief discussion of the edge shadow modifier is included here, but no separate testing or calculation of its effect is necessary when the ASHRAE 93-77 procedure is used.

## EDGE SHADOW MODIFIER

Since the absorber plate of a flat plate collector lies below the level of the outermost transparent cover, the edges of the collector which support the covers will cast shadows onto the absorber plate. This effect can be significant at high incidence angles, especially if the collector is of a deep profile.

The width of the shadows cast onto the absorber surface can be determined from Equations 2–32 and 2–33 for the usual collector orientations on tilted surfaces. The value to be used for the "height of the fence, $\Gamma$," is the height of the collector edge above the *absorber plate*.

The edge shadow modifier for absorber plates of the same dimensions as the aperture is given by:

$$K_s = 1 - \frac{L_c w_1 + W_c w_2 - w_1 w_2}{A_c} \qquad \textbf{(5-8)}$$

where:   $L_c$  is the length of the collector aperture

$W_c$  is the width of the collector aperture

$A_c = L_c W_c$  is the area of the collector aperture

$w_1$  is the shadow width from Equation 2–32

$w_2$  is the shadow width from Equation 2–33.

The edge shadow modifier is then multiplied by the direct beam solar irradiance to account for the shaded area on the absorber plate at incident angles greater than zero.

There is some benefit from the solar irradiance that strikes the inner wall of the edges opposite to those that shade the absorber plate. For this reason, though the benefit is small compared to the shadow loss, it is best for those visible inner walls to be reflecting or white in color so that the greater portion of this solar energy will be reflected onto the absorber plate. The loss due to shadows is greater than the simple geometrical loss of solar irradiance, since the forced circulation of fluid in the shaded region of the plate continues to deliver heat to this cooler portion, resulting in a negative contribution to net energy delivered.

## COLLECTOR LOSSES AND LOSS COEFFICIENT

The total losses of a flat plate collector may be represented by:

$$J_T = J_{\text{up}} + J_{\text{back}} + J_{\text{edge}} \qquad \text{Btu/hr-ft}^2 \qquad \textbf{(5-9)}$$

where $J_{\text{up}}$ is the upward loss through the cover plates given in Equation 5–11, $J_{\text{back}}$ is the loss through the back of the collector given in Equation 5–13, and $J_{\text{edge}}$ is the loss through the edges of the collector given in Equation 5–15.

The thermal loss coefficient of the collector is given by:

$$U_L = \frac{J_T}{T_i - T_a} \qquad \text{Btu/hr-ft}^2\text{-}^\circ\text{F} \qquad \textbf{(5-10)}$$

where $J_T$ is the total loss given by Equation 5–9, $T_i$ is the inlet temperature in °F and $T_a$ is the ambient temperature in °F. This thermal loss coefficient is an important quantity in evaluating the performance of a flat plate collector and should, ideally, be as small as possible.

The heat loss upward through the transparent covers of a well-constructed flat plate collector can be closely approximated. The formula was initially developed by S. A. Klein[3] and has since been modified as shown in Equation 5–11.

The standard deviation of the difference in values of $J_{\text{up}}/A\,(T_p - T_a)$ for 972 computer test trials for closure was 0.14 $W/M^2$ °K (0.0247 Btu/ft²-°R).

3. S. A. Klein. "Calculation of flat plate collector loss coefficients." *Sol. Egy.* 17, No. 1:1975.

$$J_{up} = \left[ \frac{\dfrac{T_\rho - T_a}{N}}{\dfrac{C}{T_\rho}\left(\dfrac{T_\rho - T_a}{N + f}\right)^{.33} + \dfrac{1}{h_w}} \right] +$$

$$\left[ \frac{\sigma(T_\rho^4 - T_a^4)}{\left(\dfrac{1}{\varepsilon_p + .05N(1 - \varepsilon_g)}\right) + \left(\dfrac{2N + f - 1}{\varepsilon_g}\right) - N} \right] \qquad \textbf{(5–11)}$$

where: $J_{up}$ = frontal heat loss (Btu/hr-ft²)

$T_\rho$ = plate surface temperature (°R; °R = °Rankine = °F + 460)
   $\cong (T_o + T_i)/2$ for efficient plates

$T_o$ = fluid outlet temperature (°R)

$T_i$ = fluid inlet temperature (°R)

$T_a$ = outdoor ambient temperature (°R)

$N$ = 1.00, 1.85, and 2.65 for 1, 2, and 3 cover plates[4]

$\varepsilon_p$ = emissivity of the absorber plate

$\varepsilon_g$ = emissivity of the cover glazing (0.86 for glass)

$\sigma$ = 1.7132 × 10⁻⁹ Btu/ft²-hr-°R⁴—the Stefan–Boltzman constant in English units. See Equation A–2

$h_w$ = 1 + 0.3V Btu/hr-ft²-°R where $V$ is the wind velocity in mph

$f$ = (1 − 0.227 $h_w$ + 0.0161$h_w^2$) (1 + 0.091$N$)

$C$ = 115.3 (1.0 − .00409 Σ)⁵

$\Sigma$ = tilt angle from horizontal (degrees).

The limits of the testing conditions were:

| | |
|---|---|
| 116°F ≤ $T_\rho$ ≤ 296°F | 0 ≤ $V$ ≤ 22 mph |
| 8°F ≤ $T_a$ ≤ 98°F | 1 ≤ $N$ ≤ 3 |
| 0.1 ≤ $E_p$ ≤ 0.95 | 0 ≤ Σ ≤ 90° |

With the calculated value of convection heat loss for each of the alternative cover configurations, the optimum choice of one, two, or more covers for a specific application may be made. The secondary concern associated with increasing the number of covers—an increasing reduction in solar energy admitted to the collector—may be addressed if necessary. The usual practice, however, is for collector manufacturers to supply an efficiency curve for their product and its cover configuration. The specific characteristics of the cover material, and the effect of the covers, are accommodated in the curve (which is sufficient data for an evaluation of equipment for a particular application). At the same time, be aware of the relationships from which the transmittance, $\tau$, absorptance, $\alpha$, and reflectance, $\rho$, of various cover materials may be calculated, as well as the means of quantifying the effect of internal reflections between covers in a multiple cover configuration. The equations and a discussion of their use are contained in Appendix B. As a general rule, covers should

4. W. W. S. Charters. "Solar energy utilization—liquid flat plate collectors." *Solar Energy Engineering*, edited by A. A. Sayigh. Academic Press, 1977; pages 105–135.

5. From empirical testing by AMETEK from 40° to 62°.

be constructed of materials which possess a high transmittance, $\tau$, for solar radiation, low reflectance, $\rho$, low absorptance, $\alpha$, and, of course, should be able to withstand environmental hazards, such as the impact of hailstones and high winds.

The incoming solar energy that is absorbed on the collector plate of a flat plate collector is approximately given by:

$$J_{in} = \tau_e \alpha_e I_1 = (\tau \alpha)_e I_1 \tag{5-12}$$

where: $J_{in}$ is the energy absorbed by the absorber plate in Btu/hr-ft$^2$

$\tau_e$ is the effective transmittance of the transparent covers at *solar* wavelengths, given by Equations B–3, B–5 or B–8, whichever is appropriate

$\alpha_e$ is the effective absorptance of the absorber plate at *solar* wavelengths

$I_t$ is the total solar irradiance given by Equation 3–1.

The quantity $\tau_e \alpha_e$ is referred to as the effective transmittance-absorptance product, denoted $(\tau \alpha)_e$.

Further refinement of the analysis of incoming solar energy requires that multiple reflections between a diffusely reflecting absorber plate and its covers be considered. This requires knowledge of the dependence of the reflectance of the absorber plate on incident angle and the degree to which it is specular or diffused; points which are also discussed in Appendix B. When absorber reflectance is minimized, the effects of such multiple reflections are very small.

Another refinement consists of dividing the incoming total solar irradiance into its direct, diffuse, and reflected components. In principle, Equation 5–12 should only apply to the direct beam radiation, but the distribution of diffuse and reflected radiation with various incident angles is complex; unless some assumptions are made it is difficult to determine its effective incident angle. One approach often used is to assume that diffuse radiation is isotropically distributed (equally strong from all directions) and use an average angle of 60° for its incident angle. In fact, diffuse radiation is not isotropic. Its intensity tends to increase toward the zenith (straight up) for that which emanates from clouds, while the intensity of the blue sky diffuse radiation is greatest in the direction of the sun. Moreover, diffuse and reflected radiation is polarized in varying degrees and has varying spectral characteristics.

It is often satisfactory to use values of $\alpha$ and $\tau$ that are averaged over all values of incident angles and averaged over all solar wavelengths in Equation 5–12. This is actually required for the diffuse and reflected components of $I_t$. The effects of polarization at large incident angles are usually ignored, as are the effects of variations in the index of refraction of air with temperature and humidity.

## INSULATION AND BACK LOSSES

Heat loss from the back of a flat plate collector is minimized with thermal insulation. The utility of various insulation materials can be characterized by comparing their thermal conductivity factors, $k$ (Btu per hour per square foot of insulated area per inch of thickness per F degree of temperature difference).

Precise calculations of heat loss from the back are used to quantify the characteristics of a specific system but, more importantly, provide another measure of evaluating collectors through the relationship:

$$J_{\text{back}} = \frac{T_\rho - T_a}{\dfrac{x}{k} + \dfrac{1}{h_b}} \qquad\qquad (5\text{--}13)$$

where:   $J_{\text{back}}$   is the heat loss in Btu/hr-ft$^2$ from the back of the collector

   $k$     is the insulation k-factor in Btu-in/hr-ft$^2$-°F

   $x$     is the thickness of the insulation in inches

   $T_\rho$    is the temperature of the absorber plate in °F

   $T_a$    is the ambient temperature in °F

   $h_b$    is the convection coefficient (2 to 4 Btu/hr-ft$^2$-°F)

Equation 5–13 ignores the effect of an insulation box or back pan. In order to include the effects of additional insulation and other materials with different k-factors, it is convenient to use the *R*-factor where:

$$R = \frac{x}{k} \text{ and } R_T = \frac{x_1}{k_1} + \frac{x_2}{k_2} + \cdots = R_1 + R_2 + \qquad (5\text{--}14)$$

and the equivalent values of $R$ for two separate insulation layers can be added.

## EDGE LOSSES

Edge losses may be treated in a manner similar to back losses by defining the edge area as the product of the perimeter of the collector and its thickness. If the edge is insulated with material having thermal conductivity, $k$, and thickness, $x_e$, the edge loss will be approximated by:

$$J_{\text{edge}} = \frac{k_e}{x_e} (T_\rho - T_a) \frac{\text{Edge area}}{\text{Collector area}} \qquad (5\text{--}15)$$

where:   $J_{\text{edge}}$   is the edge loss in Btu/hr-ft$^2$-°F

   $x_e$    is the thickness of the edge insulation in inches

   $k_e$    is the thermal conductivity of the edge insulation
        in Btu-in/hr-ft$^2$ °F

   $T_\rho$    is the temperature of the absorber plate, °F

   $T_a$    is the ambient temperature, °F.

Absorber plate mounting means can be similarly determined.[6]

## TRANSPARENT COVERS

The cover of a solar collector protects the absorber plate from environmental hazards, but its primary function is to reduce the heat loss upward by thermal convection. Collectors can be protected by several successive covers

6. Patented by AMETEK, Inc.

but, while each addition reduces convection loss to a greater degree, each additional cover will also reduce the amount of solar energy admitted because of the reflectance and absorptance of the added layers. The problem can be attacked by considering solar energy admission as most important, but the magnitude of convection loss is usually more critical. The rate of this heat loss increases with wind velocity, the difference between the collector plate and ambient temperatures, and the tilt angle, $\Sigma$, of the collector. But, assuming a given standard value for each of these variances, it is possible to represent the effect on collector efficiency by additional covers alone.

Figure 5–2 shows typical curves representing the efficiency of a flat plate collector as a function of the average fluid temperature for zero, one, two, and three transparent covers.

Several other general observations are noteworthy. At sufficiently low values of operating parameter, $\Delta T/I$, the performance is not improved by increasing the number of covers because the effects of the additional loss of solar transmission through the covers are greater than the decrease in the overall heat loss coefficient, $U_L$. Conversely, at high values of operating parameter, the improvement in performance as more covers are added is significant. Im-

**FIGURE 5–2** Effect at glazing covers on ASHRAE 93-77 data for black chrome and flat black paint at 40° tilt angle, 50°F ambient temperature, and $\theta = 0°$.

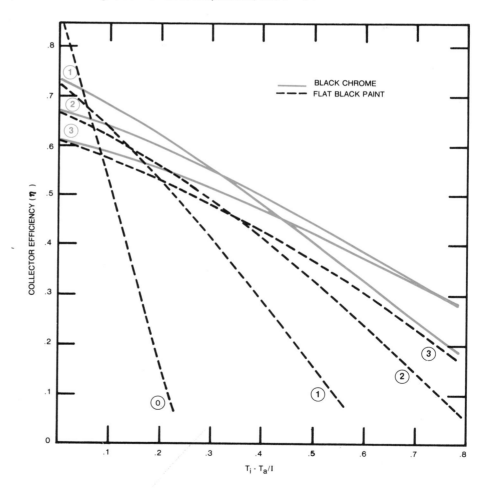

provement always results from the use of an absorber with a selective surface, provided that its absorptance of solar energy is not less than the non-selective surface. The degree of improvement increases with operating parameter but decreases with an increasing number of covers. Thus, the choice of the number of covers and whether or not to use a selective surface is dependent upon the operating parameter and the costs of various alternatives. Clearly, if plate temperatures in the neighborhood of 200°F are to be achieved, a selective surface is usually well worth its additional cost because of the marked improvement in efficiency.

## COLLECTOR PERFORMANCE MEASUREMENT

The measurements needed to determine the instantaneous efficiency of a flat plate collector through Equation 5–1 are temperature, fluid mass flow, solar irradiance, and area. Temperature measurements are required at the inlet and outlet of the collector with the difference between the two, $\Delta T$, often measured independently to improve the accuracy of the measurement. The ambient temperature surrounding the collector and the total intensity of solar radiation, including direct and indirect components incident to the collector aperture and in the plane of the collector, must be measured. Finally, the rate of mass fluid flow through the collector and the area of the collector aperture must also be measured in order to determine the instantaneous efficiency.

These measurements are the terms of the instantaneous efficiency equation, but the implementation of a test procedure for purposes of comparing collectors under equivalent conditions implicitly requires that many other quantities be known a well. For example, the tilt angle of the collector in its test setup, the specific heat of the liquid used in the collector, the wind speed and direction, and solar time should be known. Errors in the determination of these quantities can produce errors in efficiency measurement, but most experimenters previously felt that such errors were either very small, had little effect on the end result, or needed to be established only infrequently.

Since then, the errors and the magnitude of the errors were shown by the National Bureau of Standards (NBSIR 74-635) and Tech Note 899. This information was available to the ASHRAE 93 P committee near the end of its deliberations, but was without experimental verification. At that point in time, most experimenters were using single-glazed black and white pyranometers for total solar irradiance. These instruments had a tilt error which varied with irradiance intensity and the instrument. Errors as great as 13 percent were observed and data scatter was rampant. In the intervening years, these instruments have been mainly abandoned in favor of double-glazed instruments whose tilt error is smaller (0–2 percent) and is capable of being reproducibly calibrated.

The present accepted standard for solar collector testing is ASHRAE 93-77. Contained in this standard are specifications for the accuracies of the test instrumentation and descriptions of acceptable environmental test conditions. The testing consists of several mandatory operational test points, including a test near ambient conditions to determine the incident angle modifier. In addition, paragraph 8.3.2 states:

At least four different values of inlet fluid temperature shall be used to obtain the values of $\Delta T/I$. An acceptable distribution of inlet temperatures for flat-plate collectors is to set the $\Delta T$ at 10, 30, 50, and 70 percent of the stagnation temperature rise obtained at the given conditions of solar intensity and ambient air temperature.

At least four data points shall be taken for each value of $\Delta T/I$: two during the time period preceding solar noon and two in the period following solar noon, the specific periods being chosen so that the data points represent times symmetrical to solar noon. This latter requirement is made so that any transient effects that may be present will not bias the test results when they are used for design purposes. The requirement for obtaining data points equally divided between morning and afternoon is not mandatory when testing with an altazimuth mount. All test data shall be reported in addition to the fitted curve so that any difference in efficiency due solely to the operating temperature level of the collector can be discerned in the test report.

ASHRAE 93-77, ¶ 8.2.1. states:

Although a straight-line representation of the efficiency curve will suffice for many flat-plate solar collectors, some flat-plate collectors and most concentrating collectors require the use of a higher-order fit (i.e., a second order polynomial), due to variation of $U_L$ with receiver temperature.

The thought behind the 10, 30, 50 and 70 percent of stagnation points was that greater accuracy in the slope determination could be obtained by approaching the stagnation temperature. This assumes that the stagnation temperature is a constant value—in fact it is a variable according to the time of year (tilt angle, temperature, and irradiance level effects) and wind speed.

**TABLE 5–2** Stagnation Temperature

(Calculated for the steady state conditions with a Black Chrome Single Glazed Collector.)

| Normal Solar Irradiance $I$ (Btu/hr-ft²) | Ambient Temperature $T_a$ (°F) | Wind Velocity (mph) | Tilt Angle $\Sigma$ | Operating Parameter $\Delta T/I$ (hr-ft²-°F/Btu) | Stagnation Temperature (°F) | Comments |
|---|---|---|---|---|---|---|
| 370 | 50 | 0 | 40 | 1.032 | 432 | a,b,d,e |
| 370 | 50 | 5 | 40 | .902 | 384 | a,b,c,d |
| 370 | 50 | 10 | 40 | .846 | 363 | a,b,d,e |
| 370 | 100 | 0 | 40 | .978 | 462 | a,b,d,e |
| 370 | 100 | 5 | 40 | .868 | 421 | a,b,d |
| 370 | 100 | 10 | 40 | .821 | 404 | a,b,d,e |
| 330 | 50 | 0 | 40 | 1.064 | 401 | a,b,d,e |
| 330 | 50 | 5 | 40 | .924 | 355 | a |
| 330 | 50 | 10 | 40 | .863 | 335 | a,b,d |
| 330 | 100 | 0 | 40 | 1.000 | 430 | a,b,d,e |
| 330 | 100 | 5 | 40 | .885 | 392 | a,b,d |
| 330 | 100 | 10 | 40 | .836 | 376 | a,b,d,e |
| 330 | 50 | 5 | 0 | .864 | 335 | f |
| 330 | 50 | 5 | 20 | .899 | 345 | f |
| 330 | 50 | 5 | 40 | .924 | 355 | f |
| 330 | 50 | 5 | 60 | .958 | 366 | f |
| 330 | 50 | 5 | 80 | .997 | 379 | f |
| 300 | 50 | 0 | 40 | 1.09 | 376 | a,b,d,e |
| 300 | 50 | 5 | 40 | .94 | 332 | a,b,c,d |
| 300 | 50 | 10 | 40 | .88 | 314 | a,b,d,e |
| 300 | 100 | 0 | 40 | 1.03 | 409 | a,b,d,e |
| 300 | 100 | 5 | 40 | .907 | 372 | a,b,d |
| 300 | 100 | 10 | 40 | .85 | 355 | a,b,d,e |

The effects can best be understood by looking at the calculated values for commonly observed parameters on the stagnation temperature as shown in Table 5–2.

The following comments apply: (a) the wind effect on performance is very important; (b) an ambient temperature change of 50°F does not result in a 50°F increase in stagnation temperature; (c) going from 300 Btu/hr-ft$^2$ to 370 Btu/hr-ft$^2$ (a 23 percent increase) only increases the stagnation temperature above ambient by 18 percent at a 5 mph wind speed; (d) $\Delta T/I$ at the zero efficiency value decreases with increasing irradiance levels; (e) the wind speed however changes the operating parameter value, $\Delta T/I$, about 15 to 20 percent over the range 0 to 10 mph with other parameters constant; (f) stagnation temperature goes up with tilt angle.

The measurement of the wind speed is in itself no easy matter. Many weather stations have a chart recorder and the wind speed reported is the "eye-ball" mean of the pen's tracing. At the AMETEK test facility, the recording shown in Figure 5–3 was made. Shown are the actual minute-by-minute instantaneous values, the points recorded by a data logger (connected by straight lines) and the integrated value. "Eye-balling" the chart would give a mean value of 5 mph. Averaging the data logger points, gives 5.42 mph, but the integrated value is 4.67 mph. (It is interesting to note that by chance the data logger missed the peak value of 9.1 mph.)

AMETEK's conclusion was that when making wind speed measurements

**FIGURE 5–3** Actual wind speed, data logger points (connected & integrated wind speed).

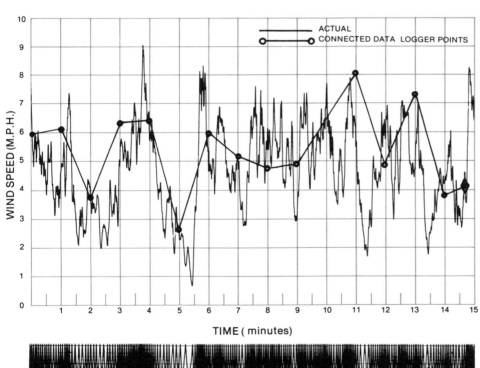

INTEGRATED WIND SPEED

**TABLE 5–3**  Test Temperatures if 10%, 30%, 50%, and 70% of the Stagnation Temperature is Used: 40° Tilt Angle, 5 MPH Wind Speed, 330 Btu/hr-ft²

(ASHRAE 93-77 Test Points)

| Collector Efficiency $\eta$ | Operating Parameter $\Delta T/I$ (hr-ft²-°F/Btu) | Incident Angle $\theta$ | Overall Heat Loss Coefficient $U_L$ (Btu/hr-ft²-°F) | Inlet Fluid Temperature $T_i$ (°F) | Outlet Fluid Temperature $T_o$ (°F) |
|---|---|---|---|---|---|
| .742 | 0 | 0 | .524 | 50 | 67 |
| .6816 | .101 | 0 | .678 | 83.4 | 100 |
| .5357 | .3036 | 0 | .784 | 150.2 | 163.5 |
| .3346 | .5061 | 0 | .842 | 217 | 226 |
| .201 | .708 | 0 | .887 | 284 | 287 |

for use in comparing performance with the theoretical model presented here, the integrated value should be used and data should be taken for at least one half hour before the test point to obtain a solid number for the average wind speed value. The cover glass of the specific collector, whose test results are shown in Table 5–3, weighed 59 pounds. Obviously it will not respond instantaneously to wind cooling effects and the half-hour time period seemed sufficient for the cover glass temperature to stabilize. From a cold start, however, it takes two hours or longer for the cover glass to get to its steady-state value.

Choosing the stagnation temperature for 40° tilt angle, wind speed of 5 mph, and solar irradiance of 330 Btu/hr-ft² resulted in the ASHRAE 93-77 test

**FIGURE 5–4**  Linear fit to 0, 10, 30, 50, and 70 percent of stagnation data points and ASHRAE 93-77 theory of parallel lines for off-angle performance.

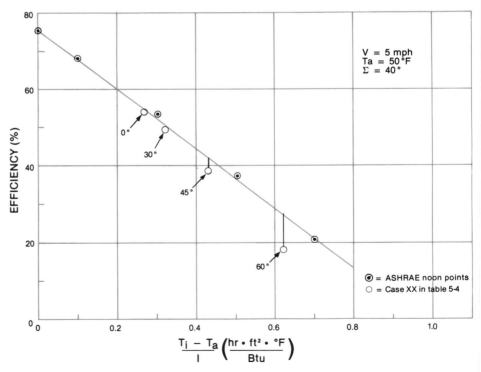

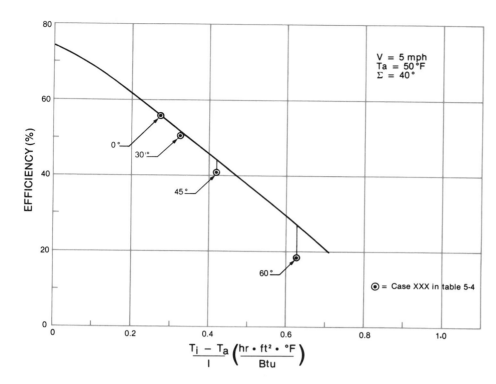

**FIGURE 5-5** Off-angle performance prediction using a perfect fit to the noon points and ASHRAE 93-77 theory of parallel lines for off-angle performance.

points shown in Table 5–3. The expected performance has been calculated from Equations 5–1 through 5–15. These points are plotted in Figures 5–4, 5–5 and 5–6.

The linear fit to the noon points results in an intercept of .755 and a slope ($U_L$) of .766 ($\eta$ = .755 − .766 $\Delta T/I$) as shown in Figure 5–4. If this equation were used to predict the performance of the same collector by making parallel lines to the noon curve for off-angle performance, the performance would be underpredicted by the amount shown in Table 5–4.

If a higher-order fit is chosen so that a perfect fit is made to the noon points, this results in Figure 5–5 (also shown in Table 5–4). The errors are still

**TABLE 5–4** Underprediction of Off-Angle Performance

| Incident Angle $\theta$ | Solar Irradiance I (Btu/hr-ft²) | Operating Parameter $\Delta T/I$ (hr-ft²-°F/Btu) | X $\eta$ | XX $\eta$ | Error XX to X % | XXX $\eta$ | Error XXX to X % |
|---|---|---|---|---|---|---|---|
| 0 | 330 | .273 | .559 | .546 | −2.4 | .559 | 0.0 |
| 30 | 277.5 | .324 | .512 | .493 | −3.9 | .507 | −1.0 |
| 45 | 215 | .419 | .428 | .396 | −8.0 | .407 | −5.2 |
| 60 | 143.8 | .626 | .241 | .185 | −31.0 | .183 | −31.7 |

X: Theory uses a variable loss at 140°F inlet
XX: Linear fit through 0, 10, 30, 50 and 70 percent of stagnation and parallel lines to this fit for off-angle performance
XXX: Higher order fit through 0, 10, 30, 50 and 70 percent of stagnation with parallel lines to this fit for off-angle performance.

greatest at the 60° point. Using higher-order fits does not greatly improve off-angle prediction accuracy. Both the linear and higher order fits underpredict expected performance at a constant temperature.

As can be seen from Equations 5–3 through 5–10, $U_L$ is not a constant, but varies in a complex way with the level of the absolute temperatures, wind speed, tilt angle, and flow rate. For a given inlet temperature and ambient temperature with constant wind speed, tilt angle, and flow rate, the loss term is a constant; varying the solar irradiance level yields a straight line plot (constant $U_L$) of efficiency versus $\Delta T/I$. This is only true for higher plate efficiency factors (.98 plus). This irradiance variation under clear skies with a collector normal to the sun will change during the course of the year from a low of 285 Btu/hr-ft² to 370 Btu/hr-ft². The sun varies in intensity as each hour passes. By means of an altazimuth mount (one that follows the sun) it is possible to have variable, normal incidence, solar irradiance during the course of one day. The effect of this was first shown by J. Hill et al: in NBS Tech Note 899 (Figure 55) reproduced here as Figure 5–7.

Another way of presenting this data is shown in Figure 5–8 as constant temperature lines with variable irradiance. The data are also shown in Table 5–5.

With a fixed mount, lower solar irradiance levels are accompanied by the beam radiation coming in at an angle. Due to reflections off the glass and absorbing surfaces, the amount of irradiance that can be absorbed is reduced. This affects the first term of Equation 5–3 but since $U_L$ is not a constant, the correct $U_L$ must be used to correctly predict off-angle performance. If an incident angle modifier ($b_0$) of 0.12 is assumed, the corrections (by direct

**FIGURE 5–6** Off-angle performance prediction using equations 5–1 through 5–15 compared to ASHRAE 93-77 normal test points.

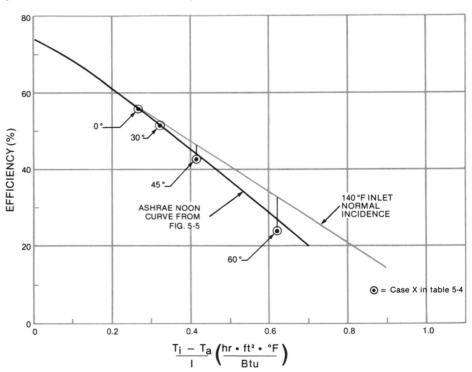

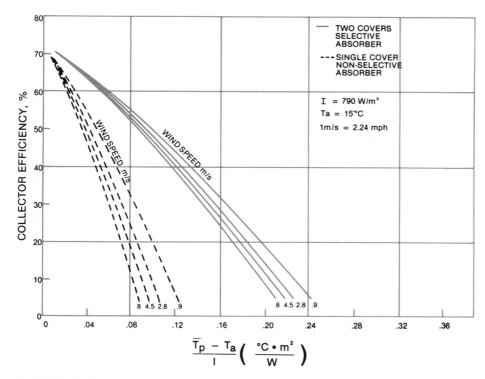

**FIGURE 5–7** The effect of variations in wind speed on the efficiency curve for single- and double-glazed flat-plate solar collectors.

**FIGURE 5–8** Calculated data for single-glazed black-chrome solar collector. Variable normal irradiance to collector with constant inlet temperature. Tilt = 40°, wind speed = 5 mph, ambient temperature = 50°F.

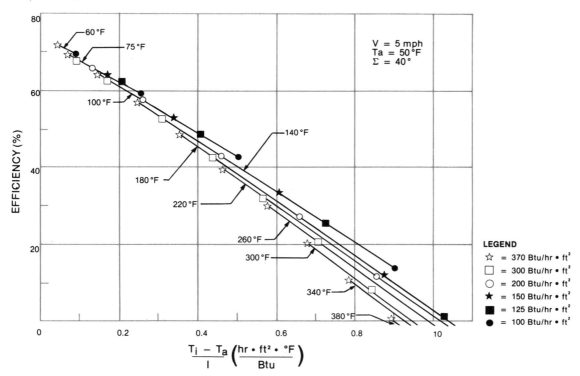

117

**TABLE 5–5**   Calculated Data for Single-Glazed Black-Chrome
Solar Collector

| Collector Efficiency $\eta$ | Operating Parameter $\Delta T/I$ (hr-ft²-°F/Btu) | Overall Heat Loss Coefficient $U_L$ (Btu/hr-ft²-°F) | Inlet Fluid Temperature $T_i$ (°F) | Solar Irradiance I (Btu/hr-ft²) |
|---|---|---|---|---|
| .742 | 0 | .542 | 50 | 370 |
| .744 | 0 | .444 | 50 | 100 |
| .703 | .068 | .647 | 75 | 370 |
| .694 | .083 | .647 | 75 | 300 |
| .672 | .125 | .647 | 75 | 200 |
| .648 | .167 | .647 | 75 | 150 |
| .630 | .2 | .647 | 75 | 125 |
| .602 | .25 | .647 | 75 | 100 |
| .658 | .135 | .708 | 100 | 370 |
| .639 | .166 | .709 | 100 | 300 |
| .588 | .25 | .708 | 100 | 200 |
| .534 | .333 | .708 | 100 | 150 |
| .497 | .4 | .708 | 100 | 125 |
| .437 | .5 | .708 | 100 | 100 |
| .579 | .243 | .769 | 140 | 370 |
| .541 | .3 | .769 | 140 | 300 |
| .443 | .45 | .769 | 140 | 200 |
| .344 | .6 | .769 | 140 | 150 |
| .265 | .72 | .769 | 140 | 125 |
| .147 | .9 | .769 | 140 | 100 |
| .496 | .351 | .809 | 180 | 370 |
| .439 | .433 | .809 | 180 | 300 |
| .289 | .65 | .809 | 180 | 200 |
| .139 | .866 | .809 | 180 | 150 |
| .019 | 1.04 | .809 | 180 | 125 |
| .407 | .459 | .843 | 220 | 370 |
| .330 | .566 | .843 | 220 | 300 |
| .127 | .850 | .843 | 220 | 200 |
| .315 | .567 | .872 | 260 | 370 |
| .217 | .7 | .871 | 260 | 300 |
| .213 | .675 | .909 | 300 | 370 |
| .092 | .833 | .908 | 300 | 300 |
| .118 | .783 | .925 | 340 | 370 |
| .012 | .891 | .953 | 380 | 370 |

subtraction) for 0°, 30°, 45°, and 60° are 0 percent, 1.4 percent, 3.8 percent, and 9.1 percent. This is shown in Figure 5–4, 5–5, and 5–6 at the appropriate points.

Measuring off-angle performance at the application temperature would be better than predicting off-angle performance from noon curves since the plot of efficiencies versus $\Delta T/I$ (with little variation in I) occurring during fixed-mount noontime tests does not yield a single straight line. This point is mainly important for rating collectors in certification programs since the Hottel-Willier-Bliss approach (one of using a linear fit through the noon points) works to within 2–10 percent for long term performance predictions from averaged solar data such as is done in the f-Chart method described in *Solar Heating Design* by Beckman, Klein and Duffie.

These approximations are workable because off-angle performance at any reasonable $\Delta T/I$ is either slightly above, below, or on the linear-fit noon curve, according to the particular collector being studied and the weather conditions.

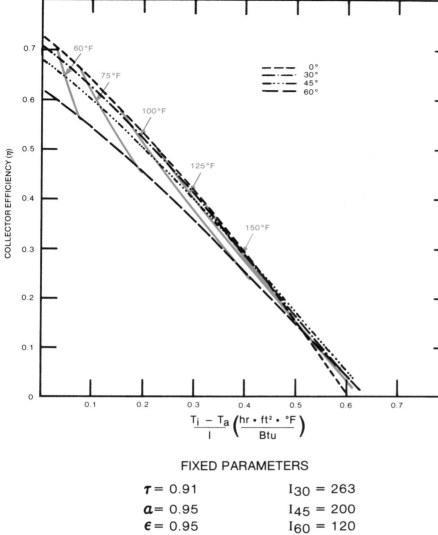

FIXED PARAMETERS

$\tau$ = 0.91           $I_{30}$ = 263

$a$ = 0.95           $I_{45}$ = 200

$\epsilon$ = 0.95           $I_{60}$ = 120

$I_0$ = 317 BTU/hr ft²   $v$ = 5 mph

Tilt = 40°           $T_a$ = 50° F

**FIGURE 5–9** Direct and off-angle performance curves for constant inlet temperature through an absorber with a flat-black painted surface.

Another way of visualizing all day performance is shown in Figures 5–9 and 5–10. The constant temperature lines show the expected performance during the course of a day within 60° to either side of noon. These lines are for a zero mass collector. Since real collectors have mass, the actual all-day performance will be to the left of these constant temperature lines in the morning and to the right during the afternoon. This is caused by the thermal lag of the absorber plate, insulation, and cover plates. What is "lost" in the morning is mostly regained in the afternoon.

The problems of accurately predicting off-angle performance were discussed, and real all-day data were compared with theory, in "The problems of

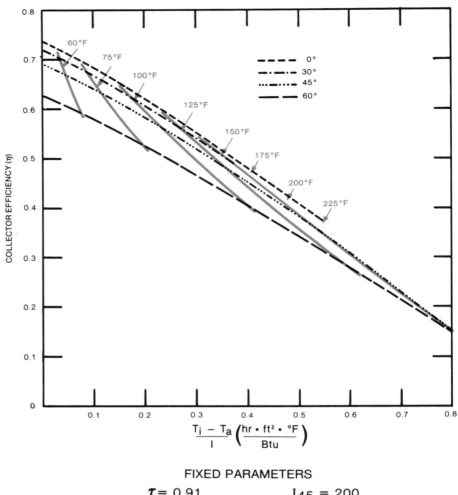

**FIGURE 5–10** Off-angle performance curves for constant inlet temperature through an absorber with a black chrome surface.

treating UL as a constant'' by John C. Bowen of AMETEK/PSG at the American section of the International Solar Energy Society, Inc.'s annual meeting in 1980.

# 6   Systems for Heating and Cooling

Systems to collect, store, and utilize solar energy may vary greatly in design because of the wide choice of equipment, the diversity of design imposed by differing needs at each installation site, and the consideration of costs versus capacity. No matter what choices are made for individual applications, however, every solar thermal system must provide for seven basic requirements: interconnection of collectors, a means of fluid circulation through the collectors, a technique to transfer thermal energy from one fluid to another, a mechanism for energy storage, a means of converting thermal energy to useful forms (such as cooling), a control system, and a technique for delivering the energy in the appropriate form to its intended application. Note that "fluid" is used above in its broadest sense, i.e., both liquids or gases. This chapter, in the main, will address systems which use a liquid as the heat transfer medium.

## SERVICE TEMPERATURE

As a first step toward a survey of solar energy equipment, solar energy systems can be categorized through one parameter of their application: the requirement for a low, medium, or high service temperature. This division indicates that a system which is cost effective in a certain temperature range is suited for that particular application only. Indeed, it is not advisable, and often not possible to use the same system for multiple applications in different temperature ranges without resorting to systems consisting of hybrid collector types and multiple storage units.

Low temperature services, ranging from below ambient to about 120°F, are typically used with solar heaters for swimming pools and for horticultural purposes; these simple systems can utilize low-cost collectors with high loss rates, $U_L$. The operating parameter ($\Delta T/I$) for such systems is usually quite low, in the vicinity of 0.1 or 0.2, so the incremental improvement in the performance of the collector through the addition of glazing covers, selective surfaces, antireflection heat mirrors, and other advanced features is usually not worth the added cost. As a result, systems that operate at low temperatures may use low-cost black plastic or rubber collectors, no glazing, and seek further economies in other components of the system as well. For example, low-cost materials such as PVC pipe may be used (if building codes permit) since it is not harmed by the low temperature fluids. It is of paramount importance, however, to observe that an efficient system at low temperatures may fail completely if operation at higher temperatures is attempted. As with any system, losses will increase rapidly with operating temperature, but this property acts as a safeguard in a low temperature application since it prevents accidental operation at temperatures that may be harmful to the other components of the system.

Medium temperatures are used in hot water service or space heating applications since the necessary temperature range of 120° to 180°F is appropriate. The design of the collectors used in these systems must be sufficiently improved to reduce the loss rate since reasonable efficiencies will be achieved only when operating parameters ($\Delta T/I$) fall in the range of 0.2 to 0.5. (As shown in Figure 5–2, additional glazings and selective surfaces, although not mandatory, do provide an increment of improvement which is significant.) Of course, system components must be able to withstand long exposure to medium temperatures.

Higher temperatures, above 160°F, are necessary for air-conditioning applications and require high performance collectors whether flat plate, the FEA classification of "special" evacuated tube collectors, or concentrating collectors. Operating requirements for these systems are much more stringent since, with the operating parameter exceeding 0.5, the use of multiple glazings and selective surfaces is not only economically effective but nearly mandatory. The system components must be able to withstand both the effects of boiling temperatures for long periods of time and extreme temperature cycling; greater skill must be exercised in controls, and safeguards against pressurization due to boiling water in stagnation must be considered.

## INTERCONNECTING COLLECTORS

In most applications, the necessary collection area is assembled through the interconnection of a number of discrete, modular, solar collection panels. The method of interconnection of these panels should recognize three needs: flow should be upward through each collector for greatest efficiency, especially when parallel banks of collectors are fed from the same header; flow should be uniform in each collector so that no panel will operate at excessive temperature and lose efficiency; and the total length of interconnection piping should be kept to a minimum, a strategy which not only contributes to material economy but increases system efficiency by keeping heat loss to a minimum.

Connecting collectors in series reduces the length of the headers required and, consequently, loss of heat from them. The flow rate through an array of collectors connected in series should be adjusted so that the temperature rise contributed by each collector in the array is equal. The temperature rise per collector can be reduced by increasing the flow rate, which slightly increases the heat transfer in the plate. If the flow rate is proportionally increased by the number of collectors in series, the losses will be equal to those of collectors connected in parallel. This is usually only practical with collectors that experienced small pressure drops at the increased flow rate.

If, on the other hand, collectors are connected in parallel, care must be taken to ensure uniform flow in each collector. This is usually accomplished by using low resistance piping for the parallel headers. Figure 6–1 illustrates some of the more economical methods of interconnecting collectors in a bank.

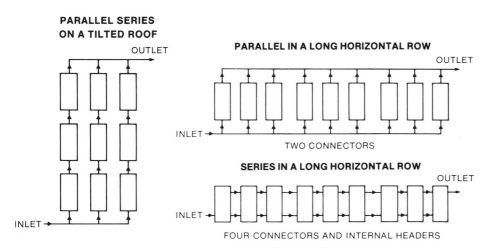

**FIGURE 6–1** Acceptable interconnection configurations.

There are two generally relevant considerations that should be observed. The first is that the number of collectors in each series of parallel rows should be equal; collector arrays containing certain large numbers of panels, such as 17 or 37 or any other prime number, are therefore difficult to arrange in convenient, equally sized groups. Secondly, the location of the inlet and outlet fluid connections on the collector will influence the interconnection geometry. Fluid connections on the top and bottom walls facilitate economical header use in the parallel configuration shown in Figure 6–1. Many collectors, to provide convenience in a series parallel configuration, use internal headers that provide fluid connections on the upper and lower ends of both *sides* of the collector. This method does economize on pipe insulation by providing headers within the collector enclosure, but it also has the dual disadvantages of requiring four connections for each collector instead of two, and more difficulty in maintenance of the internal headers.

## FLUID CIRCULATION TECHNIQUES

Solar collectors which use a liquid for heat collection may be either of the open-channel or closed-tube configuration. The open-channel type requires a circulation pump to move the required amount of fluid and to provide the head difference between the system's highest point and its lowest. Circulation in a closed-tube system is just the opposite, with the inlet at the bottom and outlet at the top, so that the closed-tube may be operated either as a thermosiphon or by forced circulation. In either case, the efficiency of the closed-tube configuration is greater than that of the open-channel as a result of one effect of siphoning on a closed system, eliminating the need for the circulation pump to work against the collector head difference. Efficiency would be reduced if a closed-tube system were connected in the same manner as an open-channel, because of the greater work required of the circulation pump and the upside-down temperature distribution on the collector absorber plate.

Fluid circulation in a closed-tube system can be accomplished without a

**FIGURE 6-2** A Thermosiphon system.

circulation pump by using the thermosiphon effect, i.e., the reduction in density of a fluid as it is heated. Thermosiphon systems can be utilized only in applications where the collector can be located below the storage or service site, since the fluid must rise from the collector to storage for use. Figure 6-2 illustrates the proper orientation of a thermosiphon system with the storage tank inlet on the top and the outlet at the bottom.

The rate of flow in a thermosiphon system depends upon the size of the tubing used, the vertical height separating the collector from the storage tank, and the temperature difference between the collector fluid and the storage fluid. The flow rate is not necessarily that which can deliver the greatest thermal energy. The use of a forced circulation pump in the system can increase the rate of energy flow to the storage tank, increase the storage tank temperature, and reduce the temperature drop across the collector, as compared to a thermosiphon system operating under the same conditions.

A thermosiphon system requires some means of preventing reverse flow; a check valve is often used. Without such a measure, heat will be removed from the storage tank at night as its fluid reverses through the collector, which then radiates thermal energy to the cold sky. The use of a circulation pump does not obviate the need because many pump designs will allow reverse flow whenever the pump is not operating.

Whether closed-tube or open-channel flow is the chosen collector configuration, fluid circulation through the entire solar thermal system may be in either a closed- or open-loop configuration. The open-loop configuration (in which the fluid, in most cases, passes through the system only once) is used in low temperature applications for agriculture and in air-heated ventilator inlets and is generally confined to the use of water or air as the means of heat transfer. The closed-loop configuration continuously recirculates the same fluid in the collector; thus, it can use special fluids, achieve higher temperatures, and is not as susceptible to the effects of contaminants in the fluid. The choice between an open- or closed-loop system will depend on many factors, including the climate, the service application, local water pressure, and building code requirements, but a few general observations are possible.

An open-loop system has several advantages over a closed-loop in hot water heating service. As in Figure 6-3, water may be taken directly from the mains and preheated by passing through the solar collectors under its own pressure before entering the hot water tank. A circulation pump is needed to maintain circulation through the collectors when hot water is not being drawn. In addition, since potable water is being circulated in the collectors, it is desirable to use a collector with a copper or steel absorber plate to prevent corrosion of dissimilar metals in contact with water. Countering these advantages is the necessity for the collector to be able to withstand street water pressures (unless a pressure reducer is used) and for a freeze drain system in climates where freezing temperatures occur. Open-loop systems are particularly advantageous in Florida, for example, where street pressure is low and freezing is rare.

In the closed-loop system, potable water does not enter the collectors so the fluid used for heat transfer may contain antifreeze and corrosion inhibitors. A heat exchanger must be added to the system to transfer heat from the closed loop to the service loop and, in fact, some building codes require a double-wall heat exchange between a toxic closed-loop and the potable service loop. This

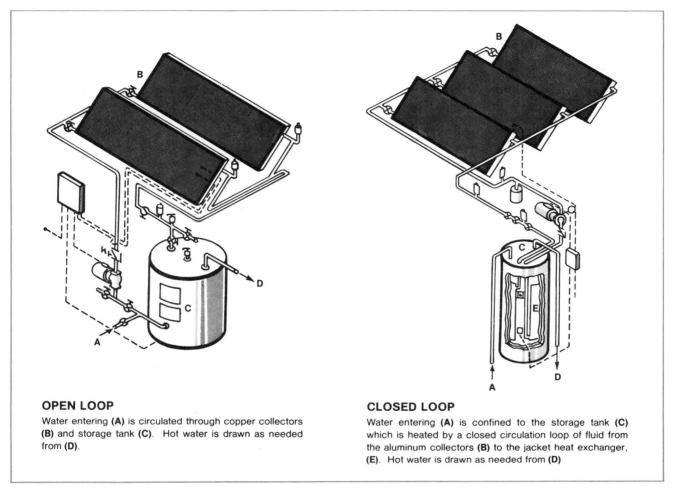

**OPEN LOOP**

Water entering **(A)** is circulated through copper collectors **(B)** and storage tank **(C)**. Hot water is drawn as needed from **(D)**.

**CLOSED LOOP**

Water entering **(A)** is confined to the storage tank **(C)** which is heated by a closed circulation loop of fluid from the aluminum collectors **(B)** to the jacket heat exchanger, **(E)**. Hot water is drawn as needed from **(D)**

**FIGURE 6–3** Open and closed loop hot-water systems.

adds to the expense of a closed-loop system but there are still some advantages. The pressure in a closed loop need not be any greater than that necessary to achieve circulation, and antifreeze can be used to provide protection. A closed-loop system would be preferred in the mountains of Colorado where there are large variations in street water pressure and many freezing nights.

The design capacity of a liquid circulation pump is proportional to the collection area and will generally fall in the range of 10 to 30 pounds per hour per square foot of collector area, or 2 to 6 gallons per minute per hundred square feet of collector area. Operating pressure in the system will be reduced through losses caused by system hardware; consideration of this potential loss can be made using the tabulated pressure loss for pipes and common fittings of various sizes contained in Figure 6–4.

## THERMAL STORAGE

The storage of thermal energy is usually accomplished by taking advantage of the capacity of many materials to retain heat. Table D–6 lists the magnitude

**FITTINGS**

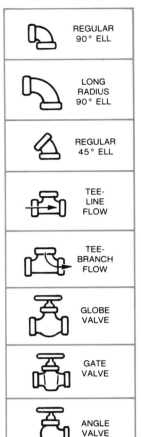

| | |
|---|---|
| | REGULAR 90° ELL |
| | LONG RADIUS 90° ELL |
| | REGULAR 45° ELL |
| | TEE-LINE FLOW |
| | TEE-BRANCH FLOW |
| | GLOBE VALVE |
| | GATE VALVE |
| | ANGLE VALVE |

**PRESSURE LOSS**

| FLOW (gpm) | COPPER OR STEEL TUBE | | |
|---|---|---|---|
| | ½" | ¾" | 1" |
| 1 | 1.8 | 0.5 | — |
| 2 | 6.4 | 1.2 | 0.5 |
| 3 | 13.1 | 2.3 | 0.7 |
| 4 | 21.7 | 4.2 | 1.2 |
| 5 | 31.9 | 6.0 | 1.6 |
| 10 | 108.0 | 19.9 | 5.8 |
| 15 | — | 40.7 | 11.6 |

Units of ft-$H_2O$ per 100 ft. of tube.

**EQUIVALENT PIPE LENGTHS OF COMMON HYDRONIC FITTINGS[a]**

| | STEEL PIPE | COPPER TUBING |
|---|---|---|
| 90° Elbow | 25 | 25 |
| 45° Elbow | 18 | 18 |
| 90° Elbow (long radius) | 13 | 13 |
| 90° Elbow (welded) | 13 | 13 |
| Reducer coupling | 10 | 10 |
| Gate valve (open) | 13 | 18 |
| Globe valve (open) | 300 | 425 |
| Angle radiator valve | 50 | 150 |
| Radiator or convector | 75 | 100 |
| Boiler | 75 | 100 |
| Tee (typical) | 100 | 100 |

Notes: a. Given in diameters of pipe.

**FIGURE 6-4** Pressure loss in pipes and fittings.

of this property, expressed as specific heat, for some of the common materials either used or considered for use in solar energy systems.

Water is often used for thermal storage with an insulated, corrosion-protected tank as its reservoir. The most common structural material for tanks is steel, protected either by rust inhibitors in the stored water or coatings on the steel surface, or both. Concrete tanks avoid the corrosion problem and are self-insulating, but they are more susceptible to leaks and are quite heavy. Alternative materials for tank construction include wood, fiberglass, and nonferrous metals, but none of these offers a significant advantage over steel or concrete. Depending upon the specific requirements at the site, compound materials (concrete mixed with plastic or fiberglass-lined wood tanks) may prove cost-effective and suitable.

Thermal storage can also be accomplished by utilizing phase changes of a solid into a liquid or a liquid into a gas. The heat of fusion (latent heat) of a solid represents the heat required to melt it without raising its temperature. Similarly, the heat of vaporization of a liquid represents the heat required to vaporize the liquid without raising its temperature. Choosing the proper substance for this method of storage is predicated on service temperature; some materials do undergo phase changes at temperatures convenient for use. Rochelle salts, metallic Gallium, Glauber salts, paraffin wax, and fluorocarbons undergo phase changes appropriate for low and medium temperature regimes. Some eutectic alloys of metals, such as solder, and Wood's metal can also be used to store latent heat, but the high temperatures at which these metals undergo phase changes are suitable only for concentrating solar cookers and not for most solar energy applications. Because of the increased costs and complexities of thermal storage by latent heat, these systems are not economical except where space limitations are severe. The latent heats of currently used or potential heat storage materials are also listed in Table D–6.

## HEAT TRANSFER

Heat is usually transferred from the closed-cycle working fluid to the storage or service fluid by means of a heat exchanger. Liquid-to-liquid heat exchange is generally accomplished by circulating one liquid through a coil of metal tubing within a tank or shell containing the other liquid, with the rate of heat transfer approximately proportional to the temperature difference between the two liquids, the area of the total surface that separates the two liquids, and the overall heat transfer coefficient.

$$J = UA\Delta T_{lm} \qquad\qquad (6-1)$$

where: $J$ = rate of heat transfer in the heat exchanger (Btu/hr)

$U$ = overall heat transfer coefficient

| Type of heat exchanger | U Free Circulation (Btu/hr-ft$^2$) | U Forced Circulation (Btu/hr-ft$^2$) | Fluid | Apparatus |
|---|---|---|---|---|
| Liquid-to-liquid | 25–60 | 150–300 | Water | Liquid-to-liquid heat exchanger |
| Liquid-to-liquid | 5–10 | 20–50 | Oil | Liquid-to-liquid heat exchanger |
| Liquid-to-gas (atm. pressure) | 1–3 | 2–10 | —— | Hot water radiators |
| Liquid-to-boiling liquid | 20–60 | 50–150 | Water | Brine coolers |
| Liquid-to-boiling liquid | 5–20 | 25–60 | Oil | Brine coolers |
| Gas-to-liquid (atm. pressure) | 1–3 | 2–10 | —— | Air coolers economizers |

$A$ = area of the heat exchanger surface (ft$^2$)

$\Delta T_{lm} = \dfrac{\Delta t_{inlet} - \Delta t_{outlet}}{\ln \dfrac{\Delta t_{inlet}}{\Delta t_{outlet}}}$ = log mean temperature difference

$\Delta t_{inlet}$ = difference between the water coming from storage and the fluid returning to the collectors (°F)

$\Delta t_{outlet}$ = difference between the water returning to storage and the fluid coming from the collectors (°F)

ln = natural logarithm

Equation 6–1 is approximate because the rate of heat transfer is also influenced by the nature of the fluid flow (laminar flow results in less heat transfer than turbulent flow) and the degree of mixing of the fluid in the shell (which may be improved with proper baffling). For most applications, the direction of flow in the two liquids should be opposite to each other (in counterflow).

Heat exchangers are also made by constructing matrices of tubes that are interwoven and, sometimes, immersed in a heat conducting fluid. This technique can be applied to situations that require heat exchange between more than two fluids or in an exchange between fluids and air. Liquid-to-air heat exchangers, like the automobile radiator, are usually constructed of finned tubes; the variety of geometries utilized in these fin and tube exchangers is too diverse to permit brief exposition, so exchange capacity is best obtained from the manufacturer. Figure 6–5 illustrates a few of the available heat exchanger geometries.

The heat pipe and related products which utilize the vaporization, transport and condensation of a heat transfer fluid in an enclosed, evacuated tube

**FIGURE 6–5** Typical heat exchange surfaces.

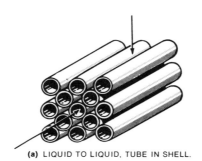

(a) LIQUID TO LIQUID, TUBE IN SHELL.

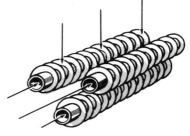

(b) LIQUID TO LIQUID OR AIR, RADIAL FIN.

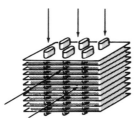

(c) LIQUID TO AIR, LAMINAR FIN.

(d) LIQUID OR SOLID TO AIR, CORRUGATED FINS.

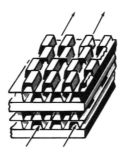

(e) LIQUID OR SOLID TO AIR, STAGGERED FINS.

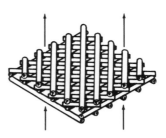

(f) LIQUID TO LIQUID OR AIR, TUBE MATRIX.

are efficient and capacious transfer devices and are used in applications ranging from cooling valves in high performance engines to home cooking utensils. Heat pipes have been proposed for many solar energy applications, but cost factors have inhibited widespread acceptance. In addition, heat pipes are most effective in conveying heat from a small area to a large one, while most solar applications require just the reverse.

## CONTROL TECHNIQUES

The automatic control of a solar thermal system is most often accomplished by controlling the fluid circulation pump in response to changes in system and service temperatures, to changes in the level of solar irradiance, and to changes in ambient temperatures. The most common type of control system consists of a control box that turns the fluid circulation pump on whenever the solar collector temperature is greater than the thermal storage temperature and turns it off when it is less. Care must be taken to adjust the temperature limits and flow rate to prevent "cycling," where the action of turning on the pump causes the collector to cool, subsequently causing the control to turn off the pump—with the process continuing in a repetitive cycle. Control boxes should have high limit controls to prevent overheating and bypass controls to allow for manual operation. Some control systems provide a circuit for sensing freezing conditions and can be used to open a drain valve. Figure 6–6 illustrates some of the more common control circuits used in thermal solar collection systems.

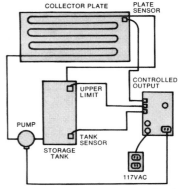

(a) PUMP CONTROL BY TEMPERATURE LIMITS

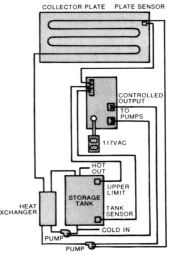

(b) DUAL PUMP CONTROL

## THERMAL ENERGY CONVERSION

The conversion of thermal energy into higher forms of energy has been one of the principal objects of the science of thermodynamics ever since the discovery of a mechanical equivalent to heat energy in the seventeenth century. The laws of thermodynamics govern the terms upon which heat energy can be converted into other forms with two important rules emerging from the application of these laws. The first is that the processes which convert heat into other forms of energy are inherently less efficient than the processes which produce heat. The second is that the ability to convert heat into other forms of energy is dependent upon the temperature difference between two sources of heat, with the corollary that heat energy cannot be extracted from one source of heat alone. The motive force for the conversion of heat to higher forms of energy is, therefore, the movement of heat from a hot object to a colder one.

The creation of mechanical energy, or work from heat, is accomplished in an engine. The expansion of a fluid upon vaporization is the basis of the Rankine cycle engine (more often known as the steam engine) and can be used to power a turbine, as well as a piston or any number of other ingenious mechanical arrangements. The efficiency of the Rankine or steam turbine cycle is usually less than that of the internal combustion engine, though efficiency does increase with an increase in steam temperature. These techniques may be useful in solar applications that involve concentrating collectors and large cen-

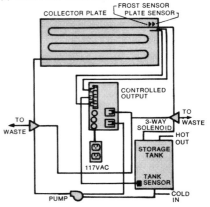

(c) CONTROL FOR FREEZE PREVENTION

**FIGURE 6–6** Typical control circuits.

tral station facilities where high temperatures and pressures are possible, but they would seem to have little application in individual solar installations.

Heat supplied by flat plate solar collectors and other medium temperature sources may be converted into mechanical power through the use of a Stirling cycle engine (Figure 6–7), which derives its power from the expansion of air as it is alternately heated and cooled.

Mechanical energy through motors powered by low-temperature heat sources can make use of thermal effects such as the Minto wheel, bimetal expansion effects, and thermoelastic effects.

The conversion of thermal energy into electricity is usually accomplished by applying rotary mechanical power to an electrical generator shaft; electricity may also be produced directly from thermal energy through the use of thermoelectric devices which consist of junctions between dissimilar metals, or thermionic devices that emit electrons when heated in a vacuum. Small amounts of electricity can be produced by the effects of mechanical stress or thermally induced mechanical stress (such as that from the thermal expansion of metals or freezing water) acting on piezoelectric devices, or by the action of condensation in an electric field like that in lightning.

## AIR CONDITIONING

As heat from hot water is ultimately conveyed to its surroundings, its energy can be utilized to produce a cooling effect. The principal methods of cooling are the mechanical vapor-compression cycle, the absorption refrigeration cycle, and the vapor jet cycle, all of which are shown diagrammatically in Figure 6–8.

**FIGURE 6–7** Stirling cycle engines: *left,* double-cylinder, two-piston; *right,* single-cylinder, piston-plus-displacer. Each has two variable-volume working spaces filled with the working fluid—one for expansion and one for compression of the gas. Spaces are at different temperatures—the extreme temperatures of the working cycle—and are connected by a duct, which holds the regenerator and heat exchangers.

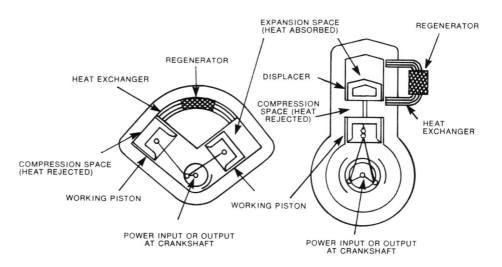

**(a)** VAPOR COMPRESSION

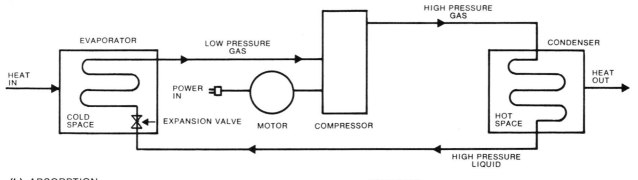

**(b)** ABSORPTION

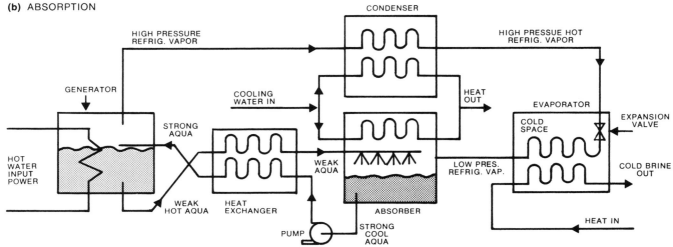

**(c)** VAPOR JET

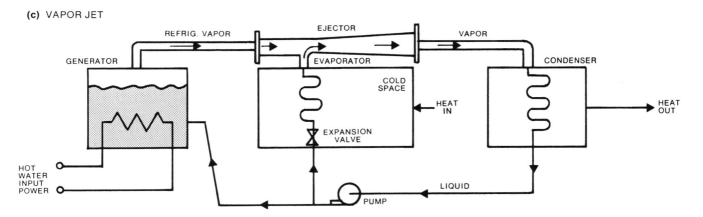

**FIGURE 6–8** Cooling cycles.

The mechanical vapor-compression cycle derives its input energy from mechanical work powering a compressor, which heats a gas by compression. The heated gas is cooled to near ambient temperature and condensed in a heat exchanger that is located outside the space to be cooled. Once cooled and condensed, the gas (or fluid) is routed through the application space where its

131

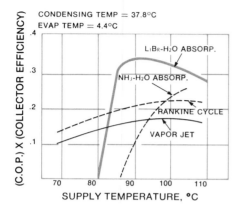

CONDENSING TEMP = 37.8°C
EVAP TEMP = 4.4°C

Li₂Br-H₂O ABSORP.

NH₃-H₂O ABSORP.

RANKINE CYCLE

VAPOR JET

**FIGURE 6-9** Comparison of cooling cycle efficiencies.

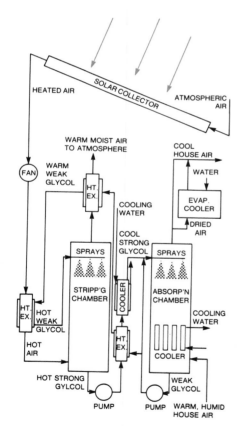

**FIGURE 6-10** Cooling by solar-powered absorption dehumidification.

expansion and vaporization absorbs heat from its surroundings to produce a cooling effect. In order for a solar energy system to supply power for a compression and expansion cycle, it would first have to convert its thermal energy into mechanical power. Because of this, the mechanical vapor-compression cycle is suitable for use in solar systems with concentrating collectors which are capable of supplying temperatures high enough to provide the necessary mechanical power.

The absorption-refrigeration cycle operates on the principle that the absorption capacity of gases in liquids is temperature and pressure dependent. For example, as a pressurized solution of ammonia in water is heated, it liberates ammonia gas at a high pressure which can then be cooled and condensed into a liquid. This refrigerant liquid ammonia is then allowed to expand into a low pressure region and vaporize, absorbing heat from the space to be cooled. The heated vapor is then reabsorbed into cool water, which is pressurized with a pump to complete the cycle. This absorption cycle is suitable for use in solar collecting systems which use high performance flat plate or concentrating collectors, but the system will usually require a cooling tower to convey heat to the environment. The lithium-bromide absorption-refrigeration cycle is similar to the ammonia cycle except that water, rather than ammonia, is the refrigerant that is absorbed into a solution with lithium bromide, not into water.

The vapor jet cycle makes use of the Venturi principle to produce a partial vacuum from an ejector, which is driven by vapor passing from a boiler generator to a condenser. The partial vacuum is used to draw vapor from an evaporator which supplies the cooling effect. The vapor jet cycle can be used at lower temperatures and offers simplicity, but the cooling tower requirement is very large and the cycle's overall efficiency is poor.

A comparison of the thermal efficiencies of the various methods of cooling is shown in Figure 6-9. In this figure, the overall thermal efficiency is the product of the solar collector efficiency and the cooling-cycle coefficient of performance (COP). The COP of a cooling cycle is the ratio of heat energy supplied to cooling effect produced, and *does not* take into account mechanical or electrical energy inputs or cooling energy losses. Figures 6-10 and 6-11 depict two possible means of interfacing a cooling system with solar collectors.

## HEAT PUMPS

A heat pump may be described as a mechanical vapor-compression cycle operating between two reversible heat sources. In the summer it cools living space by transferring heat from building interiors to the outside, and in the winter it performs the reverse by "air conditioning" (in a sense) the outside air and moving that extracted heat to living spaces.

Heat pumps have found wider acceptance in conserving energy with traditional fossil-fuel systems, but they are also useful with solar energy systems. A solar energy system incorporating a heat pump can deliver usable energy over a much wider range of storage temperatures, and therefore conserve more energy, than a system without the heat pump assist. For example, a forced air heating system operating without a heat pump requires significantly higher minimum storage temperatures. With a heat pump, storage temperatures as low as 40°F may be boosted to temperatures high enough for heating use (90°F to 130°F). When storage temperatures drop below 40°F, but ambient air temperature is above 40°F, the heat pump will use the outside air as its heat source and continue to contribute to increasing the storage temperature. The effect is equivalent to increasing the heat storage capacity and, because storage is usually at low temperatures, increasing the efficiency of the entire system. With ambient temperatures below 40°F, electric coil or oil fired burners switch on to provide heat directly to the house.

The decision whether or not to use a heat pump with a solar energy system is not a simple design choice. Analysis of the cost of the electrical energy needed to operate the heat pump versus its return in increased efficiency should be one basis for the decision.

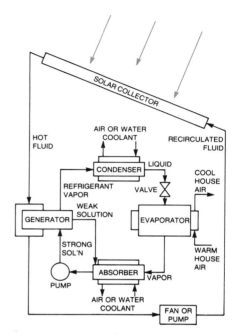

**FIGURE 6–11** Cooling by solar-powered absorption refrigeration.

## DELIVERY TECHNIQUES

The delivery of solar thermal energy usually needs to be no different than the delivery used with other sources of energy. Hot water radiators, baseboard heating, and forced air ducts are all in common use. Since large wall and roof areas are often employed for collection purposes in solar heating and cooling, some delivery systems make use of these areas more directly; these systems are, however, generally more difficult to control. Conventional hot air delivery is illustrated in Figure 6–12, a typical hot water delivery system in Figure 6–13, and passive solar delivery techniques in Figure 6–14.

In air-conditioning applications, delivery is influenced by alternative methods of storage. Since the efficiency of the absorption cycle is very sensitive to the temperature of the solar water heater (see Figure 6–9), maximum use should be made of high-temperature water directly from the collectors during those periods when it is available. At such times, the cooling apparatus is used to cool brine which is then stored. Heat storage is less efficient because it results in a lower supply temperature to the chiller than the collectors can provide directly.

For a combined heating and cooling system, two separate thermal storage tanks may be needed: one for hot water from the collectors, and another for cooled brine. If the solar system is remotely located from the serviced site, the absorption equipment should be as close as possible to the collectors.

The delivery of heated or cooled fluids in underground pipes or ducts requires thermal insulation which is protected from water, soil chemicals, and vermin. The techniques used require either special waterproof materials, pressurized conduit jackets, or pumped jacket drainage systems. Condensation of moisture can become a major problem with cooling ducts used in underground delivery unless watertight, low permeance, or vapor-barrier insulation methods are implemented.

## AUXILIARY SYSTEMS

In the vast majority of applications, a solar energy system will be designed to provide a certain percentage of a given building's energy requirement, typi-

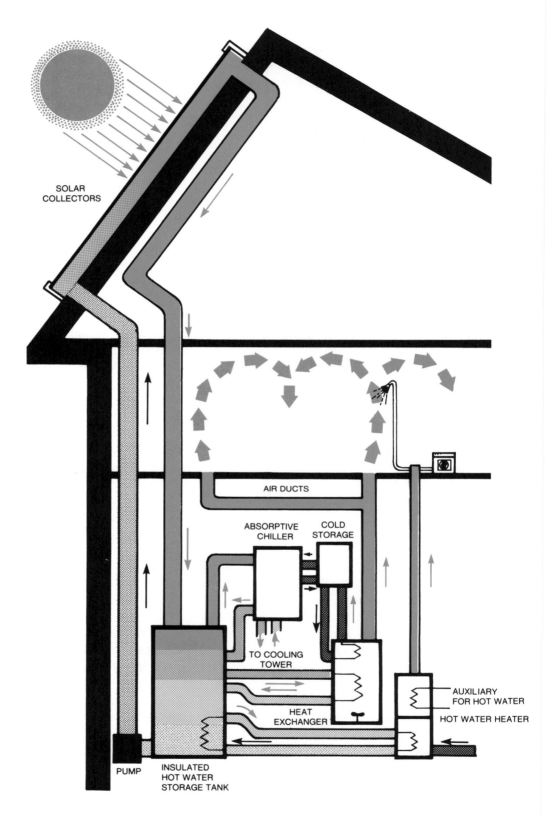

**FIGURE 6–12** Conventional delivery of solar energy for hot water, space heating, and cooling with a circulating air system.

134

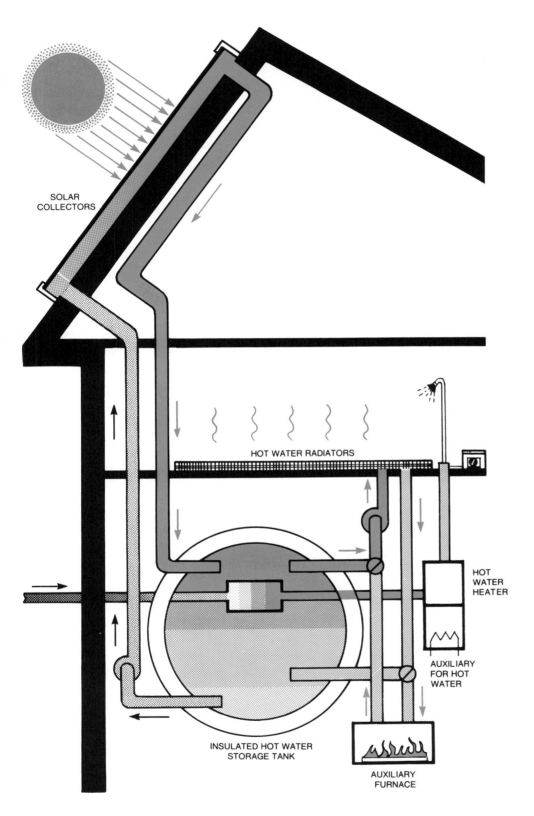

**SOLAR COLLECTORS**

**HOT WATER RADIATORS**

**HOT WATER HEATER**

**AUXILIARY FOR HOT WATER**

**INSULATED HOT WATER STORAGE TANK**

**AUXILIARY FURNACE**

**FIGURE 6–13** Conventional delivery of solar energy for hot water and space heating with circulating hot water.

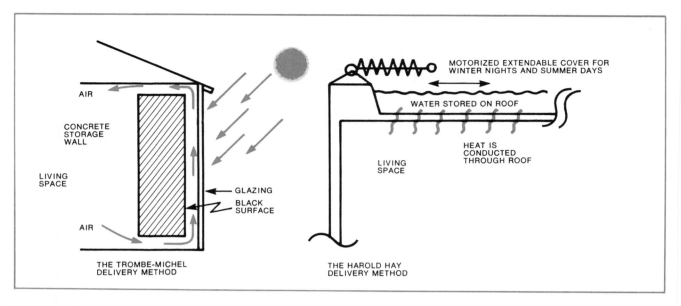

**FIGURE 6–14** Some passive solar delivery techniques.

cally in the 50–80 percent range, whether the system is intended for a minimum of providing domestic hot water or total service of space heating, cooling, and hot water. This is not to say that a solar energy system to supply 100 percent of the requirement could not be built; indeed, MIT successfully heated its first solar house with only solar energy in 1939. But the costs of added system capacity, especially in increased collector area and storage capability, sufficient to counter periods of sunless days and cold weather can become unreasonable. In addition, the system would have to be designed to meet the most extreme conditions, an infrequent occurrence, with a consequent loss in energy saved per square foot of collector area, since the system would be operating far below capacity at most times, lowering the return on the investment. As a result, an optimum solar energy system should be targeted toward that 50–80 percent provision, with an integrated auxiliary system to fill in gaps during inclement weather or when thermal storage is exhausted.

The back-up system may be chosen from a wide range of alternatives. Oil or gas furnaces, electric resistance coils, and heat pumps have all been used; in retrofit solar installations, the existing system is simply incorporated through interfacing with the solar system. Efficiency of the entire system (solar and auxiliary) can be increased by integrating the two through shared components—pumps, piping, controls—to avoid redundancy in equipment and to reduce total cost.

One precaution should be observed in the interconnection of the solar energy and auxiliary systems. In most cases, the two systems should be connected in parallel, not in series. Back-up systems connected in series inadvertently heat the solar storage, causing an extra expense in energy and less efficiency through solar heat collection since the storage temperatures will be maintained at or near usable temperatures. In addition, maintenance or storage at higher temperatures through auxiliary heating will result in increased thermal losses from storage.

136

# SOME PRACTICAL CONSIDERATIONS

Because utilization of the solar resource as an alternative energy source is a relatively new field, the body of design expertise, installation experience, and operation knowledge is not widespread. This is not to say that such knowledge and expertise cannot be found; indeed, the sum of governmental, university, and commercial sources, and the solar literature, is comprehensive. Still, an individual planning to install a solar energy system, whether to heat the water in a residential swimming pool or to cool space air in an office building, cannot always rely upon finding local expertise. By comparison with an individual considering a new oil-fired heating plant or central electric air conditioning, where the variation of efficiency and reliability in available equipment is much smaller, the telephone directory can supply a number of experienced contractors. Solar energy utilization requires more effort and investigation.

Common sense, as with most pursuits, can make its contribution to a successful solar energy system, especially in equipment installation. For example, solar collection panels are often roof-mounted, which requires consideration of related concerns. Proper orientation to the sun is the most obvious, but provision for drainage of rain water under or around the panels to avoid standing water, mounting so that the underlying roof structure is not damaged, and integrity of the mounting so that the panels stay in place for the twenty or more years of their service life are some of the ancillary considerations. On the last point, it is worth considerable care to ensure that the panels are secure, since the effects of a heavy solar panel sliding from a tilted roof to the ground can be very serious.

Other elements of system design and installation may be accomplished by any number of alternative means and, consequently, are surrounded by a larger number of related considerations. System transfer fluids containing antifreeze solutions can have potentially hazardous effects, as has already been noted. The concern is real, so much so that many communities have instituted building standards requiring a double-walled heat exchanger between the collection fluid loop and the service loop to avoid contamination of one by the other. Similar guidelines and regulatory constraints—HUD standards on the federal level, and various local and state statutes—are in effect concerning solar energy installations.

## SOLAR STANDARDS

One of the most frequently cited restrictions which limits the introduction of a new product or technology is the lack of complete standards by which to assess the value, quality and safety of the device or method. As a new technology, solar energy has, in addition to the usual institutional barriers, a unique set of complex component and system requirements which must be identified, qualified and, ultimately, certified. The problem lies in the basic fact that almost any solar device will work to one degree or another. But the answers to such questions as: How well does it work? How long will it last? and How safe will it be? require appropriate and accepted standards. Various groups have developed, or are developing, the standards required to answer these and other questions:

1. American Society of Heating, Refrigeration, and Air Conditioning Engineers (ASHRAE) has established standard testing procedures for solar collector thermal performance, thermal storage devices, solar domestic hot water systems, and solar swimming pool heaters.

2. National Bureau of Standards (NBS) has, in addition to providing valuable input to development of private consensus standards, been the primary developer of solar energy standards for the federal government. The Bureau's "Interim Performance Criteria" have been implemented in the HUD Minimum Property Standards, which include system sizing, collector descriptive information, and general installation techniques, among others.

3. American Society for Testing and Materials (ASTM) has formed a number of subcommittees to manage and administer the development of standards. The work of the subcommittee on materials performance, which is representative of other subcommittees, includes the development of testing standards for absorptive coatings, cover plates, fluids, reflective surfaces, insulation, metallic and nonmetallic containment, seals, and gaskets. Thirty standards have been adopted as of March 15, 1984, and others are coming up for a vote.

4. Sheet Metal and Air Conditioning Contractor National Association (SMACNA) has developed an Institutional Standard which includes solar energy load calculation methods, equipment selection, and suggested installation practices.

5. Underwriters Laboratory (UL) operates labeling programs for solar related equipment which qualify the product's safety, but make no attempt to certify accuracy or performance characteristics. A safety standard for solar collectors, which will include fire and casualty risks associated with commercially-built collectors, as well as safety measures for installation, maintenance and use, will be released by UL in the near future.

6. American Society of Mechanical Engineers (ASME), in its traditional role of standards development for pressure vessels, is addressing the safety of solar components and systems with respect to the high pressures and temperatures that can be developed in large arrays of flat plates, concentrating systems, and power tower installations.

## SOLAR CODES AND CERTIFICATION PROGRAMS

Certification programs, in which components or systems will be tested and their performance, maintenance, and endurance characteristics qualified, are in progress in various quarters. Once completed, these programs will assist potential solar energy equipment purchasers in choosing appropriate equipment, and provide data on which governmental bodies may establish minimum standards in their jurisdictions. Some of the principal groups involved are:

1. International Association of Plumbing and Mechanical Officials (IAPMO) has published a "Uniform Solar Energy Code" which specifies material standards for pipe and fittings, and design and test pressure criteria for components. This code has been adopted by many local jurisdictions. IAPMO also lists solar collectors and other plumbing hardware.

2. The City of Los Angeles has adopted the IAMPO as its basic code and other qualifications to it. The city maintains a program of hydrostatic testing and construction inspection for collectors and other pressurized components. Certification under this program, which addresses health and safety issues and not thermal performance or system efficiency, is required for all installations within the city.

3. The Solar Rating and Certification Corporation (SRCC) rates and publishes a list of certified solar collectors which include a schedule of fifteen thermal outputs for various operating temperatures and solar radiation conditions. The ratings are in Btu per day per collector. SRCC is sponsored by the Solar Energy Industries Association (SEIA) and Interstate Solar Coordinatory Council (ISCC).

4. The Air-conditioning and Refrigeration Institute (ARI) sponsors a solar collector certification similar to the SRCC program.

5. Florida Solar Energy Center (FSEC) operates two programs: one for listing certified solar collectors and one for listing domestic hot water systems. The solar collector listing requires ASHRAE thermal performance rating, reflecting expected Btu output for a typical Florida day for both low temperature (swimming pool) and medium temperature (domestic hot water) applications.

A laboratory certification program to qualify independent testing facilities has also been established.

These and other existing bodies of regulations and guidelines, codes and standards, undergo continual refinement or change, with new regulations added periodically. It may be reasonably assumed, therefore, that the numbers and complexity of applicable standards will increase, and that any listing of them soon becomes outdated. At the same time, and this cannot be emphasized too strongly, planning for any solar energy system must take cognizance of all federal, state and local regulations. Further, the tasks of equipment choice, system design, installation, operation, and maintenance should gain the advantages of appropriate qualification data and practical experience.

**Determining Collector Area and Storage Capacity**

As with most endeavors, there are methods of planning ranging from simple to complex which yield results coherent to the method applied, which range from ballpark to pinpoint accuracy. In this chapter, the collector area and storage capacity of a sample solar energy system are determined through a procedure that is more accurate than the procedure used in Chapter 3, but which falls short of the accuracy possible through the use of large computers. Sources and magnitudes of potential error in this example procedure are noted so that the reader may make his own choice in practical situations of whether to follow Hopkins' dictum that "Genius is an infinite capacity for taking pains," or to proceed with system design with some degree of approximation. Having said that, it is well to note that solar energy systems are more costly to purchase and install than other energy systems and that, since collector area and storage capacity are the principal design features and those most difficult to alter by afterthought, the greatest need for accuracy exists in the determination of these two components.

The capacity for heating and cooling of any system is specified by the number of Btu's per hour or per day required to achieve the desired effect. Btu requirements can be estimated by calculation or by measurement for existing buildings; for those structures still in the planning stage, a calculation must suffice. Once the Btu requirements are known, the solar system design can begin. Collector area is determined by site geography and geometry, solar irradiance statistics, collector efficiency, climatic variables, and service temperature requirements. As a general rule, collector area for spacing heating will fall in the range of one-quarter to one-half of the floor area; hot water service and hot water storage requirements fall in the range of one to two gallons for each square foot of collector area; hot air service storage requires one-third to two-thirds cubic feet of rocks for each square foot of collector area.

An important characteristic that distinguishes a solar energy system from other energy supply systems is the comparative lack of flexibility in supplying variable loads over extended time periods. The thermal storage of a solar energy system is limited in both capacity and time; it cannot supply more than its peak collection capacity for a time longer than its thermal storage will allow—rarely more than one week. The collection of solar energy for storage, which is accomplished during periods of slack demand, is similarly limited to that time period when solar irradiance is usable. By contrast, conventional fuel systems are not limited in this manner, since one may draw heavily upon them, or not at all, for an extended period. By the same token, fuel storage in a conventional system can be of long duration, neither diminishing with time nor dissipating when it is not being used. The implication within this lack of flexibility in solar systems is the possibility of an occasional, infrequent, shortfall of energy supply. It is a matter for consideration, hence the requirement for back-up conventional systems, but solar shortfalls are no cause for undue concern. Conventional fuel systems have experienced similar shortages in the past, will continue to do so more often in the future, with ever-increasing cost of new supply, until the last barrel of oil or cubic foot of natural gas is pumped from the ground. Those homes and buildings with solar energy systems will see that inevitable day arrive with less concern for the next delivery.

Another matter for consideration is the question of overdesign. It is common practice in the planning of conventional fuel energy systems to provide a significant level of overdesign, sometimes as much as 100 percent more than peak requirements. This provides a healthy margin of error in the computation of energy needs which, since the initial costs of a conventional system are lower, is affordable. In the planning of a solar energy system, overdesign is very costly because it results in poor utilization of available solar irradiance. Depending upon the control strategy used in the system, overdesign in collector area may result in increased stagnation time, or the necessity to incorporate heat wasters, for excess temperatures, in addition to higher costs for collectors, support structures, and interconnection headers. Overdesign of thermal storage capacity can lead to insufficient temperatures, as well as excessive weight, waste of living space, and cost. The question is where to draw the dividing line between too much and too little. Striking the middle ground by accurately fitting collector area and storage requirements to the average demand is the best recommendation. Back-up systems using conventional fuels or electric power should be used for overdesign capacity for those periods when solar irradiance becomes unusually weak, or demand becomes unusually large, for an extended period of time.

## CALCULATING Btu REQUIREMENTS

The most comprehensive procedure for calculating the Btu requirements for space heating and air conditioning is found in the ASHRAE *Handbook of Fundamentals*. This volume of more than 650 pages covers most of the fine details necessary to make a highly accurate computation of Btu requirements. For the purposes of this handbook some of the more important features of the detailed procedure are abstracted and simplified to yield an approximate calculation for Btu requirement. It should be noted that Btu requirements are influenced by lifestyle and variations among individuals' behavior account for substantial differences in energy consumption. The occupant of a house who sets the temperature at 80°F and leaves the windows and doors open in the winter, while consuming large amounts of hot water by frequently bathing and washing dishes and clothes, may consume more than twice the energy of one who wears sweaters at a 65°F temperature, keeps the doors and windows sealed shut, and washes dishes and clothes only once or twice a week. These examples are extreme but correspond to greatly different energy needs. Btu requirements here are calculated on the basis of average or statistical norms of behavior, and individuals must make their own adjustments to accommodate those personal factors which consume or conserve energy.

## HOT WATER

The temperature of hot water should be in the range of 110°F to 150°F, with most hot water heaters set to begin heating when the water temperature is 130°F and to shut off when 150°F is reached. The individual user then mixes hot water with cold to gain the desired temperature. Minimum hot water storage tank capacities and Btu requirements, assuming the inlet water is 60°F and that the storage tank is not quite emptied twice each day, are specified by the U.S. Department of Housing and Urban Development (HUD) and the Federal Housing Authority (FHA) for one and two-family living units as shown in Table 7-1.

## SPACE HEATING AND COOLING

The Btu requirements for heating and cooling are calculated by determining the heat loss of a structure that must be made up by heating or cooling. Heat loss is the product of the temperature difference between interior and exterior air temperatures, the area of the surface, and its $U$-factor, with loss considered outward for heating and inward for cooling. The $U$-factor is the thermal conductivity ($k$-factor) in Btu-in/hr-ft²-F° divided by the thickness (in inches) of the insulating material.

$$U = \frac{k}{t} = \left(\frac{1}{R_t}\right) \text{Btu/hr-ft}^2\text{-}°\text{F} = \frac{1}{R_1 + R_2 + R_3 + \cdots} \qquad (7\text{-}1)$$

where:   $U$   is the $U$-factor of a wall (Btu/hr-ft²-°F)

$k$   is the thermal conductivity of a material in Btu-in/hr-ft²-°F

$t$   is the thickness of the material in inches

$R_t$   is the thermal resistance of the wall.

$R_1$, $R_2$, $R_3$ are $R$-values of different materials in the wall

**TABLE 7-1**   HUD-FHA Hot Water Requirements

| | 1-1½ Baths | | | 2-2½ Baths | | | | 3-3½ Baths | | | |
|---|---|---|---|---|---|---|---|---|---|---|---|
| Number of bedrooms | 1 | 2 | 3 | 2 | 3 | 4 | 5 | 3 | 4 | 5 | 6 |
| Minimum storage | 20 | 30 | 30 | 30 | 40 | 40 | 50 | 40 | 50 | 50 | 50 |
| Btu requirement per day (thousands) | 27 | 36 | 36 | 36 | 36 | 38 | 47 | 38 | 38 | 47 | 50 |

The heat loss, $Q$, of a structure can be estimated by adding the heat losses for all exterior surfaces of the structure.

$$Q = [Q_{walls} + Q_{ceiling} + Q_{floor} + Q_{windows} + Q_{doors}](\text{Btu/hr}) \qquad (7\text{-}2)$$

$$Q_{wall} = A_{wall} \cdot U_{wall} \cdot (T_{in} - T_{out})$$
$$Q_{ceiling} = A_{ceiling} \cdot U_{ceiling} \cdot (T_{in} - T_{attic})$$
$$Q_{floor} = A_{floor} \cdot U_{floor} \cdot (T_{in} - T_{basement}) \qquad (7\text{-}3)$$
$$Q_{windows} = A_{windows} \cdot U_{window} \cdot (T_{in} - T_{out})$$
$$Q_{doors} = A_{doors} \cdot U_{door} \cdot (T_{in} - T_{out})$$

where the $Q$s are heat losses in Btus per hour, the $A$s are areas in square feet, the $U$s are $U$-factors in Btu/hr-ft$^2$-°F, and the $T$s are temperatures in °F.

$U$-factors for various building materials are sometimes provided by their manufacturers or dealers, but the $R$-factor is more common. Converting $R$-factors to $U$-factors is easily done, using Equation 7–1.

Two alternative practices in construction require an additional first step in the calculation. If walls are made of layers of different materials, the $R$s are added to obtain an $R$ for the combination. If different walls are made of materials with different $R$s, then the heat loss must be calculated for each area separately and then added.

The heat loss calculated by Equation 7–1 assumes that windows and doors are airtight and always closed. Of course, additional heat loss occurs by infiltration through open doors, windows, and cracks at a rate that depends upon wind speed and the size of the cracks. The actual heat loss from a real door is often better represented by its perimeter than by its area, since more heat is usually lost through cracks than through the area of the door.

A reasonable approximation to the total heat loss of a residential structure can be obtained by ignoring the inner wall and ceiling surface material and using only the $U$-factors of the insulation and exterior building materials in the calculation for heat loss. The result thus computed will be larger by nearly the amount that ignoring the effects of infiltration will make it smaller. For more detailed and accurate methods of calculating heat loss, references on this subject are recommended (see Bibliography).

Heat loss calculated from Equation 7–2 provides both a peak Btu requirement (when the *lowest* outdoor temperature is used) and an average Btu requirement (when the *average* outdoor temperature is used). The monthly Btu average requirement can be obtained by substituting the number of degree days from Table 3–3 in place of the temperature difference factors in Equation 7–3 and then multiplying the result by 24 hours per day. That is:

$$Q_{month} = \left[ Q \times \frac{DD}{T_{in} - T_{out}} \times 24 \right] \text{Btu/month} \qquad (7\text{-}4)$$

where:  $Q_{month}$ is the heat loss per month

   $DD$   is the number of degree days (in °F) in the month from Table 3–3

   $T_{in}$   is the desired indoor temperature (°F)

   $T_{out}$   is the *average* ambient temperature from Table 3–5 (°F).

## MEASURING Btu REQUIREMENTS

In an existing structure, the Btu requirement can be approximately measured by multiplying the electric power or fossil fuel consumed in supplying energy to the structure for a day or for an entire heating season by the appropriate conversion factor. The factor for electric heat is 1 kW = 3413 Btu/hr while that for oil heat (assuming that #2 heating oil is used) is 140,000 Btus for each gallon burned. In a typical oil heating system, however, as much as 35 percent of the heating value is lost through escape in the flue, which carries combustion products outside the structure. As a round number, it is more accurate to regard one gallon of heating oil as equivalent to the heating effect of 100,000 Btus. Each cubic foot of natural gas contributes about 1,000 Btus with more efficient combustion than oil, so that about 800 Btus are effectively provided. The recoverable heat energy from coal is about 10,000 Btus per pound or approximately 20 million Btus per ton. Table 7–2 summarizes these approximate figures.

## PROCEDURES FOR ESTIMATING AREA AND CAPACITY

### STEP 1

Locate the solar site on the maps in Figures 3–1 through 3–12 and select the nearest city among those listed in Tables 3–1, 3–3, and 3–5. Enter the values, which may be rounded to two significant figures, from these figures and tables on a monthly worksheet, as shown in Table 7–3. (The example used is based on a site near Philadelphia.)

If the site is not within 75 miles of one of the cities in the tables, select values from two other cities in opposite directions from the site and interpolate to estimate the value at the site. One should use common sense in situations where a site location is on a windy hilltop or in a misty valley, or obtain detailed climatological data when available, to make the necessary adjustments to conform to experience.

**TABLE 7–2** Approximate Btu Equivalents among Energy Sources

| Source | Unit | Approximate Heating Values in Btus | |
|---|---|---|---|
| | | Theoretical | Practical* |
| Solar power | One square foot of bright sun for one hour at solar noon | 320 | 220 |
| Electric power | One kilowatt hour | 3,413 | 3,413 |
| #2 Fuel oil | One gallon | 143,000 | 100,000 |
| Natural gas | One cubic foot | 1,000 | 800 |
| Coal | One pound | 1,300 | 1,000 |

* After system losses are considered in heating applications.

**TABLE 7–3**  Site Climate Worksheet

City:  Philadelphia
Latitude:  40° N
Longitude:  75° W
Elevation:  200 ft

| | Jan. | Feb. | Mar. | Apr. | May | June | July | Aug. | Sept. | Oct. | Nov. | Dec. | Average All Year |
|---|---|---|---|---|---|---|---|---|---|---|---|---|---|
| Solar irradiance (Ly/day) | 150 | 225 | 300 | 400 | 475 | 500 | 525 | 450 | 350 | 250 | 150 | 125 | 325 |
| (Btu/ft²-day*) (from Figs. 3–1—3–12) | 550 | 830 | 1,110 | 1,480 | 1,750 | 1,850 | 1,940 | 1,660 | 1,290 | 920 | 550 | 460 | 1,200 |
| Average ambient temperature (°F) (from Table 3–5) | 32 | 34 | 42 | 53 | 63 | 72 | 77 | 75 | 68 | 57 | 46 | 35 | 55 |
| Heating degree days (°F) | 1,000 | 870 | 720 | 370 | 120 | 0 | 0 | 0 | 40 | 250 | 560 | 920 | 404 |
| Average percent sunshine (from Table 3–1) | 50 | 53 | 57 | 56 | 58 | 63 | 62 | 63 | 60 | 59 | 52 | 50 | 57 |
| Average wind speed (mph) (from Table 3–10) | 10 | 11 | 12 | 11 | 10 | 9 | 8 | 8 | 8 | 9 | 10 | 10 | 10 |

\* NOTE: 1 Ly/day = 3.69 Btu/ft²-day.

## STEP 2

Examine the site geometry in relation to the sun. For many applications it is only necessary to select a southward facing roof or wall with exposure to the sun from 9:00 AM to 3:00 PM in the winter. Southern exposure is strongly recommended for the obvious reason, but orientation of the collector is not so critical that a minor change in direction from due south, necessitated by the realities of the site or installation error, will seriously affect the system's performance. To illustrate this point, the example used in this chapter is based upon collectors oriented at 15° east of due south, and at a 5° steeper tilt than the latitude at the Philadelphia site, even though the general rule for collectors used year-round is to orient the collector toward due south with a tilt angle equal to the site's latitude. Furthermore, the realities of any site may require a change in collector orientation; the example departs from the general rule for orientation to accommodate the influence of Philadelphia's typical weather, i.e., sun in the morning with afternoon clouds. In brief: the "optimal" orientation of collectors (facing due south at a tilt angle equal to site latitude) may not always be the optimal orientation at the location under study. The following general observations can be of assistance in determining the most appropriate collector orientation.

*Hot Water Service.* Hot water is required throughout the year and at various times during the day. Tilt the collector at approximately the angle of latitude and perhaps slightly easterly from due south to catch the morning sun.

If some seasons require awakening before dawn, tilt the collector westerly to catch the late afternoon sun and insulate the storage tank well.

*Space Heating Service.* Since heat is needed primarily in the winter, and most abundantly in the morning, tilt the collector toward the peak elevation of the sun in late January, latitude +15° (again, more or less). It may perform better if facing somewhat westerly from due south because solar collectors become more efficient as the outdoor temperature rises in the afternoon. This possibility for increased efficiency is obviously negated if late afternoon cloudiness is common at the site, an example of the manner in which theory must be combined with consideration of practical realities before decisions are made.

*Air Conditioning Service.* Air conditioning is utilized primarily in the summer and most abundantly in the afternoon and evening. Tilt the collector at the peak elevation of the sun in late July, latitude −15° (more or less). Performance may be improved if the collector faces slightly east of due south, especially in locations where mornings are usually clear, but afternoon cloudiness is frequent.

The optimum tilt orientation of a collector is neither solely nor critically determined by the tilt necessary to face the sun at its peak elevation; experience and the effects of related factors can alter the theoretical position. As an example, the convective losses from flat plate collectors tend to decrease with larger tilt angles, so greater efficiency might be pursued with greater tilt. A steep tilt angle, however, incurs a more severe wind load. Also, a smaller tilt angle in the summer will allow the sun to spend more time in front of the collector rather than behind. The final decision must, again, rest upon a combination of theoretical calculations and practical sense.

In any case, the collector angle is specified by the azimuth angle, $\phi$, the angle at which the collector tilt is directed away from due south. Due south exposure, then, is represented as $\phi = 0$. The elevation angle, $\psi$, of the collector normal is the 90° complement of the tilt angle, $\Sigma$, so $\Sigma = 90° - \psi$. Figure 2–6 contains a summary of the detailed geometric relationships. Good year-round exposure can be achieved when $\Sigma = L$, the latitude of the site and $\phi = 0$ representing exposure due south.

Once the collector normal angles, $\phi$ and $\psi$, are selected (either by design or fortuitous choice), one may proceed. A Site Geometry Worksheet like that in Table 7–4 can be used to describe the angles and the shading which form the relationship between the collector and the sun's apparent motion for each hour of the day and each month of the year. This example uses $\phi = 15°$ (facing 15° east of due south) and $\psi = 45°$ (5° steeper tilt than the latitude at Philadelphia) with the entries in the worksheet taken from Equations 2–9, 2–11, and 2–12. A programmable calculator will facilitate the calculations. The blank areas on the worksheet represent those periods before sunrise and after sunset. Zeros are entered in the worksheet whenever cos $\theta$ is negative (corresponding to those times when the sun is *behind* the collector) and for those hours or months when the collector is shaded by buildings, hills, trees, or other objects.

## STEP 3

Determine the total solar irradiance—direct, diffuse, and reflected components—incident upon the collector. All these components incident on a hori-

## TABLE 7–4  Site Geometry Worksheet

City: Philadelphia  
Latitude: 40°  
Longitude: 75°

Tilt Angles:  
From horizontal  
Σ = 45°  
From direction of due south  
φ = +15°

A = Elevation angle of the sun (Equation 2–9)  
Z = Azimuth angle of the sun (Equation 2–11)  
θ = Incident angle (Equation 2–12)

On the 21st Day of

| Local Solar Time | Dec. A | Dec. Z | Dec. Cos θ | Jan. Nov. A | Jan. Nov. Z | Jan. Nov. Cos θ | Feb. Oct. A | Feb. Oct. Z | Feb. Oct. Cos θ | Mar. Sept. A | Mar. Sept. Z | Mar. Sept. Cos θ | Apr. Aug. A | Apr. Aug. Z | Apr. Aug. Cos θ | May July A | May July Z | May July Cos θ | June A | June Z | June Cos θ |
|---|---|---|---|---|---|---|---|---|---|---|---|---|---|---|---|---|---|---|---|---|---|
| 4:00 | | | | | | | | | | | | | | | | | | | | | |
| 5:00 | | | | | | | | | | | | | | | | 2 | 115 | 0 | 4 | 117 | 0 |
| 6:00 | | | | | | | | | | | | | 8 | 99 | .17 | 13 | 106 | .15 | 15 | 108 | .14 |
| 7:00 | | | | | | | 4 | 71 | .44 | 11 | 80 | .43 | 19 | 90 | .41 | 24 | 97 | .38 | 26 | 100 | .37 |
| 8:00 | 5 | 53 | .62 | 8 | 55 | .63 | 14 | 61 | .65 | 23 | 70 | .65 | 30 | 79 | .62 | 36 | 87 | .59 | 37 | 91 | .57 |
| 9:00 | 14 | 42 | .78 | 17 | 44 | .80 | 24 | 49 | .82 | 33 | 57 | .82 | 41 | 67 | .79 | 47 | 76 | .75 | 49 | 80 | .73 |
| 10:00 | 21 | 29 | .89 | 24 | 31 | .91 | 31 | 35 | .93 | 42 | 42 | .94 | 51 | 51 | .91 | 58 | 61 | .86 | 60 | 66 | .84 |
| 11:00 | 25 | 15 | .94 | 28 | 16 | .96 | 36 | 18 | .99 | 48 | 23 | .99 | 59 | 29 | .96 | 66 | 37 | .91 | 69 | 42 | .89 |
| 12:00 | 27 | 0 | .93 | 30 | 0 | .94 | 38 | 0 | .97 | 50 | 0 | .98 | 62 | 0 | .95 | 70 | 0 | .90 | 73 | 0 | .87 |
| 1:00 | 25 | −15 | .85 | 28 | −16 | .87 | 36 | −18 | .89 | 48 | −23 | .90 | 59 | −29 | .87 | 66 | −37 | .82 | 69 | −42 | .80 |
| 2:00 | 21 | −29 | .72 | 24 | −31 | .73 | 31 | −35 | .76 | 42 | −42 | .76 | 51 | −51 | .73 | 58 | −61 | .69 | 60 | −66 | .67 |
| 3:00 | 14 | −42 | .54 | 17 | −44 | .55 | 24 | −49 | .57 | 33 | −57 | .56 | 41 | −67 | .54 | 47 | −76 | .51 | 49 | −80 | .49 |
| 4:00 | 5 | −53 | .33 | 8 | −55 | .34 | 14 | −61 | .34 | 23 | −70 | .33 | 30 | −79 | .31 | 36 | −87 | .29 | 37 | −91 | .28 |
| 5:00 | | | | | | | 4 | −71 | .09 | 11 | −80 | .08 | 19 | −90 | .06 | 24 | −97 | .05 | 26 | −100 | .04 |
| 6:00 | | | | | | | | | | | | | 8 | −99 | 0 | 13 | −106 | 0 | 15 | −108 | 0 |
| 7:00 | | | | | | | | | | | | | | | | 2 | −115 | 0 | 4 | −117 | 0 |
| 8:00 | | | | | | | | | | | | | | | | | | | | | |

zontal surface averaged over sunny and cloudy days are included in the daily solar irradiance figures on the Site Climate Worksheet from Step 1, but as a total without distinction among them. The air mass, $m$, used in the Total Irradiance Worksheet is derived by Equation 2–15, using the values of solar elevation, $A$, shown in the Site Geometry Worksheet. The direct normal solar irradiance, $I_n$, is calculated from Equation 2–14 and recorded in the Total Irradiance Worksheets (Tables 7–5 through 7–8). Next, the beam irradiance, $I_b$, on the tilted collector surface is calculated from Equation 2–13 and the values of cos θ gained from the Site Geometry Worksheet. Then, if the collector characteristics are known, the incident angle modifier is provided by multiplying $I_b$ by $K_{\tau\alpha}$ from Equation 5–6. Finally, the total irradiance, $I_t$, is determined from Equation 3–1.

The total irradiance obtained on the Total Irradiance Worksheet represents the irradiance that is obtained on a clear, sunny day. It is generally less than the solar irradiance that occurs on a sunny, though partially cloudy, day because it ignores the contribution of diffuse radiation from clouds, which can be considerable. Equation 3–6 can be used to obtain a more accurate value for diffuse radiation when cloud cover data are available but, if not, there is no

**TABLE 7-5** Total Irradiance Worksheet (Winter)

| Local Solar Time | December 21 | | | | | January 21 | | | | | February 21 | | | | |
|---|---|---|---|---|---|---|---|---|---|---|---|---|---|---|---|
| | $m$ | $I_n$ | $I_b$ | $I_b \times K_{\tau\alpha}K_s$ | $I_t$ | $m$ | $I_n$ | $I_b$ | $I_b \times K_{\tau\alpha}K_s$ | $I_t$ | $m$ | $I_n$ | $I_b$ | $I_b \times K_{\tau\alpha}K_s$ | $I_t$ |
| 6:00 | | | | | | | | | | | 14.2 | 48 | 21 | 18 | 22 |
| 7:00 | | | | | | | | | | | 4.1 | 212 | 138 | 128 | 147 |
| 8:00 | 11.4 | 79 | 49 | 45 | 52 | 7.1 | 145 | 91 | 85 | 97 | 2.4 | 270 | 222 | 216 | 239 |
| 9:00 | 4.1 | 220 | 171 | 165 | 184 | 3.4 | 244 | 195 | 189 | 210 | 1.9 | 292 | 271 | 269 | 294 |
| 10:00 | 2.8 | 265 | 236 | 232 | 255 | 2.4 | 279 | 254 | 251 | 275 | 1.7 | 302 | 299 | 299 | 325 |
| 11:00 | 2.4 | 281 | 265 | 262 | 287 | 2.1 | 292 | 281 | 279 | 304 | 1.6 | 305 | 296 | 295 | 322 |
| 12:00 | 2.2 | 288 | 268 | 265 | 290 | 2.0 | 298 | 280 | 277 | 303 | 1.7 | 302 | 269 | 265 | 291 |
| 1:00 | 2.4 | 281 | 239 | 234 | 258 | 2.1 | 292 | 254 | 249 | 275 | 1.9 | 292 | 222 | 213 | 238 |
| 2:00 | 2.8 | 265 | 191 | 182 | 204 | 2.4 | 279 | 209 | 194 | 218 | 2.9 | 270 | 154 | 140 | 163 |
| 3:00 | 4.1 | 220 | 119 | 106 | 125 | 3.4 | 244 | 139 | 121 | 142 | 4.1 | 212 | 72 | 55 | 73 |
| 4:00 | 11.4 | 79 | 26 | 19 | 26 | 7.1 | 145 | 99 | 37 | 50 | 14.2 | 48 | 4 | 0 | 3 |
| 5:00 | | | | | | | | | | | | | | | |
| 6:00 | | | | | | | | | | | | | | | |
| Total Daily Irradiance: | | | | 1681 | | | | | 1874 | | | | | 2117 | |

$m$ = air mass
$I_n$ = direct normal solar irradiance

$I_b$ = beam solar irradiance
$K_{\tau\alpha}$ = incident angle modifier

$K_s$ = edge shadow modifier
$I_t$ = total solar irradiance

**TABLE 7-6** Total Irradiance Worksheet (Spring)

| Local Solar Time | March 21 | | | | | April 21 | | | | | May 21 | | | | |
|---|---|---|---|---|---|---|---|---|---|---|---|---|---|---|---|
| | $m$ | $I_n$ | $I_b$ | $I_b \times K_{\tau\alpha}K_s$ | $I_t$ | $m$ | $I_n$ | $I_b$ | $I_b \times K_{\tau\alpha}K_s$ | $I_t$ | $m$ | $I_n$ | $I_b$ | $I_b \times K_{\tau\alpha}K_s$ | $I_t$ |
| 6:00 | | | | | | 7.1 | 103 | 17 | 7 | 19 | 4.4 | 151 | 23 | 7 | 28 |
| 7:00 | 5.2 | 163 | 70 | 59 | 74 | 3.1 | 212 | 87 | 71 | 97 | 2.4 | 221 | 84 | 67 | 98 |
| 8:00 | 2.5 | 250 | 153 | 152 | 176 | 2.0 | 256 | 159 | 147 | 177 | 1.7 | 255 | 151 | 138 | 173 |
| 9:00 | 1.8 | 281 | 230 | 224 | 251 | 1.5 | 279 | 220 | 213 | 246 | 1.4 | 272 | 204 | 196 | 234 |
| 10:00 | 1.5 | 297 | 279 | 277 | 306 | 1.3 | 291 | 264 | 261 | 296 | 1.2 | 282 | 243 | 238 | 277 |
| 11:00 | 1.3 | 304 | 301 | 301 | 330 | 1.2 | 297 | 285 | 283 | 319 | 1.1 | 287 | 261 | 258 | 298 |
| 12:00 | 1.3 | 306 | 300 | 299 | 329 | 1.1 | 299 | 284 | 282 | 317 | 1.1 | 289 | 260 | 256 | 297 |
| 1:00 | 1.3 | 304 | 274 | 270 | 299 | 1.2 | 297 | 258 | 253 | 289 | 1.1 | 287 | 235 | 229 | 269 |
| 2:00 | 1.5 | 297 | 226 | 217 | 246 | 1.3 | 291 | 212 | 202 | 237 | 1.2 | 282 | 195 | 184 | 223 |
| 3:00 | 1.8 | 281 | 157 | 142 | 169 | 1.5 | 279 | 150 | 134 | 168 | 1.4 | 272 | 139 | 122 | 160 |
| 4:00 | 2.5 | 250 | 83 | 62 | 86 | 2.0 | 256 | 79 | 57 | 88 | 1.7 | 255 | 74 | 51 | 87 |
| 5:00 | 5.2 | 163 | 13 | 0 | 16 | 3.1 | 212 | 13 | 0 | 25 | 2.4 | 221 | 11 | 0 | 31 |
| 6:00 | | | | | | 7.1 | 103 | 0 | 0 | 12 | 4.4 | 151 | 0 | 0 | 21 |
| Total Daily Irradiance: | | | | 2282 | | | | | 2290 | | | | | 2196 | |

148

## TABLE 7–7 Total Irradiance Worksheet (Summer)

| Local Solar Time | June 21 | | | | | July 21 | | | | | August 21 | | | | |
|---|---|---|---|---|---|---|---|---|---|---|---|---|---|---|---|
| | $m$ | $I_n$ | $I_b$ | $I_b \times K_{\tau\alpha}K_s$ | $I_t$ | $m$ | $I_n$ | $I_b$ | $I_b \times K_{\tau\alpha}K_s$ | $I_t$ | $m$ | $I_n$ | $I_b$ | $I_b \times K_{\tau\alpha}K_s$ | $I_t$ |
| 6:00 | 3.8 | 159 | 22 | 5 | 29 | 4.4 | 141 | 21 | 6 | 28 | 7.1 | 86 | 15 | 6 | 18 |
| 7:00 | 2.3 | 219 | 81 | 64 | 97 | 2.4 | 211 | 80 | 64 | 96 | 3.1 | 193 | 79 | 65 | 92 |
| 8:00 | 1.7 | 249 | 142 | 128 | 166 | 1.7 | 246 | 145 | 132 | 170 | 2.0 | 238 | 148 | 136 | 170 |
| 9:00 | 1.3 | 266 | 194 | 185 | 225 | 1.4 | 263 | 197 | 189 | 229 | 1.5 | 262 | 207 | 200 | 237 |
| 10:00 | 1.1 | 275 | 231 | 226 | 267 | 1.2 | 273 | 235 | 230 | 272 | 1.3 | 274 | 249 | 246 | 285 |
| 11:00 | 1.1 | 280 | 249 | 245 | 288 | 1.1 | 278 | 253 | 250 | 292 | 1.2 | 281 | 269 | 268 | 307 |
| 12:00 | 1.0 | 282 | 245 | 240 | 283 | 1.1 | 280 | 252 | 248 | 291 | 1.1 | 282 | 268 | 267 | 306 |
| 1:00 | 1.1 | 280 | 224 | 217 | 259 | 1.1 | 278 | 228 | 222 | 264 | 1.2 | 281 | 244 | 240 | 279 |
| 2:00 | 1.1 | 275 | 185 | 173 | 215 | 1.2 | 273 | 188 | 178 | 220 | 1.3 | 274 | 200 | 191 | 229 |
| 3:00 | 1.3 | 266 | 130 | 113 | 154 | 1.4 | 263 | 134 | 118 | 158 | 1.5 | 262 | 141 | 126 | 163 |
| 4:00 | 1.7 | 249 | 70 | 47 | 85 | 1.7 | 246 | 71 | 49 | 87 | 2.0 | 238 | 74 | 53 | 87 |
| 5:00 | 2.3 | 219 | 9 | 0 | 33 | 2.4 | 211 | 11 | 0 | 32 | 3.1 | 193 | 12 | 0 | 27 |
| 6:00 | 3.8 | 159 | 0 | 0 | 24 | 4.4 | 141 | 0 | 0 | 21 | 7.1 | 86 | 0 | 0 | 12 |
| *Total Daily Irradiance:* | | | | *2125* | | | | | *2160* | | | | | *2212* | |

## TABLE 7–8 Total Irradiance Worksheet (Autumn)

| Local Solar Time | September 21 | | | | | October 21 | | | | | November 21 | | | | |
|---|---|---|---|---|---|---|---|---|---|---|---|---|---|---|---|
| | $m$ | $I_n$ | $I_b$ | $I_b \times K_{\tau\alpha}K_s$ | $I_t$ | $m$ | $I_n$ | $I_b$ | $I_b \times K_{\tau\alpha}K_s$ | $I_t$ | $m$ | $I_n$ | $I_b$ | $I_b \times K_{\tau\alpha}K_s$ | $I_t$ |
| 6:00 | | | | | | | | | | | | | | | |
| 7:00 | 5.2 | 140 | 60 | 50 | 66 | 14.2 | 35 | 15 | 13 | 16 | | | | | |
| 8:00 | 2.5 | 228 | 148 | 138 | 164 | 4.1 | 189 | 123 | 115 | 134 | 7.1 | 131 | 83 | 77 | 89 |
| 9:00 | 1.8 | 260 | 213 | 207 | 237 | 2.4 | 250 | 205 | 199 | 224 | 3.4 | 231 | 185 | 179 | 200 |
| 10:00 | 1.5 | 276 | 260 | 258 | 290 | 1.9 | 272 | 253 | 251 | 277 | 2.4 | 267 | 243 | 240 | 264 |
| 11:00 | 1.3 | 284 | 281 | 281 | 313 | 1.7 | 283 | 280 | 280 | 308 | 2.1 | 280 | 269 | 268 | 293 |
| 12:00 | 1.3 | 286 | 280 | 280 | 313 | 1.6 | 287 | 278 | 277 | 305 | 2.0 | 286 | 269 | 266 | 292 |
| 1:00 | 1.3 | 284 | 256 | 252 | 285 | 1.7 | 283 | 252 | 248 | 276 | 2.1 | 280 | 244 | 239 | 265 |
| 2:00 | 1.5 | 276 | 210 | 202 | 234 | 1.9 | 272 | 207 | 199 | 225 | 2.4 | 267 | 195 | 186 | 210 |
| 3:00 | 1.8 | 260 | 145 | 131 | 161 | 2.4 | 250 | 142 | 129 | 154 | 3.4 | 231 | 127 | 114 | 135 |
| 4:00 | 2.5 | 228 | 75 | 56 | 82 | 4.1 | 189 | 64 | 49 | 67 | 7.1 | 131 | 45 | 34 | 46 |
| 5:00 | 5.2 | 140 | 11 | 0 | 16 | 14.2 | 35 | 3 | 0 | 3 | | | | | |
| 6:00 | | | | | | | | | | | | | | | |
| *Total Daily Irradiance:* | | | | *2161* | | | | | *2039* | | | | | *1794* | |

reason for concern. The performance predicted by the procedure applied here will be conservative.

In the example shown in the worksheets, an edge shadow modifier (Equation 5–1) of $K_s = 1$ was used. This is equivalent to neglecting the effect of shadows from the walls of the collector assembly on the absorber plate. Note that the values of direct normal irradiance, $I_n$, may not always agree with those tabulated by the ASHRAE method, despite the fact that similar methods are used in calculation. The differences, which are small, are due to the use of differing values for the solar constant (429 Btu/hr-ft$^2$ versus 428 Btu/hr-ft$^2$ in ASHRAE) and the use of formulas given by Equations 2–16 and 2–17 for $I_r$ and $B$ versus the use of tabular values in ASHRAE. The continuous equations avoid the step changes in values of $I_n$ as one crosses from one month to another and when considering its day to day changes.

## STEP 4

Determine the average daily solar energy available for each month of the year. This is a simple calculation which utilizes the data already collected. The Total Irradiance Worksheet shows the total hourly amount of solar energy that can be collected for each month of the year. It reflects the realities of the variations of site location and collector tilt angles and provides information on the diffuse and reflected energy portions of the energy incident to the collector surface. The daily amount of solar energy is obtained by adding the hourly contributions on the Total Irradiance Worksheet; the monthly figures are obtained by multiplying the daily amount by the number of days in each month. This is shown in the Irradiance Summary Worksheet, Table 7–9. The monthly figures are then multiplied by the average percent sunshine figures from the Site Climate Worksheet developed in Step 1. The result is the amount of solar energy that is available under those average conditions of sunny and cloudy days at the site.

Note that in this example the variation of average daily solar irradiance on the tilted surface is less than ±10 percent of 40,000 Btu/mo-ft$^2$ for the months from March through October. At the same time, the horizontal daily irradiance varies over a wide range. This indicates that the tilted surface is suited for applications that require a uniform consumption of heat during that period (as will be necessary in the supply of domestic hot water) and that during the winter months, despite the beneficial tilt angle, supplementary sources of energy may be required because of the relative reduction in available solar energy and the cooler ambient temperatures in which the collector must operate.

The value of tilting the collector can be seen by comparing the horizontal irradiance values from the Site Climate Worksheet of Step 1 (in Langleys per day) to those in the Irradiance Summary Worksheet. This is done by multiplying the horizontal irradiance values by 3.69 to convert to Btu/ft$^2$-day and then multiplying by the number of days in the month. For the example, this comparison shows that more energy is incident on the tilted surface in the winter and less in the summer than would be incident on a horizontal surface. The amount of improvement will depend on the tilt angles and the month of the year.

**TABLE 7–9**  Irradiance Summary Worksheet

City: Philadelphia
Latitude: 40°
Longitude: 75°
Elevation: 200 ft

Collector tilt angles: From horizontal, $\Sigma = 45°$
From due South, $\phi = +15°$
Reflection coefficient: $C_f = 0.25$ (Grass)

| | Jan. | Feb. | Mar. | Apr. | May | June | July | Aug. | Sept. | Oct. | Nov. | Dec. |
|---|---|---|---|---|---|---|---|---|---|---|---|---|
| Number of days | 31 | 28 | 31 | 30 | 31 | 30 | 31 | 31 | 30 | 31 | 30 | 31 |
| Total daily/irradiance Btu/ft²-day | 1874 | 2117 | 2282 | 2290 | 2196 | 2125 | 2160 | 2212 | 2161 | 2039 | 1794 | 1681 |
| Total monthly irradiance 1000 Btu/ft²-mo | 58.1 | 59.3 | 70.7 | 68.7 | 68.1 | 63.8 | 67.0 | 68.6 | 67.0 | 63.2 | 53.8 | 52.1 |
| Average % sunshine (% ÷ 100) | .50 | .53 | .57 | .56 | .58 | .63 | .62 | .63 | .60 | .59 | .52 | .50 |
| Average available solar energy 1000 Btu/ft²-mo on tilted surface | 29.1 | 31.4 | 40.3 | 38.5 | 39.5 | 40.2 | 41.5 | 43.2 | 40.2 | 37.3 | 28.0 | 26.1 |
| on horizontal surface | 17.1 | 23.2 | 34.4 | 44.4 | 54.3 | 55.5 | 60.1 | 51.5 | 38.7 | 28.5 | 16.5 | 14.3 |

## STEP 5

The performance of a flat plate collector must be evaluated. The thermal performance of solar collectors is measured and reported as efficiency (i.e., the proportion of incident solar energy which is captured and delivered to useful application).

Efficiency as a description of performance may be calculated through NBSIR 74-635 (based on aperture area and average fluid temperature) or ASHRAE 93-77 (based on gross collector area and inlet fluid temperature); regardless of the test method, the same number of collectors will be required for a given application.

But, as was shown in Chapters 4 and 5, efficiency is not a single number because it is dependent on operating conditions which vary constantly. Ambient temperature changes throughout the year, rising and falling throughout each day. Solar irradiance characteristics, both intensity and incident angle, also differ with each season and, within each day, strike the collector at a succession of decreasing, then increasing, angular values around the noon hour. The temperature in the system's thermal storage facility, another influence on collector performance, is rarely constant.

To simplify the calculations, therefore, storage temperature will be assumed constant, and the complexity of the collector's off-angle performance will be replaced by the assumption that the collector is operating only on the noon curve. The performance estimate thus calculated will put rough bounds on system size for the example site, and allow comparison of different collectors on an equal basis. Obviously, where an actual application is being sized,

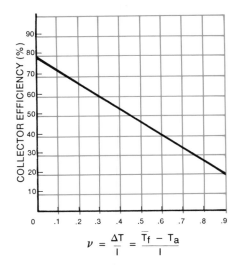

$$\nu = \frac{\Delta T}{I} = \frac{\bar{T}_f - T_a}{I}$$

**FIGURE 7–1** Collector panel of copper, with selective coating ($\alpha = 0.98$, $\varepsilon = 0.30$); double glazed with glass having a transmission $\tau = 0.90$. Collector efficiency versus operating parameter ($\bar{T}_F - T_a$)/I: Mean Fluid Temp. – Ambient Temp.

Solar Irradiance

good practice demands a thorough analysis of actual conditions with no generalized assumptions. The NBSIR 74-635 method of calculating efficiency is used here.

In this example, the AMETEK high performance collector efficiency curve, Figure 7–1, is used to determine the average collector efficiency. The curve is represented as a straight line with a y-intercept of 0.78 and a slope of −0.65.

$$\eta = 0.78 - 0.65\nu \tag{7–5}$$

where: $\eta$ = efficiency

$\nu$ = operating parameter = $\Delta T/I$

In a forced circulation system, the average fluid temperature may be selected by subtracting 5 or 10 degrees from the required service temperature. In this example, required service temperature is 145°F so that average fluid temperature circulating through the collector would be $\bar{T}_F = 140°F$. A summary of all fluid temperatures in a system operating on a clear, cloudless day would be as follows: fluid entering the collector: $T_i = 135°F$, water exiting the collector after solar heating: $T_0 = 145°F$, average fluid temperature in the collector: $\bar{T}_F = 140°F$. Note that the use of a heat exchanger between collector and storage, a common system component because of the advantages it yields, also can degrade system performance slightly because of thermal losses in the heat exchanger. Thus, in this example, $T_o$ of 145°F exiting the collector and entering one loop of the heat exchanger will yield fluid temperatures of 135°–140°F in the service loop. These temperature figures are only approximate, for purposes of example, and a well-designed system operating under good weather conditions would achieve higher temperatures. The ambient temperature for the design purposes of the example is assumed to be the average ambient temperature from the Site Climate Worksheet with total solar irradiance obtained from the Total Irradiance Worksheet. The Btus delivered is found by simply multiplying the total irradiance (from the worksheet of Step 3) by the efficiency of the collector; these data are entered in the Btus Delivered Worksheet (Table 7–10) which summarizes this information, operating parameter, and collector efficiency.

## STEP 6

Compare the Btus delivered per square foot with the service load requirement. The service load requirements and Btu requirements are approximately calculated on a monthly basis as follows:

*Hot Water.* The number of baths and bedrooms used to determine the Btu requirements in the example are taken from Table 7–11. Given a 2½ bathroom, 4 bedroom house, the minimum requirement would be 38,000 Btus per day. For hot water service of 145°F, the Btus Delivered Worksheet provides the monthly Btu per square foot of collector. The collector area may be selected to supply 100 percent of the Btu requirements in the coldest average month but this will represent the largest area to be considered. It will be more economical to use a smaller collector area and expect to use back-up heating during the coldest months. The ratio of Btus required per month to Btus delivered per square foot per month gives the area necessary to meet 100 percent of the energy requirements in that month.

*Space Heating.* The monthly space heating Btu requirement should be calculated or measured according to the ASHRAE methods outlined earlier, but a quick computation may be gained by using the following approximate technique:

Btu heating requirement =
$F \times$ (floor area in square feet) $\times$ (monthly degree days)

where: $F = 12$ for a poorly-insulated house

$F = 10$ for an average-insulated house

$F = \phantom{0}8$ for a well-insulated house.

**TABLE 7–10** Btus Delivered Worksheet

City: Philadelphia
Latitude: 40°
Longitude: 75°
Elevation: 200 ft
Service temperature: 145°F

Collector tilt angles:
 From horizontal: $\Sigma = 45°$
 From due south: $\phi = +15°$
 Reflection coefficient: $C_f = 0.25$
 Average plate temperature: $\bar{T}_F = 140°F$

| Local Solar Time | Jan. $\nu$ | Jan. $\eta$ | Jan. Btu Del. | Feb. $\nu$ | Feb. $\eta$ | Feb. Btu Del. | Mar. $\nu$ | Mar. $\eta$ | Mar. Btu Del. | Apr. $\nu$ | Apr. $\eta$ | Apr. Btu Del. | May $\nu$ | May $\eta$ | May Btu Del. | June $\nu$ | June $\eta$ | June Btu Del. |
|---|---|---|---|---|---|---|---|---|---|---|---|---|---|---|---|---|---|---|
| 6:00 | | | | | | | | | | 4.14 | — | 0 | 2.66 | — | 0 | 2.34 | — | 0 |
| 7:00 | | | | 4.61 | — | 0 | 1.26 | — | 0 | 0.86 | .22 | 22 | 0.77 | .28 | 28 | 0.70 | .32 | 31 |
| 8:00 | 1.11 | .06 | 5 | 0.71 | .32 | 47 | 0.54 | .43 | 78 | 0.48 | .47 | 85 | 0.44 | .50 | 87 | 0.41 | .51 | 85 |
| 9:00 | 0.51 | .45 | 94 | 0.44 | .50 | 120 | 0.38 | .53 | 138 | 0.35 | .55 | 139 | 0.33 | .57 | 134 | 0.30 | .58 | 131 |
| 10:00 | 0.39 | .52 | 144 | 0.36 | .55 | 163 | 0.31 | .58 | 180 | 0.29 | .59 | 178 | 0.28 | .60 | 168 | 0.26 | .61 | 163 |
| 11:00 | 0.36 | .55 | 166 | 0.32 | .57 | 187 | 0.29 | .59 | 199 | 0.27 | .61 | 196 | 0.26 | .61 | 184 | 0.24 | .63 | 179 |
| 12:00 | 0.36 | .55 | 165 | 0.33 | .57 | 185 | 0.29 | .59 | 198 | 0.27 | .60 | 195 | 0.26 | .61 | 183 | 0.24 | .62 | 176 |
| 1:00 | 0.39 | .52 | 144 | 0.36 | .55 | 160 | 0.32 | .57 | 175 | 0.30 | .59 | 173 | 0.28 | .60 | 161 | 0.26 | .61 | 157 |
| 2:00 | 0.50 | .46 | 100 | 0.44 | .50 | 120 | 0.39 | .53 | 133 | 0.36 | .55 | 133 | 0.34 | .56 | 126 | 0.32 | .57 | 123 |
| 3:00 | 0.77 | .28 | 40 | 0.63 | .37 | 61 | 0.56 | .42 | 74 | 0.50 | .45 | 78 | 0.47 | .47 | 77 | 0.44 | .49 | 75 |
| 4:00 | 2.16 | — | 0 | 1.39 | — | 0 | 1.07 | .09 | 8 | 0.94 | .17 | 16 | 0.86 | .22 | 20 | 0.81 | .25 | 21 |
| 5:00 | | | | | | | 5.16 | — | 0 | 3.00 | — | 0 | 2.41 | — | 0 | 2.13 | — | 0 |
| 6:00 | | | | | | | | | | | | | | | | | | |
| $\bar{T}_A$ | | 32 | | | 34 | | | 42 | | | 53 | | | 63 | | | 72 | |
| $\bar{T}_F - \bar{T}_A$ | | 108 | | | 106 | | | 98 | | | 87 | | | 77 | | | 68 | |

| Btus delivered per square foot per day | | | | | | |
|---|---|---|---|---|---|---|
| Clear day | 858 | 1043 | 1183 | 1215 | 1168 | 1141 |
| Per month | 26,600 | 29,200 | 36,700 | 36,500 | 36,200 | 34,200 |
| × % sunshine | 13,300 | 15,500 | 20,900 | 20,400 | 21,000 | 21,600 |

$\nu$ = operating parameter = $\Delta T/I$
$\eta$ = efficiency

**TABLE 7–10** BTUs Delivered Worksheet (continued)

City: Philadelphia  
Latitude: 40°  
Longitude: 75°  
Elevation: 200 ft  
Service temperature: 145°F

Collector tilt angles:  
From horizontal: $\Sigma = 45°$  
From due south: $\phi = +15°$  
Reflection coefficient: $C_f = 0.25$  
Average plate temperature: $\bar{T}_F = 140°F$

| Local Solar Time | July $\nu$ | $\eta$ | Btu Del. | Aug. $\nu$ | $\eta$ | Btu Del. | Sept. $\nu$ | $\eta$ | Btu Del. | Oct. $\nu$ | $\eta$ | Btu Del. | Nov. $\nu$ | $\eta$ | Btu Del. | Dec. $\nu$ | $\eta$ | Btu Del. |
|---|---|---|---|---|---|---|---|---|---|---|---|---|---|---|---|---|---|---|
| 6:00 | 2.42 | — | 0 | 3.82 | — | 0 | | | | | | | | | | | | |
| 7:00 | 0.68 | .34 | 32 | 0.73 | .31 | 27 | 1.09 | .07 | 5 | 4.88 | — | 0 | | | | | | |
| 8:00 | 0.38 | .53 | 89 | 0.39 | .53 | 87 | 0.44 | .49 | 81 | 0.61 | .38 | 51 | 1.04 | .10 | 9 | 1.98 | — | 0 |
| 9:00 | 0.28 | .60 | 135 | 0.28 | .60 | 139 | 0.31 | .58 | 137 | 0.37 | .54 | 122 | 0.46 | .48 | 97 | 0.63 | .37 | 61 |
| 10:00 | 0.24 | .63 | 167 | 0.23 | .63 | 176 | 0.25 | .62 | 179 | 0.30 | .59 | 164 | 0.35 | .55 | 147 | 0.41 | .52 | 134 |
| 11:00 | 0.22 | .64 | 184 | 0.22 | .64 | 193 | 0.23 | .63 | 197 | 0.27 | .61 | 188 | 0.32 | .57 | 170 | 0.36 | .55 | 159 |
| 12:00 | 0.22 | .64 | 182 | 0.22 | .64 | 193 | 0.23 | .63 | 197 | 0.27 | .60 | 186 | 0.32 | .57 | 170 | 0.36 | .55 | 162 |
| 1:00 | 0.24 | .62 | 162 | 0.24 | .63 | 171 | 0.25 | .62 | 175 | 0.30 | .59 | 163 | 0.35 | .55 | 148 | 0.40 | .52 | 137 |
| 2:00 | 0.29 | .59 | 127 | 0.29 | .59 | 133 | 0.31 | .58 | 135 | 0.37 | .54 | 123 | 0.44 | .49 | 105 | 0.50 | .45 | 95 |
| 3:00 | 0.41 | .51 | 79 | 0.41 | .51 | 82 | 0.45 | .49 | 79 | 0.53 | .43 | 68 | 0.68 | .34 | 47 | 0.81 | .25 | 32 |
| 4:00 | 0.75 | .29 | 25 | 0.78 | .27 | 22 | 0.88 | .21 | 17 | 1.20 | 0 | 0 | 1.96 | — | 0 | 3.75 | — | 0 |
| 5:00 | 2.17 | — | 0 | 2.71 | — | 0 | 4.50 | — | 0 | | | | | | | | | |
| 6:00 | | | | | | | | | | | | | | | | | | |
| $\bar{T}_A$ | | 77 | | | 75 | | | 68 | | | 57 | | | 46 | | | 35 | |
| $\bar{T}_F - \bar{T}_A$ | | 63 | | | 65 | | | 72 | | | 83 | | | 94 | | | 105 | |

Btus delivered per square foot per day

| | July | Aug. | Sept. | Oct. | Nov. | Dec. |
|---|---|---|---|---|---|---|
| Clear day | 1182 | 1223 | 1202 | 1065 | 893 | 780 |
| Per month | 36,600 | 37,900 | 36,100 | 33,000 | 26,800 | 24,200 |
| × % sunshine | 22,700 | 23,900 | 21,600 | 19,500 | 13,900 | 12,100 |

In the example, $F = 10$ and a 2000-square-foot floor area is given in the determination of the monthly Btu requirements for space heating (Table 7–12).

*Air Conditioning.* At the present time, air conditioning requires a service temperature of approximately 190°F. The lithium-bromide absorption system is the most efficient air-conditioning technique with input energy available from thermal sources now known. Alternatively, heat pumps may use mechanical energy to drive a compressor, but the efficiency of a conversion of solar-heated fluids to mechanical energy is poor without the use of concentrating collectors.

For those air-conditioning applications with a 200°F temperature requirement, Btus delivered must be recomputed for a 200 degree service during the summer months, as has been done on the Air Conditioning Worksheet (Table 7–13). The value of $\bar{T}_F$ used in the calculation is $200° - 5° = 195°F$.

In practice, air conditioning loads should be determined using the ASHRAE procedure, since the load depends on humidity as well as outside

154

## TABLE 7–11 Hot Water Worksheet

City: Philadelphia

Collector angles: $\phi = 15°$
$\Psi = 45°$
$\Sigma = 45°$

Reflection coefficient: $C_f = 0.25$

House: 2000 ft², insulated, 2½ baths, 4 bedrooms

Collector: AMETEK, double-glazed

| | Jan. | Feb. | Mar. | Apr. | May | June | July | Aug. | Sept. | Oct. | Nov. | Dec. | Average All Year |
|---|---|---|---|---|---|---|---|---|---|---|---|---|---|
| Hot water Btus required ($10^6$ Btu/mo) | 1.2 | 1.1 | 1.2 | 1.1 | 1.2 | 1.1 | 1.2 | 1.2 | 1.1 | 1.2 | 1.1 | 1.2 | 1.1 |
| Btus delivered ($10^3$ Btu/ft²) | 13 | 16 | 21 | 20 | 21 | 22 | 23 | 24 | 22 | 20 | 14 | 12 | 18.9 |
| Ratio (ft²) | 90 | 71 | 57 | 54 | 57 | 51 | 53 | 50 | 51 | 62 | 79 | 99 | — |
| % of needs met by 3 AMETEK collectors (48 ft²) | 53 | 68 | 84 | 89 | 84 | 94 | 91 | 96 | 94 | 77 | 61 | 48 | 78 |
| % of needs met by 4 AMETEK collectors (64 ft²) | 71 | 90 | 100 | 100 | 100 | 100 | 100 | 100 | 100 | 100 | 81 | 65 | 92 |

Note: The effects of storage size and certain other considerations (p. 151, step 5) on collector performance are ignored here for the sake of simplification. This calculation technique is primarily aimed at comparing collector performance. More exact techniques are needed to properly size systems.

## TABLE 7–12 Space Heating Worksheet

City: Philadelphia

Collector angles: $\phi = 15°$
$\Psi = 45°$
$\Sigma = 45°$

Reflection coefficient: $C_f = 0.25$

House: 2000 ft², insulated, 2½ baths, 4 bedrooms

Collector: AMETEK, double-glazed

| | Jan. | Feb. | Mar. | Apr. | May | June | July | Aug. | Sept. | Oct. | Nov. | Dec. | Average All Year |
|---|---|---|---|---|---|---|---|---|---|---|---|---|---|
| Degree days | 1000 | 870 | 770 | 370 | 120 | 0 | 0 | 0 | 40 | 250 | 560 | 920 | 408 |
| Btu heating requirement ($10^6$ Btu/mo) | 20 | 17 | 14 | 7.4 | 2.4 | 0 | 0 | 0 | 0.8 | 5.0 | 11 | 18 | 8.0 |
| Btus delivered ($10^3$ Btu/ft²) | 13 | 16 | 21 | 20 | 21 | 22 | 23 | 24 | 22 | 20 | 14 | 12 | 19 |
| Ratio (ft²) | 1504 | 1096 | 670 | 362 | 114 | 0 | 0 | 0 | 37 | 256 | 791 | 1488 | — |
| % of needs met by 30 AMETEK collectors (480 ft²) | 32 | 44 | 72 | 100 | 100 | 100 | 100 | 100 | 100 | 100 | 61 | 32 | 78 |
| % of needs met by 40 AMETEK collectors (640 ft²) | 43 | 58 | 96 | 100 | 100 | 100 | 100 | 100 | 100 | 100 | 81 | 43 | 85 |
| % of needs met by 60 AMETEK collectors (960 ft²) | 64 | 88 | 100 | 100 | 100 | 100 | 100 | 100 | 100 | 100 | 100 | 65 | 93 |

155

## TABLE 7–13 Air Conditioning Worksheet

City: Philadelphia

Collector Angles: $\phi = 15°$
$\Psi = 60°$
$\Sigma = 30°$

Reflection coefficient: $C_f = 0.25$

House: 2000 ft², insulated,
2½ baths, 4 bedrooms

Collector: AMETEK, double-glazed

| | Jan. | Feb. | Mar. | Apr. | May | June | July | Aug. | Sept. | Oct. | Nov. | Dec. | Average All Year |
|---|---|---|---|---|---|---|---|---|---|---|---|---|---|
| Normal maximum temperature | 40 | 42 | 51 | 64 | 74 | 83 | 87 | 85 | 78 | 68 | 56 | 43 | 64 |
| Degrees exceeding 65°F | — | — | — | — | 9 | 18 | 22 | 20 | 13 | 3 | — | — | — |
| Days per month | 31 | 28 | 31 | 30 | 31 | 30 | 31 | 31 | 30 | 31 | 30 | 31 | — |
| Cooling degree days | 0 | 0 | 0 | 0 | 279 | 540 | 682 | 620 | 390 | 93 | 0 | 0 | — |
| Relative humidity | .59 | .57 | .53 | .49 | .52 | .54 | .55 | .54 | .56 | .53 | .56 | .61 | .55 |
| A/C load ($10^6$ Btu) | 0 | 0 | 0 | 0 | 8.3 | 16.7 | 21.4 | 19.1 | 12.5 | 2.8 | 0 | 0 | — |
| A/C Btus delivered ($10^3$ Btu/ft²) | — | — | — | — | 17.4 | 19.2 | 20.0 | 20.3 | 17.6 | 14.7 | — | — | — |
| Ratio (ft²) | — | — | — | — | 480 | 870 | 1070 | 940 | 710 | 190 | — | — | — |
| % of needs met by 30 AMETEK collectors (480 ft²) | — | — | — | — | 100 | 55 | 45 | 51 | 68 | 100 | — | — | — |
| % of needs met by 40 AMETEK collectors (640 ft²) | — | — | — | — | 100 | 74 | 60 | 68 | 90 | 100 | — | — | — |
| % of needs met by 60 AMETEK collectors (960 ft²) | — | — | — | — | 100 | 100 | 90 | 100 | 100 | 100 | — | — | — |

temperature. An approximate method, useful for estimating air conditioning load, is:

$$\text{A/C load} = F \frac{2(\text{relative humidity})}{COP} (\text{floor area, ft}^2)$$

(monthly cooling degree days)

where: A/C load is the air conditioning hot water requirement (Btu)

$$F = \begin{cases} 12 \text{ for poorly-insulated house} \\ 10 \text{ for average-insulated house} \\ 8 \text{ for well-insulated house} \end{cases}$$

COP is the coefficient of performance for the air-conditioning system used.

The monthly cooling degree days are obtained by multiplying the number of days in a month by the amount by which the normal monthly average temperature (Table 3–5) exceeds 65°F. The average percent humidity data necessary for the calculation is contained in Table 3–4. In the example, based on a site near Philadelphia with a 2000-square-foot home with $F = 10$ and an absorption unit with $COP = 0.7$, the relevant figures are contained in the Air Conditioning Worksheet.

The number of solar collectors needed is summarized in the Collector Area Worksheet (Table 7–14), which consolidates the information obtained on previous worksheets.

**TABLE 7–14**  Collector Area Worksheet

City: Philadelphia  
Collector angles:  $\phi = 15°$  
 $\Psi = 45°$  
 $\Sigma = 45°$  
Reflection coefficient:  $C_f = 0.25$

House:  2000 ft², insulated  
2½ baths, 4 bedrooms  
Collector:  AMETEK, double-glazed

| | Area (ft²) | | Percent of Needs Met During the Year | | |
|---|---|---|---|---|---|
| Number of Collectors | Effective Solar Collection | Physical* | Hot Water | Space Heating | Air Conditioning |
| 3 | 48 | 54 | 74 | — | — |
| 6 | 96 | 108 | 92 | — | — |
| 30 | 480 | 540 | 100 | 53 | 60 |
| 40 | 640 | 720 | 100 | 65 | 75 |
| 60 | 960 | 1080 | 100 | 82 | 97 |

* The physical area is the area of the collector, including its box or frame. Additional area is usually required for fluid connection headers and access paths for maintenance.

## STEP 7

*Determination of storage capacity.* The storage capacity for a solar heating system should range from one to two gallons for each square foot of collection area. If it is less than one gallon, it will be in danger of overheating; storage of more than two gallons per square foot of collector may result in lower storage temperature levels than are desirable during any one day.

Storage capacity for hot water heating may be calculated at 1.5 gallons per square foot. In the example, using three AMETEK collectors, this amounts to 72 gallons (1.5 gal/ft² × 48 ft²); the use of six AMETEK collectors would require 144 gallons of storage capacity.

The three-collector example supplies 74 percent of the total hot water heating Btu requirement with fulfillment of the remaining 26 percent to be achieved with a back-up water heater. In this example, two 40-gallon insulated hot water tanks with heat exchange loops would provide adequate service. The use of two exchange loops is necessary not only because of building code demands, but to prevent contamination of drinking water by the collector antifreeze, should leaks develop inside either tank. The use of two storage tanks has other advantages as well. It allows for safe temperature control through control of the intertank circulation pump without disturbing the collector circulation pump. It also allows for large separation distances between storage tanks to take advantage of bulk transfer between the tanks which reduces the thermal loss in the connecting pipes.

In the six-collector example, the solar storage tank will require a capacity of about 100 gallons, but it will supply 92 percent of hot water heating for the location and tilt angles used.

In space heating, the storage requirements are also calculated at 1.5 gal-

**TABLE 7–15** Solar Storage Capacity Factors (Gallons Per Square Foot of Collector Area)

| Sunshine | Average Annual Percent Sunshine (from Table 3–1) | Storage Factor (Gal/ft²) | | |
|---|---|---|---|---|
| Cloudy | 0–55 ⟶ | 1.9 | 1.7 | 1.5 |
| Normal | 56–69 ⟶ | 1.7 | 1.5 | 1.3 |
| Sunny | 70–100 ⟶ | 1.5 | 1.3 | 1.1 |

| Temperature Extremes | Difference Between Annual Extremes (from Tables 3–6, 3–7) |
|---|---|
| High | 120–150 |
| Normal | 90–120 |
| Low | 0–90 |

### Example Storage Capacity Factors

| 1.9 | 1.7 | 1.5 |
|---|---|---|
| Cleveland | Seattle | |
| Detroit | Portland, OR | |
| | Wilmington, DE | |

| 1.7 | 1.5 | 1.3 |
|---|---|---|
| New York | Boston | Miami |
| Chicago | Philadelphia | San Francisco |
| Washington, DC | Providence | Honolulu |
| Omaha | Baltimore | |
| Salt Lake City | Atlanta | |
| Portland, ME | Houston | |
| Minneapolis | Richmond | |
| Nashville | Charlotte | |

| 1.5 | 1.3 | 1.1 |
|---|---|---|
| Denver | Phoenix | Los Angeles |
| Albuquerque | Reno | |

lons per square foot of collector area, while for air conditioning it may be more desirable to use 1.2 gallons per square foot for more rapid achievement of the high storage water temperatures necessary for the supply of the absorption apparatus. The factors used to determine the storage capacity requirements are listed in Table 7–15.

## OTHER METHODS

Work at the University of Wisconsin has resulted in several different solar performance calculation programs. TRNSYS (Klein *et al.,* 1973) is a general simulation computer program, but it takes a fair amount of expertise to set up and run. *Solar Heating Design by the f-CHART Method,* written by Beckman, Klein and Duffie, simplified this general method and their approach was produced as a computer program which has had a number of updated versions. The user is asked to provide a list of descriptive numbers pertaining to the size and ASHRAE "linear fit" performance of the collectors, the nearest city about which weather data is known, and the various loads put upon the system. The program will then compute the expected performance of the system and give an economic analysis if the cost and interest rates are known. This program is also available through several data base services. DOE support for the computer program work makes it fairly inexpensive to access.

A shortcut method of providing average annual solar fraction for domestic hot water systems based on the f-CHART method is to be found in Appendix E. This work was done with support from DOE and HUD by David E. Cassel at Mueller Associates, Baltimore. The original metric nomograph and tables were published by ASHRAE in its *Solar Domestic and Service Hot Water Manual* in 1983 and in its 1984 Systems volume. The values of the location factor *K* are slightly in error in these publications, as they were corrected to correspond exactly with the f-CHART results at a later date. The corrected values are provided in Table E–2.

# PART III

## Solar
## Electrical
## Technology

# 8 Physics and Chemistry: An Overview of Photovoltaics

Electricity is one of the most adaptable forms of power and the ability to directly convert solar energy into electricity adds to its versatility. The most successful methods devised for this conversion are those which apply the photoelectric effect. Described on the atomic level, the photoelectric effect is the excitation of a conducting electron from a stable orbit by the direct action of light of sufficiently high energy. A portion of the spectrum of sunlight is capable of energizing electrons to forceful conduction as it interacts with certain materials, especially those junctions between oppositely doped semiconductors.

There are other phenomena that can result in the production of an electric current from the action of light. When two dissimilar metals are joined at a junction which is heated by the action of light, an electric potential results from the *thermoelectric effect*. When light of sufficient intensity strikes any metal which is sufficiently heated in a vacuum, electrons will be released by the *thermionic effect*. Both phenomena, although they have been investigated for solar applications, are one step removed from direct conversion to electricity, since they depend upon the conversion of sunlight to heat before producing electricity. Thus, they tend to be less efficient than the photoelectric effect in producing electricity. In addition, there are other methods involving the differences in ionization potentials characteristic of various phases of some materials or, most common of all, the conversion of mechanical energy into electricity through an engine- or turbine-driven generator. However, in large-scale solar applications, especially those using concentration techniques, there is no reason to expect that photovoltaic converters based on the photoelectric effect will soon be more cost effective in producing electricity than the steam cycle.

Photovoltaic cells do offer some unique advantages. They have no moving parts which, in principle, translates into less maintenance; they may be used advantageously with concentration but do not require it; and, since operation does not involve the use of heated fluids or other materials, insulation is unnecessary. Further, the flexibilities of transmission and storage techniques are greater for electric power than for thermal power, photovoltaic cells are light in weight with greater portability, and they may be used in very small areas, to power wristwatches or calculators.

163

The principal impediment to the widespread use of photovoltaic cells is their current high cost, which is substantially higher than the level at which power generation under peak conditions would be competitive with existing conventional electric power rates. There are, however, many important developments that have contributed to successive cost reductions in the recent past. The proponents of solar cells extrapolate future cost reductions from past efforts based upon anticipated economies of scale and programmed technical development efforts, but their role will probably be confined to specialty applications for several years to come.

The study of photovoltaic power generation is fundamentally a study of various forms of energy: the kinetic and potential energies of an electron in motion, the quantum energy of light, the energy contained in an electric current, and so forth. As is the case with numerous disciplines underlying the collection and utilization of solar thermal energy, a comprehensive treatment of all theory relevant to photovoltaic processes is beyond the scope and available space of this handbook. The objective here, a qualitative understanding of photovoltaic processes and devices, as well as their practical application, will be enhanced by some familiarity with basic electron physics.

It is important, therefore, to briefly review the manner in which electrons are distributed around the nucleus of an isolated atom, and then consider the changes in that distribution when atoms come together to form a molecule. Each general class of molecules exhibits a characteristic distribution of electrons and, as will be seen, a characteristic distribution of energy levels. This energy level configuration is an important determinant of the properties of the material—the type of bonding formed with other atoms, and the degree to which the material will conduct an electric charge, are two examples—and so it is of interest here. Of particular importance to an understanding of photovoltaic materials and processes is the class of solids used to fabricate photovoltaic cells, i.e., semiconductors, and their electrical properties, e.g., resistivity and conductivity, among others. With this foundation established, succeeding chapters will consider more specific topics: the structure of a photovoltaic cell, the materials and means used in their fabrication, the method by which they generate an electric current, and the utilization of these cells in systems to produce electricity from solar radiation.

An additional note of caution is important. In both the preceding section on solar thermal technology and in the solar photovoltaic section to follow, the review of theoretical knowledge underlying each discipline has been purposefully abbreviated and greatly simplified. This approach has been used for two reasons. First, comprehensive and rigorous treatments of all of these concepts—thermodynamics, semiconductor physics, chemical bonding, etc.—are readily available in numerous textbooks which makes their repetition here unnecessary. Secondly, summary presentation of theory serves the objective of providing sufficient familiarity with the concepts to better understand practical applications without exceeding the space limitations appropriate to this handbook. Summarizations of theory relevant to the photovoltaic concepts are especially vulnerable to any implied or apparent accuracy because of the nature and relatively early developmental stage of the discipline. The field of photovoltaics is based upon theories of the solid state, which are still undergoing refinement.

## ATOMIC STRUCTURE

The structure of the atom of each of the elements—from hydrogen, which is the simplest in nature, to the most complex man-made elements—follows a common pattern. Sometimes likened to a planetary system (see Figure 8–1), an isolated atom is comprised of a nucleus occupying a central position like that of a sun, and electrons moving in a series of orbits about the nucleus like planets. The nucleus is comprised of protons and neutrons bound together. The number of electrons surrounding the nucleus varies according to the element, but an isolated atom is electrically neutral since the positive charges in the nucleus are balanced by a like number of negatively charged electrons. The planetary model of atomic structure is greatly over-simplified, but it serves to focus attention on an atom's electrons and electron behavior, which is crucial to an understanding of photovoltaic electricity. As will be seen, photovoltaic cells absorb the energy contained within incident sunlight to provide free electrons (and their positively charged counterparts), collect these charged particles, and channel their flow through an external circuit to do useful work.

## ENERGY LEVELS

Before proceeding, distinction of terminology is appropriate. Research has shown that it is impossible to simultaneously describe properties of an electron such as its position and momentum. Instead, one may only specify the average (or expected) values for these dimensions. Therefore, an electron occupies a shell of probability around the nucleus, rather than moving through a series of successive points which describe a specific geometric form such as a circle or ellipse. Rather than speak of orbits, one must envision "electron clouds" around the nucleus. One such electron cloud is shown for the lithium atom in Figure 8–2. The energies available to an electron consist of discrete levels which can be bundled together into "shells." This terminology is only approximate. Only shells, and not subshells, will be discussed.[1] Words such as "orbit" and "location" are necessary even though electrons do not occupy specific orbits nor can electrons be located in the usual sense of the word.

The atoms of each element can be characterized by a specific number of protons: the electrons are distributed in a characteristic number and configuration of shells. Hydrogen has a single electron in a single shell, lithium has 3 electrons distributed among 2 shells, germanium has 32 electrons and 4 shells, and so on. An example of such energy levels for the lithium atom is shown in Figure 8–3. Each energy level shown represents the total energy of the electron occupying that shell (a total comprised of kinetic energy of its motion and the potential energy due to electrostatic forces acting upon it) and identifies the shell's energy level. These allowable energy levels are separated by energy gaps which represent energies not accessible to any electron. An electron within a given atomic system may shift to an orbit with a higher energy level, but only if that electron gains sufficient additional energy to make the transition

1. The shells we will talk about are characterized by the fact that all electrons in a given shell will have the same principal quantum number, n.

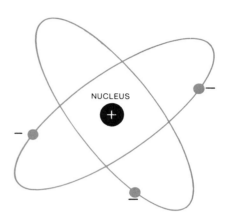

**FIGURE 8–1** Planetary model of the lithium atom.

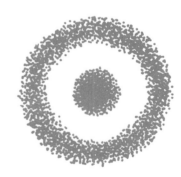

**FIGURE 8–2** Electron cloud for the 2s state in the lithium atom.

**FIGURE 8–3** Energy levels available to the electrons in a lithium atom. (Note: the exact meaning of the labels 1s, 2s, etc., is not essential to our discussion.)

across the gap to the higher level. Similarly, an electron may shift to a lower energy level, but in doing so must lose by emission energy equivalent to the difference between the levels.

## PERIODICITY

Each shell has space to accommodate only a certain number of electrons (the maximum number depends upon the shell) such that additional shells are established at a greater distance from the nucleus as needed. Hydrogen, the lightest element, has a single electron in a shell which can accept two electrons. Lithium, slightly heavier, has three electrons distributed among two shells; two electrons fill both available sites in the first shell. The third electron requires the establishment of a new shell which has sites available for eight electrons.

In progression through the elements, from the lightest to the heaviest in atomic weight, the number of electrons necessary for each of the elements progressively increases. In order to accommodate all of these electrons the capacity of the outer shells also increases. What is of importance here is that all of the elements may be arranged into groups based upon the configuration and the number of occupied and unoccupied electron sites in the outer shells of each. First tabulated in the familiar *Periodic Table of the Elements* by the Russian chemist D. E. Mendeleev (a simplified version is shown in Figure 8–4),

**FIGURE 8–4** A simplified version of the Periodic Table of the Elements. The numbers under the abbreviation for any element represent the number of electrons in each shell. (These numbers are provided only up to element 54).

| IA | IIA | IIIB | IVB | VB | VIB | VIIB | VIII | VIII | VIII | IB | IIB | IIIA | IVA | VA | VIA | VIIA | VIIIA |
|---|---|---|---|---|---|---|---|---|---|---|---|---|---|---|---|---|---|
| 1 H 1 | | | | | | | | | | | | | | | | | 2 He 2 |
| 3 Li 2, 1 | 4 Be 2, 2 | | | | | | | | | | | 5 B 2, 3 | 6 C 2, 4 | 7 N 2, 5 | 8 O 2, 6 | 9 F 2, 7 | 10 Ne 2, 8 |
| 11 Na 2, 8, 1 | 12 Mg 2, 8, 2 | | | | | | | | | | | 13 Al 2, 8, 3 | 14 Si 2, 8, 4 | 15 P 2, 8, 5 | 16 S 2, 8, 6 | 17 Cl 2, 8, 7 | 18 Ar 2, 8, 8 |
| 19 K 2, 8, 8, 1 | 20 Ca 2, 8, 8, 2 | 21 Se 2, 8, 9, 2 | 22 Ti 2, 8, 10, 2 | 23 V 2, 8, 11, 2 | 24 Cr 2, 8, 12, 2 | 25 Mn 2, 8, 13, 2 | 26 Fe 2, 8, 14, 2 | 27 Co 2, 8, 15, 2 | 28 Ni 2, 8, 16, 2 | 29 Cu 2, 8, 17, 2 | 30 Zn 2, 8, 18, 2 | 31 Ga 2, 8, 18, 3 | 32 Ge 2, 8, 18, 4 | 33 As 2, 8, 18, 5 | 34 Se 2, 8, 18, 6 | 35 Br 2, 8, 18, 7 | 36 Kr 2, 8, 18, 8 |
| 37 Rb 2, 8, 18, 8, 1 | 38 Sr 2, 8, 18, 8, 2 | 39 Y 2, 8, 18, 9, 2 | 40 Zr 2, 8, 18, 10, 2 | 41 Nb 2, 8, 18, 12, 1 | 42 Mo 2, 8, 18, 13, 1 | 43 Te 2, 8, 18, 14, 1 | 44 Ru 2, 8, 18, 15, 1 | 45 Rh 2, 8, 18, 16, 1 | 46 Pd 2, 8, 18, 18 | 47 Ag 2, 8, 18, 18, 1 | 48 Cd 2, 8, 18, 18, 2 | 49 In 2, 8, 18, 18, 3 | 50 Sn 2, 8, 18, 18, 4 | 51 Sb 2, 8, 18, 18, 5 | 52 Te 2, 8, 18, 18, 6 | 53 I 2, 8, 18, 18, 7 | 54 Xe 2, 8, 18, 18, 8 |
| 55 Cs | 56 Ba | 57 La | 72 Hf | 73 Ta | 74 W | 75 Re | 76 Os | 77 Ir | 78 Pt | 79 Au | 80 Hg | 81 Te | 82 Pb | 83 Bi | 84 Po | 85 At | 86 Rn |
| 87 Fr | 88 Ra | 89 Ac | | | | | | | | | | | | | | | |

| | | | | | | | | | | | | | |
|---|---|---|---|---|---|---|---|---|---|---|---|---|---|
| 58 Ce | 59 Pr | 60 Nd | 61 Pm | 62 Sm | 63 Eu | 64 Gd | 65 Tb | 66 Dy | 67 Ho | 68 Er | 69 Tm | 70 Yb | 71 Lu |
| 90 Th | 91 Pa | 92 U | 93 Np | 94 Pu | 95 Am | 96 Cm | 97 Bk | 98 Cf | 99 Es | 100 Fm | 101 Md | 102 No | 103 Lw |

each of the vertical groups of elements, whether lighter-weight elements at the top of each column or the heavier, more complex atoms near a column's bottom, exhibit common characteristics. The table's horizontal arrangement, also based on electronic configuration in the outer subshells, is also meaningful. Generally, elements positioned on the far right side of each row are characterized by outer shells which have their electron sites filled. Elements on the far left side of each row are characterized by outer shells with one or two electrons occupying sites in a shell with additional sites available. An orderly progression of the ratio of unoccupied to occupied electron sites in the outer shells occurs in reading left-to-right across each row.

Atoms "prefer" to exist with completed outer shells, i.e., an electron in every available site, and, if possible, completely filled shells. Atoms in the far right column of the Periodic Table (the Noble gases) are configured in precisely this manner (completely filled shells). With completely filled outer shells, these elements offer little reactivity to other atoms. Elements in the first column exhibit the opposite state-of-affairs: one electron in a shell with other available sites. These atoms are highly reactive as they seek to contrive completed outer shells. There are two general ways of doing so. The atom may fill the open sites with electrons "borrowed" from other atoms and, thus, complete the shell. Alternatively, an atom such as sodium may seek to "lend away" its electron to other atoms to gain the illusion that the uncompleted shell is gone and that the next innermost shell, which is filled, has become a completed, outer shell.

## ATOMIC BONDS

The outer electrons of neighboring atoms in a solid interact in the formation of bonding forces which hold solids together. There are three types of bonds. The first type is exhibited by the bonding between sodium (Na) atoms and chlorine (Cl) atoms. Bonding between the sodium and chlorine atoms occurs when the sodium atom "donates" its single outermost electron to an open position in the outer shell of the chlorine-forming sodium chloride (NaCl), or table salt. There is an electrical charge transferred in the process. As independent atoms, both the Na and the Cl start out being electrically neutral. Donation of the Na electron (with its negative charge) removes the atom's electrical neutrality and results in a $+1$ charge to the Na atom. Acceptance of the electron by the Cl atom, similarly results in a $-1$ charge. The electrostatic attraction between the positive sodium ion and the negatively charged chlorine ion produces an *ionic bond* (Figure 8–5). Since all of the electrons are closely bound in completed shells, there are no electrons available to participate in conduction of current. Compounds like NaCl, thus, have the high resistivity of an insulator.

A second type of bond is formed by metals because these elements will easily give up the one, two, or three electrons present in the outer shell to form positively charged ions. In their pure metallic form, solids of these elements form a crystal lattice of ionic atoms with the freed outer electrons contributed not to another specific atom but to the crystal lattice as a whole. The lattice is held together by electrostatic attraction between the positive ions and the free electrons distributed through the crystal. Solids such as metals therefore ex-

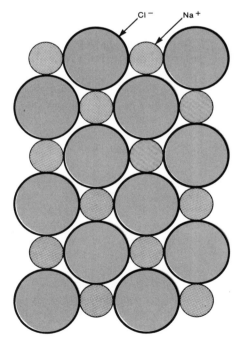

**FIGURE 8–5** An example of ionic bonding: the NaCl lattice.

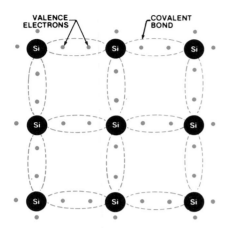

**FIGURE 8–6** Covalent bonds in silicon.

hibit *metallic bonding;* the wealth of free electrons in the crystal lattice is available to carry a current producing the high conductivity which one expects of a pure metal.

A third type of bond, *covalent bonding* (Figure 8–6), also serves the objective of allowing each atom in the solid to pursue a complete outer shell by filling as many electron sites as possible. Covalent-bonded atoms "share" electrons. As noted earlier, hydrogen has one electron and one open site available in its single shell. Two hydrogen atoms, thus, pool their single electrons to form a hydrogen molecule in a type of joint venture which allows to each atom a "time-sharing" of two electrons. The effect is that each hydrogen atom has its shell completed. Neither of the shared electrons can be identified as belonging to a particular atom; instead they belong to the bond. (It should be noted that these bonding types are not mutually exclusive; many compounds—the gallium arsenide used in photovoltaic cells is one example—are characterized by a mixture of both ionic and covalent bonding.)

## SOLIDS

When atoms form a solid, the electrons within that solid are still restricted to certain allowable energy levels. There is, however, a significant difference between the energy levels available to an electron in an isolated atom, and those energy levels available to the electrons within a solid. As millions of such atoms come together to form solids, electrons in the outer shells of proximate atoms interact, with the result that the formerly discrete energy levels of each atom split into new levels located just above and just below the original level (Figure 8–7). In a sense, the proliferation of energy levels occurs to provide sufficient "elbow room" for the outer shell electrons crowded together. The new levels are located so closely to one another and to the original discrete levels, that they effectively form a level with breadth, i.e., an *energy band*. The energy bands characteristic of a solid are still separated by energy gaps within which electrons may not be located, but the existence of widened bands in a solid produces an important result. Electrons located in each atom's more distant shells gain more latitude of location and movement within the widened bands of the solid as compared to the discrete levels of an isolated atom. Electrons located in the innermost orbits

**FIGURE 8–7** The energy levels available to (a) a single isolated lithium atom, (b) many isolated lithium atoms, (c) many interacting lithium atoms ("many" means a number approximating the number of atoms in a solid).

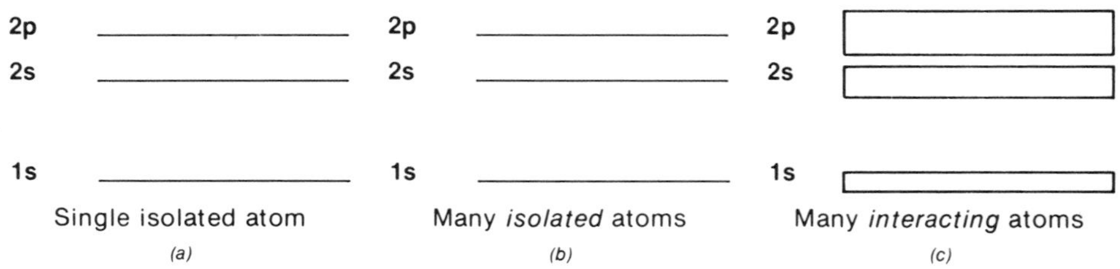

| 2p ———————— | 2p ———————— | 2p ☐ |
| 2s ———————— | 2s ———————— | 2s ☐ |
| | | |
| 1s ———————— | 1s ———————— | 1s ☐ |
| Single isolated atom | Many *isolated* atoms | Many *interacting* atoms |
| (a) | (b) | (c) |

of the atoms in a solid are more strongly held by the electrostatic forces exerted by the nucleus, and continue to occupy discrete energy levels.

## CONDUCTORS AND INSULATORS

The alteration in the configuration of energy levels when atoms form a solid may be regarded as a two-stage process. The first has already been noted as the splitting of discrete energy levels into a series of new levels as the atoms move closer and their outer shells begin to overlap. As the atoms move even closer, and reach the equilibrium spacing characteristic of each solid (where the forces of attraction and repulsion between atoms are in balance), two clearly distinguishable outer bands, sometimes separated by an energy gap, are formed. The lower energy-level-band is called the *valence band*, the higher level is the *conduction band*. Electrons may move from one band to the other but, as was the case for discrete energy levels, the process requires either a sufficient energy input (to move up from valence to conduction), or energy emission (to move in the reverse direction). Dependent upon the type of solid (Figure 8–8), both bands may have occupied and/or available electron sites. The ratio of occupied to available sites is an important determinant of the electrical properties of the material.

Conductivity, the ability to carry a current when a potential is applied, and resistivity, the reciprocal of conductivity, are two properties of interest. Current flow requires the transport of charge within the material, electrons capable of spatial movement. A general classification of the conductivity/resistivity character of solids may be made as follows:

1. Those solids characterized by a band containing both electrons to trans-

**FIGURE 8–8** Band diagrams for (a) conductor, (b) insulator, and (c) semiconductor at low temperatures. Bands occupied by electrons are shaded. The energy band gap is indicated by $E_g$.

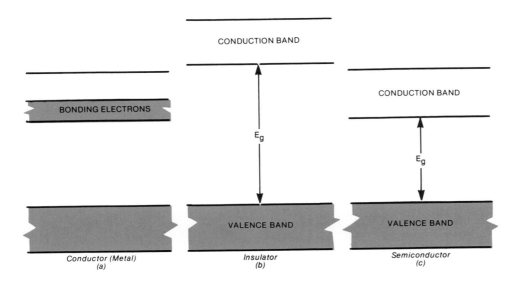

port a charge, as well as open band sites to which those electrons may move when a potential is applied, will conduct current.

2. Materials with a relatively narrow band gap across which electrons may be excited from the valence band to enter the conduction band will also exhibit conductivity.

3. In metals, the conduction and valence bands overlap. This lack of energy gap provides an obvious benefit to conductivity.

4. An insulator is a material characterized by a completely-filled valence band and a completely-empty conduction band, as well as a large energy gap; the valence band has no open sites to which its electrons may move, the conduction band has open sites but no electrons to carry a charge, and the breadth of the energy gap makes traversal difficult. Though other factors (material purity, temperature, etc.) influence a given material's electrical properties, its energy band structure is the prime determinant.

## SEMICONDUCTORS

Since photovoltaic cells are fabricated of semiconducting materials, the properties of this general class of solids are of importance. Semiconductors have a band structure much like that of insulators. At low temperatures (0°K is a standard comparison point), the valence band is completely filled, and the conduction band is completely empty. The energy gap, $E_g$, between the two bands, however, is significantly smaller in a semiconductor than that found in insulators. The smaller size of the energy gap is an important property of semiconductors. Relatively modest amounts of energy input are sufficient to excite some valence band electrons over the gap to the conduction band. Once in the conduction band, with its abundance of open sites, the electrons can move freely to conduct an electric current.

Various forms of energy can raise an electron from the valence band to the conduction band. Even at room temperature there is sufficient thermal energy to accomplish this. Another form of energy is of obvious importance: the energy of incident light (photons) can also be sufficient to excite some electrons over the gap to the conduction band.

### DIRECT AND INDIRECT SEMICONDUCTORS

The characteristic band structure of a crystalline solid may be quantified by calculating the energy levels and wave functions of the solid's electronic structure. Without detailing the specifics of the calculation, it is worth noting that electrons can be thought of as having a propagation constant, K, which depends upon the direction of travel of the electrons in the lattice. A precise calculation of variations in energy, E, and propagation constant, K, must therefore accommodate a three dimensional environment.

As an approximation of their relationship, the E and K of silicon and gallium arsenide crystals are plotted (Figure 8–9) in one crystal dimension to illustrate the shape of the valence and conduction bands as well as the energy gap, $E_g$, which they define.

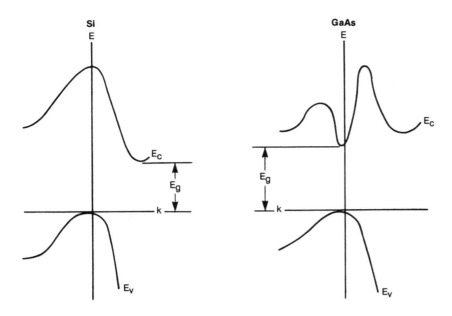

**FIGURE 8–9**  Simplified energy bands for silicon (left), and gallium arsenide.

Note that for the gallium arsenide (GaAs) crystal, the electron energy has a conduction band minimum and a valence band maximum at the same value of K. By contrast, silicon exhibits a conduction minimum and valence maximum at different K values. The consequence is that an electron in GaAs may be excited from valence to conduction band without a change in K. This is known as a *direct transition*. A similar electron transition in silicon necessitates a change in K which, in turn, requires a change in electron momentum. The latter is called an *indirect transition* because the electron must undergo a change in both momentum and energy to accomplish a successful transition of the energy gap $E_g$.

## CHARGE CARRIERS IN SEMICONDUCTORS

As has been noted, the high conductance of metals is due to an abundance of freely moving conduction band electrons in a crystal lattice. Insulators exhibit high resistance because of a wide energy gap and the consequent scarcity of conduction band electrons. Semiconductors are neither good conductors nor good insulators (hence, their name), but are materials with an electron-filled valence band and an empty conduction band at 0°K. Conduction of current by a semiconductor, then, requires excitation of electrons across the band gap; each electron transit alters the relative populations of filled and unfilled sites in both the valence and conduction bands. The mechanism may be illustrated with a simplified model.

As excitation energy (thermal, optical, etc.) is supplied to the semiconductor, some valence band electrons absorb sufficient energy to jump across the energy gap to an available site in the once-empty conduction band (Figure

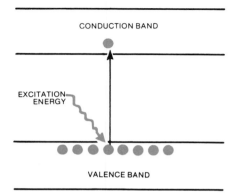

**FIGURE 8–10** Excitation of an electron from the valence band to the conduction band in a semiconductor.

8–10). Note that with the transfer of the first electron across the gap, its former position in the valence band becomes open and available. The now-open electron site is termed a "hole." The process of electron excitation to the conduction band, to use the actual terminology, is the creation of an electron-hole pair (EHP). One EHP increases the conductance of the semiconductor in two ways: under the application of a voltage the electron which has been excited across the gap may move from site to site in the otherwise empty conduction band while the opening of one available site, the hole, in the densely populated valence band allows its electrons some latitude for movement. Semiconductor technology refers to the valence band activity as a movement of the hole; in actuality it is the movement of electrons changing the position of the hole. The flow of current in a semiconductor, then, is accounted for by the motion of these two charge carriers: the negatively charged electrons and the positively charged holes.

The excitation of an electron is not a rare event and neither is the inverse process in which an electron encounters a hole, thereby annihilating each other. The process is known as *recombination*. Since each recombination removes two charge carriers from current conduction (the negative electron and positive hole), one objective in the fabrication of a photovoltaic cell is to reduce the frequency of recombination.

## INTRINSIC AND EXTRINSIC SEMICONDUCTORS

An *intrinsic semiconductor* is defined as a pure and perfect crystalline material, i.e., no impurities and no lattice structure defects. At $0°K$, an intrinsic semiconductor has a completely-filled valence band, an empty conduction band, and exhibits no conduction. The band diagram of Figure 8–8c illustrates an intrinsic semiconductor. An energy input equivalent to the size of the band gap, $E_g$ (thermal energy can be sufficient) will generate electron-hole pairs as electrons are excited from the valence band to the conduction band; the number of electrons in the conduction band, of course, must equal the number of holes in the valence band. At equilibrium, and under steady-state conditions, the formation of new electron-hole pairs will be precisely balanced by the rate of recombination.

The introduction of impurities or defects into the lattice structure of a semiconductor (a process called *doping*), provides the ability to vary the conductance of the material by engendering either a predominance of electrons, or of holes. When a crystal is doped so that the equilibrium concentration of charge carriers is different from that of the undoped material, the semiconductor is termed *extrinsic*.

Additional levels in the energy band structure, located within the band gap, are created in doped semiconductors or those with lattice defects. There are two general types of additional levels, each serving a specific objective (Figure 8–11). The first type is known as a *donor level:* a level filled with electrons located in the energy gap just under the conduction band. The energy differential between the donor level and conduction band is small enough that the input of small amounts of energy will excite the electrons and, thus, build a substantial concentration of negative charge carriers in the conduction band. The

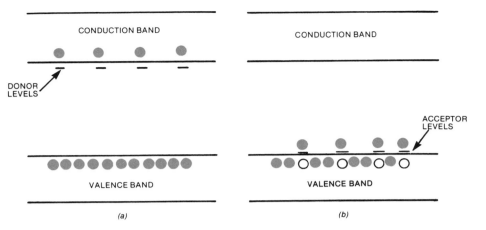

**FIGURE 8–11** Doping in semiconductors: (a) Donor levels in an *n*-type material; (b) acceptor levels in a *p*-type material.

location of the donor level, and the ease with which its electrons can be transferred to the conduction band is a function of the impurity dopant used. Phosphorus, and other elements from column V of the periodic table will introduce a properly situated donor level within a silicon lattice from which the conduction band may be enriched with electrons. Materials with a relative abundance of electrons gained in this manner, i.e., relative to the intrinsic concentration of the EHP, are called *n-type semiconductors*.

Alternative impurity dopants for silicon, boron for example, form a new level just above the valence band, which at 0°K is empty of electrons. Input of energy here causes excitation of valence band electrons into the new level which, because it accepts these electrons, is called an *acceptor level*. The process produces a significantly higher concentration of positive-charge-carrying holes in the valence band compared to the concentration of negative electrons in the conduction band. The material is therefore called a *p-type semiconductor*.

The origin of the additional electrons which produce an *n*-type semiconductor may be visualized as shown in Figure 8–12. Assume this is a covalently bonded silicon-crystal lattice with the silicon atoms located at the characteristic crystal-lattice points. Assume there is an introduction of an antimony (Sb) atom as a dopant to replace a Si atom. The Sb atom brings with it five outer electrons: four of these electrons are utilized in forming four covalent bonds with neighboring silicon atoms. The fifth electron, however, has no bonding duties to perform and is free to travel through the crystal lattice.

In a similar way, *p*-type semiconductors can be formed by doping with boron (B). The boron atom has an outer shell with only three electrons. All three are utilized in bonding with neighboring silicon atoms—the integrity of the lattice has first call on electrons—leaving one of the four necessary bonds incomplete, i.e., a hole (Figure 8–13).

In semiconductor manufacture, doping purposefully produces either a predominance of electrons or a predominance of holes; for example, for semiconductors belonging to Group IV of the Periodic Table, dopants from Group V (phosphorus, arsenic, antimony) produce *n*-type semiconductors, and Group

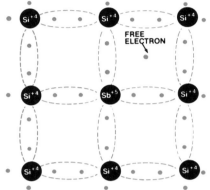

**FIGURE 8–12** A silicon lattice in which one of the Si atoms has been replaced by an antimony atom. Since Si has a valence of 4, while Sb has a valence of 5, an extra electron is available for conducting electric current.

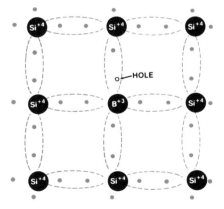

**FIGURE 8–13** A silicon lattice in which one of the atoms has been replaced by a boron (B) atom. Since Si has a valence of 4, and B a valence of 3, there is one electron missing. This missing electron behaves like a hole and can conduct electric current through the lattice.

III elements (boron, aluminum, gallium) produce *p*-type. The utilization of two types for photovoltaic purposes is explained in succeeding chapters. Also, dopant atoms are substituted into the crystal lattice by the billions, not one at a time, producing a corresponding increase in the number of electrons or holes.

Note the free electrons that characterize the *n*-type material and the holes that characterize the *p*-type material. They are produced solely by doping during fabrication. No electron hole pairs have been formed. When EHP's are created through the input and absorption of energy, the additional electrons or holes contribute even more of the desired charged particles to participate in conduction. Table 8–1 summarizes the parameters of the impurity doping process for silicon and germanium.

**TABLE 8–1**  Summary of Features of Impurity Conduction: Silicon and Germanium

| Conductivity Type | *n*-Type | *p*-Type |
|---|---|---|
| Conduction by | Electrons | Holes |
| Energy band in which carrier moves | Conduction | Valence |
| Sign of carrier | Negative | Positive |
| Valence of impurity atom | 5 | 3 |
| Name for impurity atom | Donor | Acceptor |
| Typical impurities | Elements of Group V: Phosphorus (P) Arsenic (As) Antimony (Sb) | Elements of Group III: Boron (B) Aluminum (Al) Gallium (Ga) Indium (In) |

# 9   Structural Makeup

The photovoltaic effect—the generation of voltage and current by positive and negative charge carriers produced through the absorption of photon energy—has been known for more than a century. Alexandre Becquerel first recorded the phenomenon in 1839 when he detected a photovoltage in an electrolyte in which silver chloride and platinum electrodes were immersed. Other materials capable of producing an electromotive force of small magnitude were described by a succession of investigators who followed and whose work produced the first true solid-state device, a selenium photocell, in the late 1800s. Research broadened and accelerated in this century, especially since the 1930s, but the generation of significant amounts of electrical energy from solar light is a relatively recent development.

Bell Laboratories' introduction in 1954 of a workable solar cell marked the first significant stage. That cell was a single crystal, planar-junction silicon device which proved to be the forerunner of contemporary silicon-based photovoltaic cells. Its development was a direct result of substantial advances made in the late 1940s and early 1950s that included improved methods for purification of silicon and the development of the Czochralski method of growing silicon crystals. The Bell Laboratories' cell was workable, but not economically practical because of its high cost per unit of energy produced as compared to conventionally generated power.

Development work during the last 25 years has pursued a number of avenues including techniques to refine the existing technology and to discover new methods and materials. Silicon-based technology was developed first, and made substantial progress early, chiefly due to two factors. Because the Bell Laboratories' cell was silicon, the technology received the continued focus of investigators. Secondly, the impetus of the space program, beginning in 1958 with the launching of Vanguard I and its requirement for a source of solar generated electricity as primary power, had a highly beneficial effect on research. Costs per unit of generated power were still high, even into the 1970s, but substantial progress was made in increasing cell conversion efficiency.

A number of other semiconducting materials—cadmium sulfide, gallium arsenide, and cadmium telluride—were known to have significant potential, and development efforts were addressed to these alternative materials as well. One of the potential advantages of these alternative materials was the possibility of alleviating material and fabrication difficulties of silicon-based cells. Some of the alternatives proved to be more troublesome than silicon and went by the wayside. Several others fulfilled the early promise and were refined by numerous public and private development efforts. At the present time, though much work remains to be done on each of the competing technologies, alternative photovoltaic materials have reached satisfactory levels of development maturity. The question of ultimate superiority—whether silicon or one of the alternative materials will prove to be clearly superior—is yet to be answered. Silicon cells provide relatively high conversion efficiencies at the moment, 12–16 percent depending upon the cell, but the final cost per unit of energy produced by an operating system is the critical measure. When prepared by electrodeposition, cadmium telluride uses comparatively inexpensive materials and consumes substantially lower energy during the fabrication of the cell. It can produce electricity at a lower cost per electrical unit, even at half the conversion efficiency of silicon. Further resolution of the question may show one system to be superior for commercial generation of solar power, another system more appropriate for residential use, and a third for remote power applications.

As might be expected, choosing a semiconducting material to fabricate a solar photovoltaic cell brings with it an associated set of inherent advantages and disadvantages, processing concerns, costs, and operational characteristics. As a first step toward the specifics of solar cell operation, an overview of general cell types, by methods and materials, is appropriate.

## SOLAR CELL JUNCTION TYPES AND MATERIALS

A photovoltaic solar cell is made of a semiconductor material. Incorporated into this material is an electrostatic inhomogeneity, called a *junction*, which serves to separate photogenerated electrons and holes. There are three general types of junctions:

*Homojunctions*. These cells are comprised of a single semiconducting material which has been doped during fabrication to yield at least two layers of opposite conductivity types. The cells may be constructed as *n*-on-*p* or *p*-on-*n*, with the first letter indicating the layer near the cell's light sensitive surface, and the second letter the cell's base material.

*Heterojunctions*. An *n*-on-*p* or *p*-on-*n* junction is formed between different semiconductor materials, e.g., cadmium sulfide/cuprous sulfide.

*Schottky Barrier*. The junction formation occurs here due to a readjustment of electric charges which provide an electrostatic barrier between the semiconductor and a deposited thin metal layer. In some cells, an extremely thin insulating layer is inserted between the metal and semiconductor; the configuration is referred to as an MIS device (Metal-Insulator-Semiconductor). If the insulating layer is an oxide, the cell is termed an MOS device (Metal-Oxide-Semiconductor).

Each of these junctions may be constructed of a *single crystal, polycrystalline,* or *amorphous* material. Single crystals may be cut from ingots doped to provide the proper base polarity, with the opposite polarity formed through diffusion, ion implantation, or epitaxial growth (growth of single crystal layers from a gaseous or liquid phase).

Polycrystalline materials can be formed through a casting process or by depositing thin films onto a substrate (e.g., by vacuum deposition, chemical vapor deposition, or electrodeposition).

Amorphous materials can also be formed by deposition of their films on a substrate, usually by means of an electrical discharge. One other means of general classification, by cell construction, might also be noted. There are two main types of junctions (see Figure 9–1):

*Planar.* The junction is in a plane parallel to the top surface of the cell.

*Vertical.* The junction has vertical components to aid in the collection of charge carriers.

However the junction is formed between the opposite polarities of the cell, its purpose is to create a permanent electric field within the cell. In the case of the *p-n* junction, the field results from the diffusion of electrons from the *n*-type region across the junction into the "electron deficient" *p*-type region. Similarly, holes from the *p*-type region diffuse across the junction into the "hole-deficient" *n*-type region. This diffusion of charges produces a permanent built-in voltage equal to the sum of diffusion potentials of the holes and electrons.

This summary defines the essence of the junction formation and the process by which an inherent potential is formed. But because the junction and its properties are critical to the operation of a photovoltaic cell, it is worthwhile to expand the discussion in greater detail.

**FIGURE 9–1** *Top,* a typical planar junction *n-on-p* solar cell (shown without its usual glass cover). *Below,* a vertical-junction solar cell.

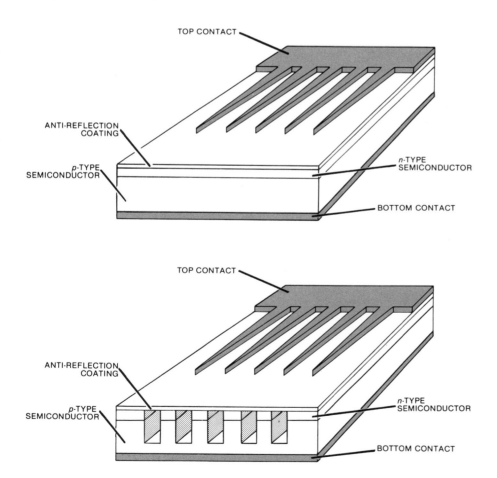

## CARRIER CONCENTRATIONS

Recalling that the location of an electron cannot be precisely described but must be stated in terms of a probability, the distribution of electrons in solids, including the semiconductors in photovoltaic cells, is governed by a special branch of probability theory known as Fermi-Dirac statistics. Though the underlying statistical arguments are complex, their result, the Fermi-Dirac Distribution Function, is straightforward and is shown graphically in Figure 9–2.

The function of f(E) yields the probability of electron occupation of an available energy site at some absolute temperature T. Figure 9–2 depicts the shape of the distribution function for several temperatures including 0°K.

A quantative use of the function is not important, but several qualitative points should be noted. The Energy $E_f$ is called the *Fermi Level* and is an important quantity in semiconductor behavior analysis. When the energy level under consideration is equal to the Fermi Level, the probability of electron occupation is precisely one half. As can be seen from the figure, at 0°K every available site up to $E_f$ is filled by electrons, all states above $E_f$ are empty. (Note that f(E) gives the probability of occupancy for an *available* energy state; if there are no available states—as is the case within the band gap of a semiconductor—there can be no electrons located there.)

Adding the Fermi Level to the same simplified representation of a semiconductor's energy band structure used in the preceding chapter illustrates why the Fermi Level is an important concept. Recall that for an intrinsic semiconductor, the concentration of holes in the valence band is equal to the concentration of electrons in the conduction band. The Fermi Level, thus, lies approximately at the midpoint of the band gap (actual, but slight, displacement occurs because of varying densities of available states in the two bands). In *n*-type semiconductors, the concentration of conduction-band electrons is considerably higher than the hole concentration of the valence band, and the Fermi Level lies much closer to the conduction band. Conversely, in a *p*-type material the higher concentration of the valence-band holes versus conduction-band electrons gives rise to a Fermi Level near the valence band. An example of Fermi Level locations is given in Figure 9–3.

## ENERGY BAND STRUCTURE

Each of these concepts—junction types, charge carrier concentration, and Fermi Level—may now be combined to convey a more accurate representation

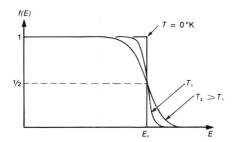

**FIGURE 9–2** The Fermi-Dirac distribution function.

**FIGURE 9–3** Simplified energy band diagrams for intrinsic, *n*-type, and *p*-type semiconductors.

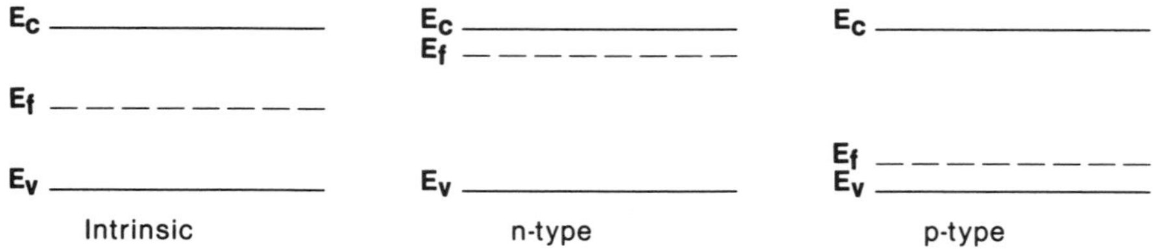

| $E_C$ ———————— | $E_C$ ————————<br>$E_f$ — — — — — — — | $E_C$ ———————— |
|---|---|---|
| $E_f$ — — — — — — — | | |
| $E_V$ ———————— | $E_V$ ———————— | $E_f$ — — — — — — —<br>$E_V$ ———————— |
| Intrinsic | n-type | p-type |

of the structure of a photovoltaic cell. A useful and common tool is the energy band diagram, which in its most elementary form has already been encountered in Figure 9–3.

Figure 9–4 illustrates a general energy band diagram for a *p-n* junction. The Fermi Level is equalized in the *n* and *p* regions due to the flow of electrons to the *p* region and holes to the *n* region. This leaves behind a stationary space charge, across which there is a potential difference referred to as the *diffusion potential*, $V_D$.

With the integral junction field in place, any excess charge carriers which appear (i.e., those electrons and holes generated by absorption of photon or thermal energy) will be influenced by the field. Minority carriers (electrons which are generated in the *p*-region, holes which are generated in the *n*-region) which appear within the integral junction field, or depletion region, will be swept across the junction to that region where each is a majority carrier. Electrons will be driven toward a lower energy state in the *n*-region, holes toward their lower energy state in the *p*-region. Some minority carriers will be generated in regions of the crystal outside the influence of the field; a fraction of these will recombine and make no contribution. Another fraction, however, will diffuse into the depletion region and also be swept across the junction. This accumulation of majority charge carriers on both sides of the junction creates a voltage across it and, with an external circuit provided, will allow a current to flow.

A more precise depiction of the energy band structure for each of the three general types of junctions is now possible. As shown in Figure 9–5, the band diagram for the homojunction device is the same as that in Figure 9–4. The three regions represented include the depletion region where the permanent junction field is high enough to reduce the probability of the loss of minority carriers through recombination. Methods have been developed (and are available in appropriate texts) to analyze the operation of this and other junction types. In essence, these equations determine the separate contribution to the photocurrent by carriers generated in the surface, depletion, and base regions

**FIGURE 9–4**  Energy band diagram for a *p–n* junction. The drift of electrons and holes under the influence of the diffusion potential is shown.

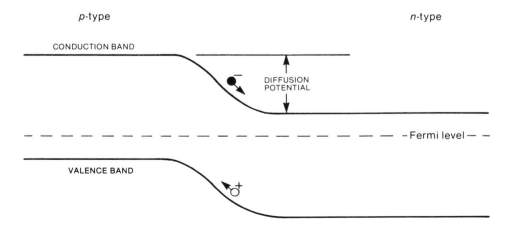

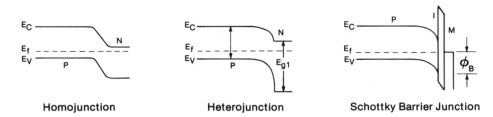

Homojunction  Heterojunction  Schottky Barrier Junction

**FIGURE 9–5**  Energy band diagram for the three junction types.

of the device, considering parameters such as wavelength of incident light, carrier mobility and lifetime, recombination, doping densities, and so forth. By varying these parameters, and with the precise determination of the effect of these variations on device performance made possible by these methods, designers may optimize performance through material and processing modifications. These modifications constitute engineering on an atomic scale.

The same three regions are valid for heterojunctions, but utilization of different semiconductor materials in these devices produces discontinuities in the band edges at the interface between components. Note that the magnitude of the band gap differs with the large band gap in the surface region and the small band gap in the base region. This configuration achieves the so-called "window" effect (i.e., superior response over a large range of wavelengths of incident sunlight). High energy photons, which are absorbed near the surface of a homojunction cell and, thus, are susceptible to a loss of charge collection efficiency through recombination, are absorbed at the junction region of many heterojunction devices because of the wider band gap in the surface region. This allows the field in the heterojunction device to sweep the photogenerated minority carriers across the junction, and thus may provide greater collection efficiency. But there is a price to be paid for this property. Discontinuities in the energy band may be introduced. However, the device may be engineered to minimize the size of the discontinuities to avoid both excessive impedance and recombination of minority carrier flow across the junction.

Not every heterojunction device is superior in performance to a homojunction. As just one example, the collection efficiency of most heterojunction cells made with silicon is lower than that of a silicon homojunction, principally because of the introduction of interface states which promote recombination. A comparison of the major material properties, including optical properties such as the absorption coefficient, is contained in Chapter 10.

The Schottky barrier cell is usually considered to be a two-layered device; its surface region is a very thin metal layer which is partially transparent. Collection efficiency is improved in these devices because the field region extends to the surface metal. The insulating layer of the MIS device does not impede carrier collection; the layer is so thin, carriers can tunnel[1] through it. The barrier height, $\phi_B$, shown in Figure 9–5, is an important design parameter in these cells. It is determined by the *work function* (the amount of energy required to remove an electron) of the metal, and by the *interface states* which exist in the band gap. In some MIS devices the electric field in the junction is intense enough to form an *inversion layer region* (i.e., a region in the *n*-type

---

1. Tunneling is the mechanism through which an electron or hole, which encounters a barrier height $\phi_B$ greater than its total energy, can "penetrate" the barrier rather than having to surmount it. There is no classical (i.e., non-quantum mechanical) analogue for the phenomenon.

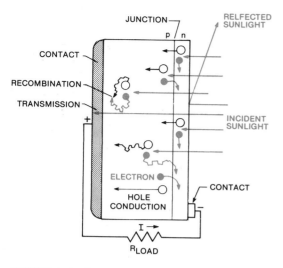

**FIGURE 9–6** Schematic representation of light interaction and current flow in a photovoltaic cell.

semiconductor which provides a junction behaving very much like the *p-n* junction).

As is the case with other junction types, the optimal device design for a Schottky Barrier cell must recognize and accommodate the individual and interactive impact of a wide range of variables. As an example, the photocurrent delivered by these devices decreases rapidly with decreasing transmission of light through the metal layer, but is not strongly affected by the magnitude of the barrier height. In polycrystalline cells, passivating treatments may be necessary in order to eliminate the effects of grain boundaries[2] and interface states.

## THE SOLAR CELL

A generalized solar photovoltaic cell with an external load resistance may be represented as shown in Figure 9–6. The circuit's negative contact is made at the cell's surface and the positive contact connected to a layer on the back of the cell. (The polarity would be reversed for a *p*-on-*n* cell.)

With sunlight incident on the cell's surface, the varying energy content of photons will cause them to penetrate the crystal lattice to varying depths. Absorption of the photon energy by the lattice atoms creates electron-hole pairs. Both the negative and positive charge carriers thus produced move through the crystal lattice; those formed within the junction region are swept across by its electric field; those formed outside the region move randomly; some to enter the field region and its influence, some to be lost through recombination outside the field. Those electrons which are collected are channeled through the external circuit, leaving the cell at the negative contact, performing work, then re-entering at the positive contact where each recombines with a hole.

2. The use of polycrystalline semiconductors, while reducing material and processing costs associated with monocrystalline structure introduces the problem of grain structure, boundaries, and size. In the worst possible case, grain boundaries can act as a minority carrier sink and majority carrier barrier such that with randomly oriented grains, only the topmost contribute to the photocurrent; those below are effectively isolated from this junction by the boundaries. Passivation of grain boundaries can reduce these effects, but a fibrous or columnar structure is preferable.

The essential interaction is the creation of charge carriers by incident optical energy. In the *n*-type material, the created electron is a *majority carrier* while the created hole is a *minority carrier,* and conversely for a *p*-type material. Those minority carriers which can diffuse to the junction are swept across it by the junction field.

The concentration of minority carriers decreases very rapidly (exponentially) with the distance from the cell's transition region because of recombination. Improvements in minority carrier properties—such as *lifetime* (the time between generation and recombination) and *diffusion length* (distance traveled before recombination)—will benefit cell efficiency by raising the probability that the carrier will reach the junction and contribute to current flow.

## LOSS FACTORS

An ideal photovoltaic cell would convert all of the photon energy incident on its surface to electrical energy. As is the case with solar thermal energy collectors, however, there are a number of loss factors which reduce conversion efficiency.

*Solar Spectrum.* Those photons in the solar spectrum having energies less than the band gap cannot generate electron-hole pairs and will not contribute to the photovoltaic output. Any photons having energies in excess of the band gap will not be fully utilized in photovoltaic conversion; the energy by which the photon energy exceeds the band-gap energy is converted to heat and thereby not available for photovoltaic output.

*Reflection.* A fraction of incident light will be reflected from the surface of the cell. As before, the coefficient of reflectance represents the fraction of light lost in this manner, and the coefficient will vary with temperature, wavelength, angle of incidence, and polarization. Anti-reflection coatings are used to reduce this loss.

*Absorptance.* The absorption properties of a photovoltaic cell (represented by its absorption coefficient) are of paramount importance in determining how much of the incoming solar energy is converted to electricity. Figure 9–7 illustrates the absorption coefficient of a number of photovoltaic materials as a function of the wavelength of incident light. Several points should be noted.

Semiconductors, with a high absorption coefficient (such as cadmium telluride), are capable of absorbing a large fraction of the incident sunlight in very thin layers. For example, less than 2 micrometers of CdTe is necessary to absorb more than 90 percent of the usable sunlight while more than 100 micrometers of silicon is required. Therefore, in CdTe, minority carriers are generated very near the junction and have less chance for recombination.

*Cell Series Resistance.* The inherent resistance of the cell's materials can reduce the cell's efficiency. Most of this resistance loss occurs in conducting electricity from the junction, through the thin upper layer, and to the front electrical contact. The type and location of contact (grid, screen, or point contact), and the junction depth impact on the magnitude of this resistance; the use of alternative front contacts, and varying junction depths to reduce this resistance loss, must also consider the loss of incident light caused by surface masking.

*Cell Shunt Resistance.* Any conductive path within the cell which parallels

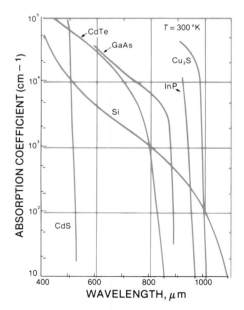

**FIGURE 9–7** Absorption spectra of various photovoltaic materials. (Reprinted with permission of G. Pouvesle, Ed., *Acta Electronica,* Ref. 11.)

that of the load resistor will tend to partially short out the cell and thereby reduce its current output.

*Junction Loss.* Because the voltage developed by a cell is dependent on the number of excess minority carriers produced, cell voltage will increase as illumination intensity increases. There is, however, a maximum voltage—theoretically the voltage of the band gap—which any cell can produce. This limitation results from the fact that photogenerated charge carriers counterbalance the junction potential; were photocarriers generated in numbers sufficient to produce a voltage equivalent to the junction potential, the cell's internal field would cease to exist. In reality, a cell's maximum voltage never approaches the band gap energy very closely. This maximum value, the cell's *open circuit voltage,* is limited by junction losses, and in many junctions is proportional to the size of the band gap.

*Collection Efficiency.* As noted earlier, some fraction of the minority carriers produced through the absorption of photon energy are not collected. Most of these are carriers produced outside the influence of the junction field, which recombine either in the bulk of the cell or at its surface.

The magnitude of each of these loss factors, as well as the effect of their interactions, is of obvious importance to designers and manufacturers attempting to maximize the operation of their photovoltaic cells. Of primary interest to those considering the installation of these cells to provide useful electric power, is the overall performance and cost of a system needed for a specific application. These points are discussed in the following chapters.

## EQUIVALENT CIRCUIT OF A SOLAR CELL

Some of the previous sections have dealt with the details of solar cells on a microscopic level. However, once a solar cell is designed and fabricated, the final product has a very simple form: it is a flat plate with two wires attached to it. If light is allowed to shine on the cell, then an electric current can be drawn from the wires.

In order to visualize what is inside the solar cell, we can construct an equivalent current (see Figure 9–8) which includes, in a simplified manner, the various loss mechanisms that can occur.

**FIGURE 9–8** Equivalent circuit of a solar cell.

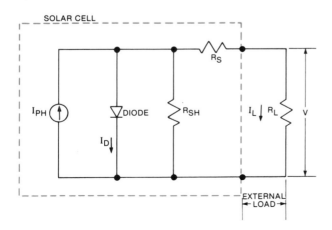

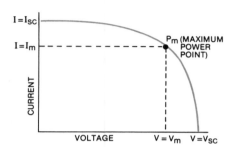

**FIGURE 9-9** *I-V curve for a solar cell.*

The equivalent circuit contains a current source which generates a photo-current $I_{PH}$. Across the current source is a diode which represents the rectifying nature of the junction. A current $I_D$ flows through this diode (this is sometimes referred to as the *dark current*).

Resistive losses are lumped into a shunt resistance, $R_{SH}$, and a series resistance, $R_S$. The external load, $R_L$, has a current, $I_L$, flowing through it to produce an external voltage, $V_L$.

If the external load resistor $R_L$ connected across an illuminated cell is varied from zero to very large values, an I-V curve—such as shown in Figure 9-9—is obtained. When the load resistance is zero, no voltage is developed across the load and the current is just equal to its *short circuit* value, $I_{sc}$. On the other hand, when the load resistance is so large that essentially no current flows through it, the voltage is equal to its *open circuit* value, $V_{oc}$.

The power output of the solar cell is given by:

$$P_{\text{out}} = I_L V_L$$

At some point on the I-V curve this power output has its highest value, $P_m$. At this point, the corresponding current and voltage are $I_m$ and $V_m$, respectively.

We can now define the efficiency of a solar cell as:

$$\text{Efficiency} = \frac{\text{Electrical Power Output}}{\text{Light Power Input}}$$

or using the symbol, $\eta$, for efficiency:

$$\eta = \frac{P_m}{F_i A} = \frac{I_m V_m}{F_i A}$$

where $F_i$ is the incident energy in the sunlight per unit area and $A$ is the area of the cell.

An alternate way of writing the efficiency is in terms of the open circuit voltage, $V_{oc}$, and the short circuit current, $I_{sc}$:

$$\eta = \frac{I_{sc} V_{oc} FF}{F_i A}$$

where $FF$ is a number called the fill factor, which describes the "squareness" of the I-V curve. The "squarer" the curve (i.e., as the "knee" in the curve becomes sharper) the larger will be the area of the rectangle which can be drawn within, and the greater the maximum power output for a given $I_{sc}$ and $V_{oc}$.

The fill factor is always less than unity[3] and is strongly affected by the values of series and shunt resistance. An increase in series resistance or a decrease in shunt resistance will always lower the fill factor.

The literature for most commercially available photovoltaic cells and modules, which is the usual source of performance data, will provide an overall power rating and specific values for the five significant points on the curve. $I_{sc}$, $V_{oc}$, $I_m$, $V_m$, and $P_m$ for certain specified operating conditions. Some also include I-V curves so that the effect of variations in cell temperature and

3. Typical commercially available cells have fill-factor values between 0.60 and 0.80.

illumination intensity on the peak power performance of the array may be deduced. These data are helpful in comparing performance among a number of modules from different manufacturers.

## THEORETICAL LIMITS TO SOLAR CELL EFFICIENCIES

As mentioned in previous sections, a solar cell cannot convert all of the solar energy that falls on it. The maximum efficiency obtainable from a solar cell is not merely due to losses such as reflection or series resistance, but is the result of more fundamental limitations. The most important of these is the size of the energy-band gap of the semiconductor.

Figure 9–10 shows the variation in efficiency with band gap for ideal homojunctions. The curves were obtained by assuming specific values for mobilities, carriers, lifetimes, carrier densities, and effective masses, and then calculating the efficiency for each temperature. Such curves should only be used in a qualitative sense in comparing different semiconductors.

An even more fundamental limit, independent of material properties, can be obtained by a detailed balancing calculation, and gives efficiencies several percentage points higher than those in Figure 9–10.

An ideal solar cell can operate at very high efficiencies for photons having an energy equal to the band gap. Therefore, it is reasonable to expect a stack of solar cells—each one with a slightly different band gap—to perform very well. Such a stack of cells is referred to as a multijunction solar cell. The theoretical efficiency of an infinite number of such cells, with band gaps varying from zero to infinity, has been calculated by Vos and Pauwels to be 87 percent at 300°K.

While such a structure is impractical, it shows the effectiveness of combining cells with different band gaps. Even two cells with different band gaps can provide significant increases in efficiency over a single band gap structure.

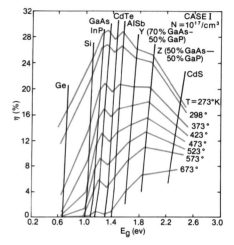

**FIGURE 9–10** Efficiency versus energy gap.

# 10 Comparing Solar Cell Technologies

A general comparison of the advantages and disadvantages of the semiconducting materials used in the manufacture of photovoltaic cells is not without pitfalls. What may be considered a disadvantage (e.g., the need for ultra-pure silicon produced by energy intensive methods) may be alleviated by new technologies. Conversely, a theoretical advantage (e.g., the strong optical absorption coefficient of a material such as gallium arsenide) can bring with it a disadvantage (significantly higher potential for surface losses of charge carriers through recombination). The breadth and pace of photovoltaic research and development, and the increasing variety of materials, processing refinements, and cell configurations which result, can make any comparative listing quickly outdated.

A companion listing of the state-of-the-art of each of these material technologies can be useful, but the same questions of accuracy and timeliness intrude. Progress is announced continuously in a wide range of activities. Engineering refinement, especially in the manufacturing process, includes efforts to increase junction potential and collection efficiencies, reduce the incidence of defects in the cell, and minimize the electrical resistance of ancillary elements such as cell-to-cell connections. Perhaps the single most important objective is the pursuit of higher efficiencies at lower costs. This concern is manifest in a trend toward thin-film configurations, rather than monocrystalline cells. While the majority of commercially available solar cells are monocrystalline and silicon-based (partially due to their developmental head-start), the cost and difficulty of producing large single crystals of virtually any semiconducting material in the required purity are becoming less economically acceptable. Two alternatives to monocrystalline cells—polycrystalline or amorphous thin-film semiconductors—are presently less efficient but are much less expensive. Progress in the development of these thin film cells is rapidly improving their cost effectiveness.

It is important to reiterate, as a preface to the comparative survey of various materials and technologies which follows, that conclusive statements are impossible. The interplay of inherent material properties, the continuing accumulation of developmental advances and new directions, and engineering accomplishments (both real and potential), bring constant change to the technology.

## SILICON-BASED CELLS

Silicon, as has been noted, was the material used for the original Bell Laboratories' cell and has benefitted from three decades of research into its use in photovoltaic applications. As a result, single crystal *p-n* type silicon cells comprise the largest class of photovoltaic cells which are commercially available. Though silicon (as $SiO_2$) is the second most abundant element in the earth's crust, production of a silicon cell from this inexpensive raw material is a complex and costly process. About 3,500 kilocalories per kilogram (kcal/kg) is needed to break the silicon-oxygen bond, and considerable further energy is needed to purify raw silicon into the electronic-grade material used to produce large, single crystal ingots (Figure 10–1) from which the silicon wafers are cut. Some of the additional expensive processing steps are: doping, etching, ion implementation, anti-reflection coatings, encapsulation, and so forth.

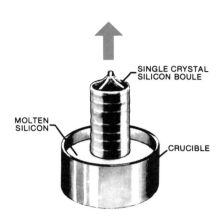

**FIGURE 10–1** Growth of a single crystal boule of silicon by the Czochralski method.

Two methods have been used to reduce the total cost of a finished silicon-based array. The first is to begin with less expensive starting materials such as metallurgical-grade silicon or solar-grade silicon. The second is to develop less expensive methods of producing the requisite large blocks of the material. In either approach, of course, the semiconducting properties necessary for photovoltaic use must be preserved.

Two systems under development can serve as examples of combining both cost reduction approaches. The heat exchange method (HEM) starts with metallurgical-grade silicon and utilizes a helium gas heat exchanger to solidify directionally a single crystal silicon cube in a silicon crucible. A second system, which also utilizes less expensive metallurgical-grade silicon, casts semicrystalline blocks with a proprietary process which sweeps impurities in the molten silicon to one end of the block as it cools. That portion of the block may then be removed before the ingot is sawed into wafers. Because the block is semicrystalline, this method has the potential problem of crystal grain size and grain boundaries, but blocks can be produced with crsytal grains so large that the cells produced exhibit properties similar to single crystal cells.

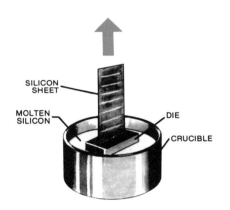

**FIGURE 10–2** Silicon sheet pulled through a die.

These methods offer economy in materials, yet both still require that an ingot be cut into wafers which is a costly, time-consuming, and wasteful operation. (About 50 percent of the silicon in an ingot is lost in sawing.) A number of research efforts have addressed this problem by developing methods to produce silicon in the form of a ribbon of appropriate thickness.

Methods used to produce the silicon ribbons vary. Often the ribbon is pulled horizontally or vertically through a die which controls the ribbon's physical dimensions and solidification rate (Figure 10–2). These methods can produce impurities in the ribbon, however, if the die reacts with the molten silicon. A second ribbon method produces thin silicon sheets from a shallow quartz crucible of molten silicon without the use of a die. A third process, called the dendritic web method (Figure 10–3), utilizes the characteristics of the crystals at the edges of the ribbon to control the shape and thickness of the product as it is being pulled.

These methods are ingenious, but each requires high energy usage in melting the silicon raw material. To reduce that cost, other investigators have been investigating the use of thin films of silicon on a supporting substrate to gain the material's photovoltaic benefits while conserving material and energy. One

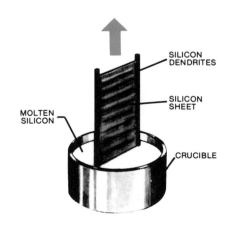

**FIGURE 10–3** Silicon sheet produced by dendritic web method.

such method is vapor deposition from various silicon compounds, to produce a polycrystalline sheet which is then rapidly melted with a scanning laser to form a single crystal ribbon. Other methods produce the thin silicon film on ceramic or graphite substrates.

All of these approaches produce a silicon material which is characterized by its crystalline configuration (i.e., single crystal, semicrystalline, or polycrystalline). Another class of materials, which includes silicon, are produced as *amorphous semiconductors*. These amorphous materials have no crystalline structure yet may be developed with the properties necessary for photovoltaic use.

Methods developed for amorphous silicon involve the deposition of a thin film of silicon on a substrate by decomposing silane ($SiH_4$) or other silicon compounds in a glow discharge produced in a vacuum system. The critical issue in these methods is to incorporate a controlled amount of impurity during the vacuum deposition to optimize the photovoltaic properties of the thin film. In fact, amorphous silicon has a band gap slightly larger than crystalline silicon and better absorption of visible light. These advantages have yet to be fully exploited in practice. The thin-film methods hold promise, however, even at lower than theoretical efficiency and even with some question concerning the long-term stability of amorphous films under operating conditions. The materials and fabrication costs are projected to be much lower than those of the crystalline methods.

The cost of all silicon-based cells could be reduced substantially if the purification method for silicon could be made less costly. Research into several such purification schemes has been supported by the Department of Energy. One such approach, proposed by the Union Carbide Corporation, is now in a pilot production and has been projected to eventually decrease the cost of semiconductor-grade poly-silicon to $10/kg.

## CUPROUS SULFIDE/CADMIUM SULFIDE

The attraction of relatively inexpensive manufacturing has been associated with the $Cu_2S/CdS$ photovoltaic technology since its initial development in the early 1950's. The highest efficiency cells have been produced by evaporating cadmium sulfide under vacuum onto a heated conducting substrate. The topmost region of the CdS is converted to $Cu_2S$ by dipping into a solution of copper chloride. Such cells have achieved an efficiency of 9 percent in the laboratory. (If zinc is added to the cadmium sulfide to form CdZnS, an efficiency of 10 percent can be obtained.)

Cells of this type have been produced and marketed for several years, but have been nagged with problems of instability. $Cu_2S/CdS$ cells have been found to react with moisture in the air; even the miniscule amounts of atmospheric moisture that are able to infiltrate an encapsulated array can have a deleterious effect on the cell junctions. Changes in the composition of the semiconducting material, new processing, or new encapsulation methods may resolve this difficulty.

## II-VI SEMICONDUCTORS

A number of other semiconducting materials related to cadmium sulfide and cuprous sulfide (related by virtue of their placement in the second and sixth columns of the Periodic Table) are also being investigated. Chief among them are cadmium selenide, copper indium selenide, and cadmium telluride. Many of these compounds are direct semiconductors and thus very strong absorbers of light. The result is the conservation of material by making the cell extremely thin. In some cases, less expensive polycrystalline materials can be used since the grain boundaries can be passivated to avoid charge carrier recombination at those sites. For example, a heterojunction solar cell made of copper indium selenide/cadmium sulfide has attained an efficiency of over 10 percent in thin-film form. The method used requires careful control of three separately-heated sources for the vacuum deposition of the copper indium selenide.

High quality thin films have been produced by inexpensive electrodeposition. The AMETEK photovoltaic cell is of this type; an electrodeposited thin film of cadmium telluride configured as a Schottky barrier cell.

## III-V SEMICONDUCTORS

Materials from the third and fifth column of the Periodic Table have been used to produce semiconductors which, with further refinement, hold promise for photovoltaic applications. Perhaps the best known of these compounds is gallium arsenide (GaAs), but other compounds such as indium phosphide, gallium phosphide, and aluminum gallium arsenide are also being investigated.

The properties of these materials are well-known and some, such as gallium arsenide, have almost ideal characteristics for photovoltaic use, (e.g., good thermal stability and charge collection efficiency). Unfortunately, all of the compounds used to date incorporate either gallium (Ga) or indium (In), elements which are very rare and expensive. These elements are by-products of the smelting of other metals, and neither has heretofore had any significant commercial use. Their cost would further increase with increased demand which would result from their use in full-scale production of photovoltaic cells. Currently gallium costs 550 times more than silicon and indium costs approximately 300 times more than silicon. Moreover, the processes used to produce the monocrystalline configuration of these materials (e.g., epitaxial crystal growth), are as expensive and energy-intensive as the monocrystalline silicon methods. Research into countering these problems by utilizing thin films to reduce material costs is being pursued.

## ORGANIC SEMICONDUCTORS

Research has recently shown that certain organic materials can act like semiconductors. In particular, polyacetylene can be doped both $p$ and $n$-type, and rectifying junctions made between them.

Compounds of phthalocyanine and merocyanine have been used to form solar cells. However, efficiencies are generally below one percent and degradation effects are a problem.

## OTHER APPROACHES

An alternative to making an all solid-state solar cell is to use a semiconductor in conjunction with a liquid electrolyte. A properly chosen electrolyte can act as a Schottky barrier. The rectifying barrier is thus formed by the simple act of dipping a semiconductor into an electrolyte.

Such a *photoelectrochemical* (PEC) cell is useful mainly because it can also be used for storage. For example, the photocurrent can be used to decompose water to hydrogen and oxygen, which can then be stored for later use in a fuel cell.

Alternatively, solar energy can be stored in the electrolyte by proper choice of a *redox* couple. In this case, both energy conversion and storage occur in a single cell.

The main problem in all of these systems has been chemical instability. Generally, the rectifying junction occurs between an *n*-type semiconductor which acts like an anode in the electrolyte. Therefore, anodic dissolution of the semiconductor is a source of instability.

However, recently a *p*-type semiconductor (indium phosphate, InP) has been used which should be much more stable since under illumination it is cathodically protected.

The use of liquid electrolytes together with silicon has been proposed by Texas Instruments as a low cost approach to both energy conversion and storage. In order to avoid the high cost of silicon solar cells, Texas Instruments uses tiny beads of silicon which are formed on a flat surface similar to sandpaper. These beads are then immersed in an electrolyte (hydrogen bromine, HBr) as shown in Figure 10-4. Upon illumination, the electricity generated is either used immediately or energy is stored in the liquid bromine generated at the

**FIGURE 10-4** A photovoltaic array using tiny silicon beads. Energy can be stored in the hydrogen and bromine for use at a later time.

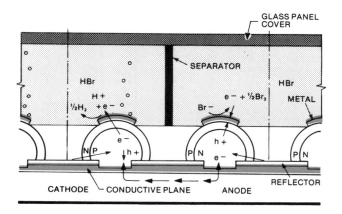

anode and hydrogen gas at the cathode. The bromine and hydrogen are stored separately and can be combined in a fuel cell to produce electricity when the sun is not out.

## SUMMARY

Though this survey of alternative photovoltaic technologies is far from exhaustive, it should be apparent that at the current state-of-the-art, each of the alternatives incorporates an inherent mix of advantages and disadvantages both real and potential. The technologies which will prove to be commercially acceptable will be those which achieve satisfactory progress in the following areas:

1. Reduced energy consumption of the fabrication process.
2. Less rigorous material-purity requirements.
3. Reduced raw material costs.
4. Reduced costs in processing cells into a finished array.

Those factors, in turn, are distilled into a single measure: the cost per watt of electrical energy produced by the photovoltaic array. As discussed in Chapter 11, the point at which the cost per watt becomes competitive with electricity generated by other means varies with the application.

# 11 Photovoltaic Systems

Photovoltaic power applications may be divided into two general types. The first are those *stand-alone* systems used to supply a relatively small electrical load of predictable demand. The majority of these installations are made at remote facilities such as broadcasting or weather-monitoring stations, navigation aids, electrochemical corrosion protection, or power supply, in regions where connection to the conventional power grid is difficult or impossible. The variation in solar irradiance at the site is the critical design parameter with the system usually sized for worst-case conditions. The cost per unit of energy delivered is of less concern with these systems because the alternative—gasoline or diesel powered generators—has always entailed substantially higher than utility grid costs per unit of electricity delivered. As a result, photovoltaic power systems for remote locations are increasingly becoming competitive. A number of such systems have been in use worldwide, and have performed reliably and cost effectively.

The second general application is where systems are used to supply residential, commercial, or industrial loads. The electrical demand of these systems is characteristically much larger and more variable so that photovoltaic power cannot be used as the sole source of electrical energy. Two separate but interconnected supply routes are used. As in the case with stand-alone systems, the photovoltaic system is connected, through its power conditioner, to the load and to the storage. An additional supply route is also used—connection of the non-solar power grid to both the load and to storage. The arrangement is represented in the block diagram of Figure 11–1.

The objective of the dual system is continuity of power supply under all conditions. When photovoltaic power is generated in excess of immediate demand, the surplus is fed first into storage until its capacity is reached, then into the conventional grid to earn credits by running the utility meter backward. During those times when solar generated electricity is not equal to the load demand (e.g., at night, under unfavorable weather conditions, or when storage is depleted), electricity is supplied directly from the utility grid to the load.

As is the case with solar thermal energy systems, design of a given photovoltaic system varies with solar irradiance, climate, load, and other factors which are specific to each installation site.

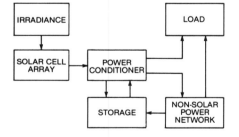

**FIGURE 11–1** Block diagram of the elements in a large-scale photovoltaic power system.

## MODULES AND ARRAYS

The photovoltaic *module* is the basic unit of the solar collection component of the system; a number of modules are connected together to form the *array*. Each module is an assembly of interconnected photovoltaic cells (from two to as many as one hundred, depending on the design) mounted on some supporting surface and covered and/or encapsulated.

Series and parallel connections are used both among the individual cells within each module and among the modules which compose the array. For example, cells may be interconnected as shown in Figure 11–2. Total voltage output of the module is determined by the output voltage of each individual cell multiplied by the number of cells connected in series. The strings of series-connected cells are then interconnected in parallel to provide the necessary current. The current output of the module is determined by multiplying the current output of each individual cell by the number of cells connected in parallel. The string terminations feed their output into a power collecting conductor called a *bus*. Like the interconnections between the cells, bus connectors may consist of wires, but more often the bus connections are flat conductors.

The same strategy of appropriate series and parallel connections is also used with modules so that the voltage and current output of the array may be tailored to the demand of the load.

The relationships for the array (or, with the attention to subscript changes, a module) are summarized in Table 11–1.

There is another advantage to the strategy of an appropriate combination of series and parallel interconnections, apart from the ability to match array voltage and current output to the application. Were all of the cells within a module (or all of the modules in an array), connected in series only, the shadowing of some portion of the array or the failure of a cell(s) would immediately limit current output to the low value in the non-illuminated or failed component. Similarly, were all connections made in parallel, array output could be reduced not only because shadowing reduces energy input to the array, but also because of an increase in internal energy losses within the non-illuminated portion of the array.

A second strategy used to conserve array power output is the use of *blocking* (sometimes termed *isolation*) diodes. Blocking diodes are connected in the array circuit, between the string terminations and the power bus, to act as

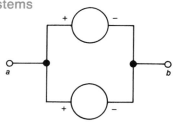

a. TWO CELLS IN PARALLEL

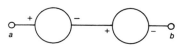

b. TWO CELLS IN SERIES

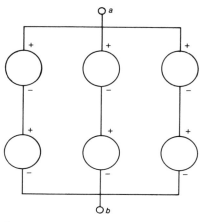

c. THREE STRINGS IN PARALLEL (EACH STRING HAS TWO CELLS IN SERIES)

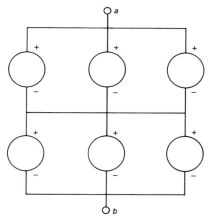

d. TWO SUBMODULES IN SERIES (EACH SUBMODULE HAS THREE CELLS IN PARALLEL)

**FIGURE 11–2** Parallel and series connection of solar cells (also applicable to solar cell modules, panels, and electrochemical batteries, but not to ideal voltage or current sources).

**TABLE 11–1** Determining Array Output: Voltage, Current, and Power

$$V_a = N_s V_c$$
$$I_a = N_p I_c$$
$$P_a = N_t P_c$$
$$P_a = V_a I_a$$
$$P_c = V_c I_c$$
$$N_t = N_s N_p$$

where:
$N_s$ = number of cells connected in series
$N_p$ = number of cells connected in parallel
$N_t$ = total number of cells in the array
$V_c$ = individual cell output voltage
$I_c$ = individual cell output current
$I_a$ = array output current
$V_a$ = array output voltage
$P_c$ = cell output power
$P_a$ = array output power

"one-way" signs controlling the electrical traffic. Current produced by illumi-nated cells is allowed to flow to collection in the power bus, but current flow in the reverse direction from the bus to non-illuminated or defective cells (which would occur whenever the string output voltage is less than bus voltage) is blocked. The addition of blocking diodes to the circuit exacts a price—a small voltage drop across the diode which lowers the output voltage—but a properly designed system will optimize the trade-off between energy lost through shad-owed cells and energy lost within the diodes.

Blocking diodes also contribute protection from serious subsystem failure resulting from short circuits anywhere in the circuit (e.g., cell failure within a string, faults in connectors, defects in terminals, and so forth). Should such a fault develop, the blocking diode isolates the malfunctioning components and prevents the drainage of power from the cells still operating.

Whatever the specific geometry and construction of the individual solar cells, and the pattern of interconnection, there are two general types of photo-voltaic arrays: Flat plate and concentrating.

## FLAT PLATE ARRAYS

Contemporary flat plate arrays used in unconcentrated sunlight consist of four elements: the cells and associated connections, an encapsulating cover, a frame to hold the module, and a supporting structure for the array.

The structural element to which the solar cells, inter-connections, and associated wiring are affixed may be either a *substrate* (upon which the back of the cell is mounted) or a *supersubstrate* (on which the cell's upper or active surface is affixed). Supersubstrates and the adhesives used to mount the cells must be transparent. A variety of materials have been used as substrates, but it is desirable to use materials with some degree of thermal conductance to help dissipate the heat. Some module designs utilize white paint to reflect sunlight from the exposed yet inactive interstitial substrate area between cells.

Encapsulation or covering of the module is necessary to protect the solar cells and interconnections from mechanical damage, environmental hazards (like those discussed at length in the section on thermal collectors), and mois-ture intrusion. A variety of glass and polymeric materials are used in the assem-bly of modules (Figures 11–3 and 11–4). The following properties of encapsula-tion materials are of concern:

1. *Weather Resistance:* The material's ability to resist deterioration due to moisture, UV radiation, and ozone or other pollutants.
2. *Transparency:* Loss of transparency reduces optical transmittance and, thus, energy input to the cells.
3. *Reflection:* Loss of incoming light because of reflection. Reflection can occur at any interface between two materials having different indices of refraction. Because any photovoltaic cell is an assembly of such inter-faces between materials, (e.g., air to cover, cover to cell surface, cell active region to cell bulk material), losses by reflection can be substan-tial. These losses can be reduced by using anti-reflective coatings.
4. *Thermal Expansion:* Daily temperature cycling, which causes expan-

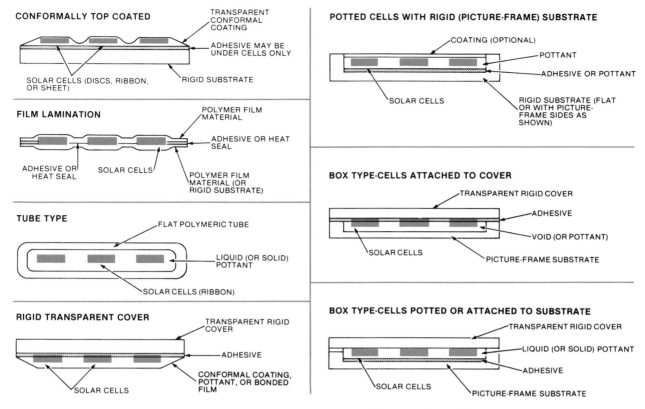

**FIGURE 11–3** Polymeric windows for terrestrial modules (Reprinted with permission of the IEEE).

sion and contraction of encapsulating materials, substrates and other components, can cause deterioration of the module and its sealing.

5. *Thermal Conductivity:* Transfer of heat from the module is desirable for the reasons noted in (4) and to minimize the operating temperature of the photovoltaic cells.

6. *Durability:* Materials must be resistant to potential damage from environmental hazards such as hailstones, surface abrasion from windblown grit, as well as vandalism.

7. *Soil and Fungal Accumulation:* Reduction in energy input to the cells results from the accumulation of static electric charges on the cover, promoting the attraction of airborne dust, and from the growth of fungus nourished by some additives used in the manufacture of polymeric materials.

Substrate, cells, and covering are usually held in assembly by frames which provide the greatest dissipation of heat. The completed modules are mounted on a supporting structure, sited to face in a generally southern direction, and tilted at an angle which approximates the local latitude. A supporting structure with an adjustable tilt angle can allow for seasonal tilt adjustments.

Many photovoltaic arrays are mounted not on free-standing structures, but on a building's roof. Roofs are generally good thermal insulators; an array

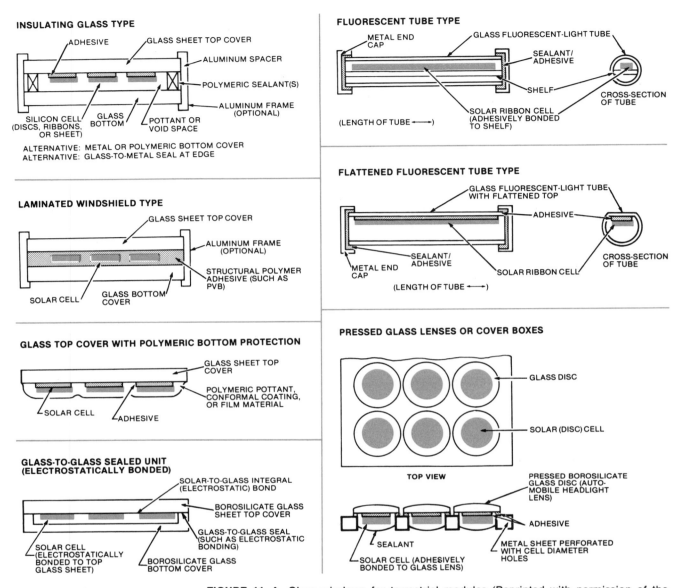

**FIGURE 11-4** Glass windows for terrestrial modules (Reprinted with permission of the IEEE).

mounted too close to the roof surface may suffer a loss of the beneficial cooling by wind and convection currents. As shown in Figure 11–5, the cooling action provided by these currents is dependent upon the spacing between adjacent modules, the height of the modules, and the length of the array. In roof mounting, it is advisable to provide the largest gap that is practical between the roof and the array so that air may move upward to remove heat. This will also prevent debris build-up behind the collectors and subsequent damming of water. Spacing between modules is also helpful in allowing hot air beneath the array to escape.

Arrays mounted on a free-standing support structure may have provision for an adjustable tilt angle; but whether tilt is adjustable or not, most flat plate

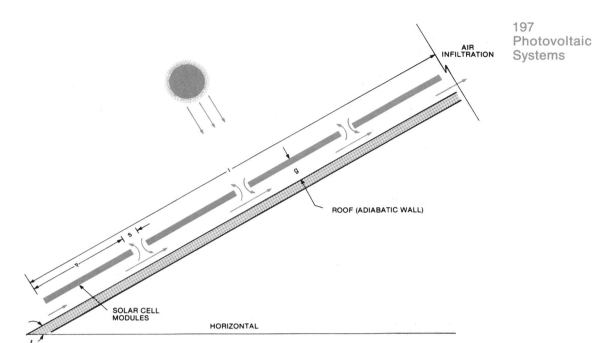

**FIGURE 11-5** Cross section of a rooftop installation of modules.

photovoltaic arrays are not equipped to track the sun in its daily motion. The gain in power output from a flat plate array made possible by a tracking system is offset by the power requirement of the tracking drive. The additional costs of purchasing, installing, and maintaining the orientation system make the comparison even less favorable. Flat plate systems built for large power output, and attended by personnel, may benefit from orientation systems, but it is the concentrating array which offers a substantial benefit from solar tracking.

## CONCENTRATING ARRAYS

One approach to reducing the cost per photovoltaic watt is to increase the level of solar irradiance to those cells. Like solar thermal collectors, there are a number of methods to concentrate solar irradiance ranging from relatively inexpensive reflectors to complex optical concentrators; the operational benefit of a tracking system increases with concentration. Low concentration arrays that use sheet reflectors (side mirrors, yielding concentration ratios of two or less) can be installed in a fixed position, but their tilt angle should be adjusted seasonally. Concentration ratios in excess of two must use a tracking system to achieve the full benefit of the additional components. Concentrating systems may be reflective (mirrors) or refractive (lenses), focus the sun in one plane or two, provide concentration ratios of 1.5 to 1000, require no, one, or two axis tracking, and provide a line or area of focusing.

Before discussing the general types of concentrating systems, it is worthwhile to briefly review the relationship between photovoltaic cell conversion efficiency and the temperature surrounding the operating cell. Recall that, for a given cell, conversion efficiency increases with increasing illumination intensity, but only up to a certain point. It also causes an increase in cell temperature which results in degradation of cell performance.

Concentrating systems require some provision for cooling. Systems operating under relatively small concentration ratios (<10) may use passive arrangements such as finned heat sinks. Systems operating with higher concentration ratios must have some active cooling such as a jacket of circulating water.

A variety of concentrating methods are shown in Figure 11–6 and 11–7, their operation characteristics summarized in Table 11–2.

*Linear Concentrators.* Linear concentrators focus sunlight in one plane and have been used with and without tracking systems. If an orientation system is used, tracking need only be done in one direction, which is an advantage of these linear or trough concentrators. The axis of rotation may be placed in a north-south direction (daily tracking) or in an east-west direction (seasonal tracking). Inclining the axis of rotation of the former to the local latitude angle (a combination known as *polar axis tracking*) will increase the array's performance, but care must be taken to prevent shadowing of some modules by adjacent modules during the early and late hours of the day.

The least complex linear concentrators use sheet reflectors (side mirrors) on either side of the modules. Concentration ratios of approximately two can be achieved in this way so that tracking is not required.

Another linear concentrator which can achieve much higher concentration ratios utilizes a line-focusing parabolic trough. The photovoltaic cells are mounted along the focal line of the parabola on either a square or round tube, or channel. This type of mount allows for passage of water or other liquid coolant (Figure 11–8) while tracking can be simplified by keeping the reflector fixed and orientated only on the photovoltaic receiver. One such system has been designed to produce 360 kW at peak output from 600 modules. A liquid cooling

**FIGURE 11–6** Illustrative examples of reflective concentrators. (From *Solar Cell Array Design Handbook* by Hans Rauschenbach. Copyright 1980 by Van Nostrand Reinhold Company. Reprinted by permission of the publisher.)

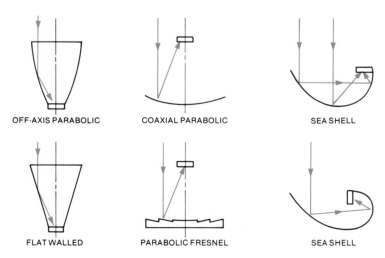

OFF-AXIS PARABOLIC    COAXIAL PARABOLIC    SEA SHELL

FLAT WALLED    PARABOLIC FRESNEL    SEA SHELL

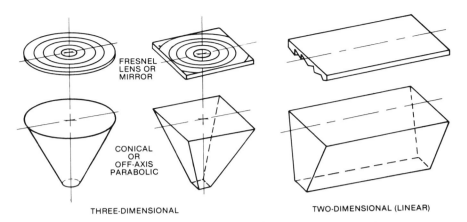

FRESNEL LENS OR MIRROR

CONICAL OR OFF-AXIS PARABOLIC

THREE-DIMENSIONAL

TWO-DIMENSIONAL (LINEAR)

**FIGURE 11–7** Illustrative examples of two-dimensional and three-dimensional concentrators. (From *Solar Cell Array Design Handbook* by Hans Rauschenbach. Copyright 1980 by Van Nostrand Reinhold Company. Reprinted by permission of the publisher.)

system is expected to maintain cell operating temperature at approximately 55°C under a concentration ratio of 20.

Daily and seasonal tracking can be maintained through computer control. Each module consists of three linear reflectors, an extruded aluminum heat sink, and the cells which are mounted downward. The primary reflector at the bottom of each module reflects incoming light upward into a parabolic trough in

**TABLE 11–2** Typical Solar Concentrators

| Name | Method of Concentration | | Usual Range of Concentration Ratio | Type of Tracking Required | | | Focal Zone | | |
|---|---|---|---|---|---|---|---|---|---|
| | Reflective | Refractive | | None | One Axis | Two Axis | Point | Line | Area |
| *Flat Reflectors* | | | | | | | | | |
| Side mirrors (north side of the absorber, noon reversible mirrors, "V" troughs, etc.) | ■ | | 1.5–3.0 | ■ | | | | | ■ |
| Fixed flat mirrors, movable focus | ■ | | 20–50 | | ■ (absorb) | | | ■ | |
| Multiple heliostats, redirecting to a central absorber | ■ | | 100–1000 | | | ■ | ■ | | |
| *Single Curvature Reflectors* | | | | | | | | | |
| Truncated cones | ■ | | 1.5–5 | ■ | ■ | | ■ | | |
| Compound parabolic concentrator | ■ | | 3–10 | ■ | ■ | | ■ | ■ | |
| Parabolic cylinder (E-W, N-S, or tilted axis) | ■ | | 10–30 | | ■ | | | ■ | |
| Reflecting linear fresnel | ■ | | 10–30 | | ■ | | | ■ | |
| *Double Curvature Reflectors* | | | | | | | | | |
| Paraboloids | ■ | | 50–1000 | | | ■ | ■ | | |
| Hemispheres | ■ | | 25–500 | | | ■ | ■ | | |
| Reflecting circular fresnel | ■ | | 50–1000 | | | ■ | ■ | | |
| *Refracting Lenses* | | | | | | | | | |
| Linear fresnel | | ■ | 3–50 | ■ | ■ | | ■ | ■ | |
| Circular fresnel | | ■ | 50–1000 | | | ■ | ■ | | |

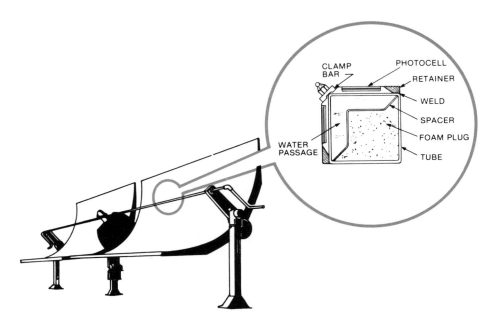

**FIGURE 11-8**   Line-focused hybrid system.

which the cells are mounted. All of the modules are linked together for daily elevation tracking, and the entire array of modules is mounted on a platform which rotates around a central pivot for seasonal tracking. One potential advantage to this arrangement is the inverted mounting of the cells which forestalls problems of climatic hazards such as hail damage and accumulation of dirt.

These examples in no way exhaust the methods used to provide linear concentration of the solar irradiance into a photovoltaic receiver. Methods similar to those shown for solar thermal collectors, and any number of other designs, have been proposed or built. The same diversity of designs is true for the axial concentrators which are summarized below.

*Axial Concentrators.* These concentrators are constructed so that the concentrating component is located on the same optical axis as the photovoltaic cell. The concentrator may be a Fresnel lens or a parabolic mirror (Figure 11-9). These systems utilize both daily and seasonal tracking and achieve high concentration ratios which, in turn, often require an active cooling system.

In a typical Fresnel lens concentrating system, each module consists of cells mounted on a substrate with passive cooling fins. Fresnel lenses, cast in groups of four, are hermetically sealed to the substrate to produce a concentration ratio of 40 and to isolate the concentrated solar irradiance from people and inflammables.

Typical of reflecting axial concentrators are two parabolic designs engineered to provide different features. The Cassegrain type shown in Figure 11-10 allows compactness through a shorter module length for the same focal length-to-entrance aperture ratio than would be obtained with an equivalent conventional parabolic reflector. Two mirrors are used, reflecting the incoming beam twice with cooling fins attached to both mirrors. In addition, the secondary mirror fins are usually treated with some highly reflective paint to assist in heat sinking because this side of the module faces the sun. Wavelength-selec-

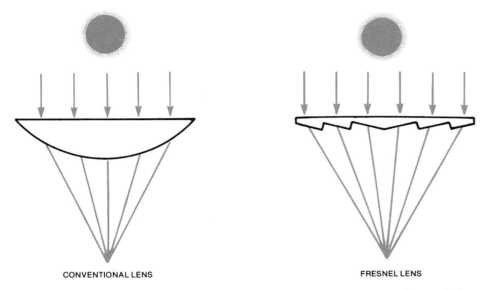

**FIGURE 11–9** Illustrative examples of refractive concentrators. (From *Solar Cell Array Design Handbook*. Copyright 1980 by Van Nostrand Reinhold Company. Reprinted by permission of the publisher.)

tive reflectors can also be used to remove those parts of the solar spectrum which contribute primarily to heat generation, rather than the photovoltaic process. The entire module can be hermetically sealed to protect the cells and the mirrors.

An orientation system for the Cassegrain type is expensive as well as complex. Some less expensive, lightweight designs utilize a rigid parabolic reflector (or a thin aluminized sheet reflector), mounted against a parabolic cavity to concentrate sunlight on the photovoltaic receiver mounted at the focal point. The entire assembly may be contained in a spherical plastic enclosure (Figure 11–11) to provide protection from the environment and to absorb wind

**FIGURE 11–10** Illustrative examples of two-stage concentrators. (From *Solar Cell Array Design Handbook*. Copyright 1980 by Van Nostrand Reinhold Company. Reprinted by permission of the publisher.)

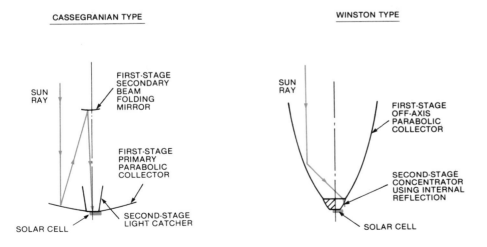

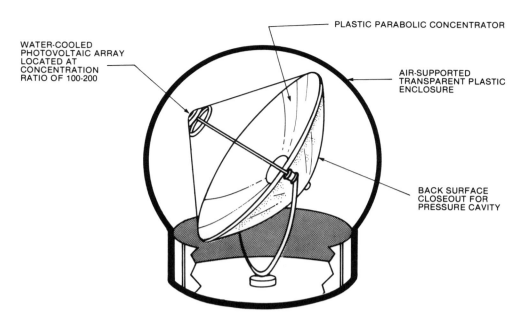

WATER-COOLED
PHOTOVOLTAIC ARRAY
LOCATED AT
CONCENTRATION
RATIO OF 100-200

PLASTIC PARABOLIC CONCENTRATOR

AIR-SUPPORTED
TRANSPARENT PLASTIC
ENCLOSURE

BACK SURFACE
CLOSEOUT FOR
PRESSURE CAVITY

**FIGURE 11–11**   Lightweight parabolic concentrator module.

loading. Provisions for solar tracking are simplified because of the unit's light weight, and methods for active cooling of the receiver are simplified because of its compact size.

One specialized concentration system, similar to multiple heliostat concentrators, is the mirror-field system. As shown in the module of Figure 11–12, multiple parabolic mirrors focus sunlight onto an actively-cooled receiver. The mirrors are oriented in groups of four in the y-axis with each of the individual mirrors mounted on a pivot and linked for x-axis orientation (Figure 11–13). Though this type of multiple mirror system has been primarily applied to solar thermal and furnace systems, it is receiving attention for photovoltaic use.

## POWER CONDITIONING AND CONTROL

The power conditioning component of a photovoltaic system is necessary for two main reasons. First, variations in the environment under which each system operates cause a significant variation in the DC output voltage produced. Random variations in solar irradiance because of cloud cover, and the more gradual variations resulting from diurnal and seasonal changes, require power conditioning and regulation before the power is applied to the load and/ or storage. Secondly, an inverter to convert DC to AC is usually needed because the majority of residential loads are AC, and to match the solar-generated output to the characteristics of the utility grid input. A mixed AC/DC supply is also possible, but conditioning and control of the power in such a system, certainly in one used for a residential application, would appear to be too complex and expensive for serious consideration.

The precise configuration of the power conditioning and control system

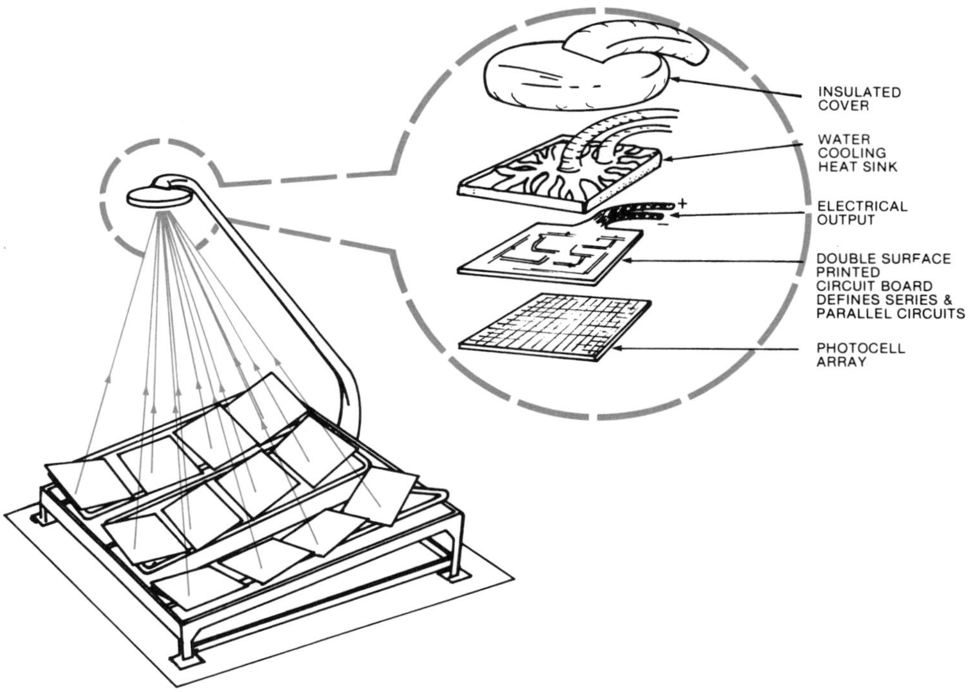

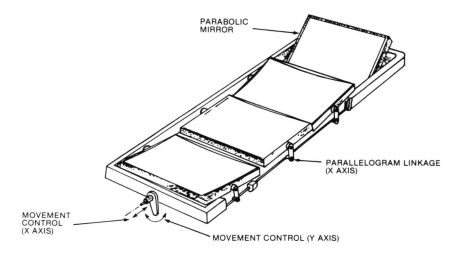

INSULATED
COVER

WATER
COOLING
HEAT SINK

ELECTRICAL
OUTPUT

DOUBLE SURFACE
PRINTED
CIRCUIT BOARD
DEFINES SERIES &
PARALLEL CIRCUITS

PHOTOCELL
ARRAY

**FIGURE 11–12**  Multiple parabolic mirrors focus sunlight onto an actively cooled receiver.

will vary. The majority of photovoltaic system power conditioners will require provision for each of three functions:

1. Power conversion in one of the following categories:
   *DC-DC:* Both input and output are DC.
   *DC-AC:* Input is DC from the photovoltaic array, output is AC.
   *DC-AC-DC:* Input is DC with output either AC or DC. The unit is

**FIGURE 11–13**  Mirror control.

PARABOLIC
MIRROR

PARALLELOGRAM LINKAGE
(X AXIS)

MOVEMENT
CONTROL
(X AXIS)

MOVEMENT CONTROL (Y AXIS)

sometimes called a *bidirectional* inverter, with inversion of the regulated AC back to DC for battery charging.

2. Power control of the entire system to include configuration control, automatic system start and shutdown, monitoring of storage status with regulation for overload protection, and switching between photovoltaic and non-solar supply.

3. An element to monitor array output and to match circuit input impedance to it so that system operation is maintained at the maximum power point at all irradiance levels.

The components used in power conditioning systems for photovoltaics have, until recently, been modified versions of those used with computers, aircraft generators, and wind turbines. As a result, the technology and utilization of these units is well-understood. With the exception of efficiency, performance of power conditioning systems has been generally satisfactory.

Efficiency concerns arise because each step of power conditioning in any electrical circuit—whether inversion, regulation, or other—has inherent losses. Losses in power conditioning equipment connected to the conventional utility grid were of small consequence in the past because of the "limitless" nature and low cost of power. As a result, available conventional equipment with efficiencies of an average of 20 percent at partial load and 80 percent at full load have been acceptable for uses in computer power supply and so forth; but when applied to photovoltaic systems, conventional equipment with conventional efficiencies are less tolerable. Development work has therefore proceeded on power conditioning equipment more attuned to the needs of photovoltaic systems. Fortunately, concurrent advances in solid-state electronics have made possible the replacement of rotary inverting equipment with solid-state switching devices. First used with wind turbines (an application with design requirements similar to photovoltaics), these power conditioning systems can provide efficiencies approaching 95 percent at full load as well as solid-state reliability.

## STORAGE

The inherently cyclic nature of solar thermal and photovoltaic energy systems (i.e., full operation possible only during good weather and daylight hours with no energy collection at night), requires provision for storage and/or connection to the utility grid. Virtually all point-of-use photovoltaic systems will incorporate a storage element to provide power during poor weather conditions and at night. A number of methods are currently in use.

### BATTERIES

The most immediately useful means of storing surplus energy produced by small-to-medium size photovoltaic systems are batteries. Most systems now in operation use nickel-cadmium or lead-acid batteries, chiefly because of their availability and well-understood technology. Though heavy and requiring periodic maintenance, batteries can be combined much like the photovoltaic cells themselves to obtain the desired voltage and current.

Of special importance in choosing battery storage for a photovoltaic system are the specifications for the unit's depth of discharge without damage, the number of charge/discharge cycles during its life, and storage capacity. The most useful lead-acid type is the lead-calcium battery capable of operating to 80 percent of maximum depth of discharge, with 1500–2000 cycles/life, and an energy density of 11–22 watt hours per kilogram (Wh/kg). Nickel-cadmium batteries, by comparison, are capable of 100 percent of maximum rated depth of discharge, for thousands of cycles, and an energy density of 33–40 Wh/kg. It is important to note that generalized comparisons between battery types are difficult to make; each photovoltaic system and its application must determine which of these batteries will be more suitable. Some applications will require little or no storage; others may require storage of several days to several weeks. Operating temperature, the required rate of discharge, and other system design parameters will also influence the choice of battery. Table 11–3 lists relevant data for major battery types.

Many of the types listed are undergoing development, but to date, most have exhibited substantial disadvantages. Sodium-sulfur and lithium-sulfur systems promise a substantial improvement in energy density with a possibility of tolerable cycles/life, but their requirement of elevated operating temperatures (in excess of 300°C) calls into question their utility for many systems. The metal-air battery can be manufactured at lower material costs, but the lifetime of its secondary electrode (like that in the nickel-zinc battery), is too limited for photovoltaic use.

At the current time, then, the lead-acid batteries are perhaps the best choice for storage of excess electrical energy. An acceptable cycle life, a proven technology, and the capability of assembling as many or as few of the units as required, make them the most practical alternative for many contemporary systems. Batteries add complexity to the system, require voltage regulation to prevent overcharging, require additional maintenance and, because battery lifetime is shorter than cell lifetime, more frequent replacement.

**TABLE 11–3**  Storage Battery Data

| Battery Type | Energy Density[a] Wh/kg | Life Cycles[b] | Cost $/kWh |
|---|---|---|---|
| Silver–Zinc | 100–120 | 100–300 | 900 |
| Nickel–Cadmium | 33–40 | 300–2000 | 600 |
| Nickel–Iron | 22–33 | 3000 | 400 |
| Lead–Acid[c] | 11–22 | 1500–2000 | 50 |
| Nickel–Zinc[d] | 66–88 | 250–350 | 20–25 |
| Zinc–Chlorine[d] | 66 | 500 | 10–20 |
| Sodium–Sulfur[d] | 170–220 | 1000 | 15–20 |
| Lithium–Sulfur[d] | 130–170 | 1000 | 15–20 |
| Zinc–Oxygen[d] | 160 | ? | ? |
| Aluminum–Air[d] | 240 | ? | ? |

(a) Values shown are for 1 hr and 6 hr rate of discharge respectively.
(b) Range shown is from severe to modest duty.
(c) Typical automotive battery.
(d) Projected figures–batteries still at experimental stage.

## FUEL CELLS

Electrolysis can be used to simultaneously evolve hydrogen gas (which may be stored as a gas, liquid, or metal hydride) and oxygen gas (which may be stored or vented to the atmosphere). The hydrogen and oxygen may then be recombined in a fuel cell to produce electricity. Commercial electrolyzers with an efficiency of 70 percent have been constructed; some experimental systems operating at higher pressures and temperatures (200°C) have produced efficiencies of 85 percent.

Widescale use of fuel cells for energy storage within photovoltaic systems must await further development. Present technological shortcomings include a requirement for platinum as an electrocatalyst, short electrode lifetime, and difficulties in manufacturing the multicell units necessary for large scale electricity storage and reconversion. Storage of the evolved hydrogen as a gas or liquid requires expensive containment and refrigeration, which add complexity—especially in a small, point-of-use system. Conversion of the hydrogen to a hydride, by reaction with a suitable metal, has attractions since storage would be relatively compact, require no refrigeration, and would release hydrogen at near room temperature. But costs are still high.

Proposals have been made to locate large photovoltaic arrays/electrolyzers/fuel cells at dispersed locations under utility management. Another option is central location of the array and electrolyzers but with the fuel cells located at the point-of-use. The hydrogen fuel generated at the central location would then be supplied to residential fuel cells by pipeline or truck. Each of these proposals, however, is predicated on further development and cost reduction of electrolyzer/fuel cell units.

## FLYWHEELS

The use of a flywheel as a means of storing mechanical or kinetic energy is an old technique which has been suggested for adaption to photovoltaic system use. Photovoltaically generated electricity would be used to drive a motor and thus store kinetic energy in a flywheel. This stored energy can then be converted back to electricity by having the flywheel drive a generator.

The method has a number of disadvantages for point-of-use systems, chiefly its low energy density and energy efficiency resulting from bearing friction and windage. Large, central utility flywheel systems appear to hold slightly better promise because of recent advances in flywheel operation in vacuum on hydrostatic or magnetic bearings that could reduce energy losses. Even in such a system, however, the energy density problem appears less amenable to improvement. A 1975 report on a 500 MW flywheel system utilizing state-of-the-art technology, projected a maximum density of 26 Wh/kg for the system, compared to a required density of 88 Wh/kg for cost effectiveness. Flywheel systems could be useful for certain specialized tasks such as rapid following of small load variations of temporary supply while switching between primary supply sources.

## PUMPED HYDRO

Pumping water into an elevated reservoir (from which a hydroelectric turbine is then powered) is the only large-scale storage method currently in use. The United States has approximately 10,000 MWh of conventionally pumped storage capacity. The method is relatively simple and proven and could be adapted to photovoltaic storage, but expansion is limited by a lack of suitable sites. Proposals have been made to locate the pumps and the lower reservoir in natural or excavated caverns, with the higher reservoir on the surface. This method would allow the installation of a floating photovoltaic array on the surface of the upper reservoir to conserve space. Improved equipment (such as reversible pump/turbines or impulse turbines and multistage high-pressure pumps powered by a common shaft) is under development to provide greater efficiency. Even with expansion to new sites and improved equipment, however, the sheer size of such an installation and the slowness of its response precludes significant central utility use. Expanded storage capacity at topographically suitable sites will probably continue, but the construction of "artificial" differential-head reservoir-pairs is very much dependent upon the investment/tax treatment accorded.

## SEMICONDUCTING MAGNETS

The storage method with the highest conversion efficiency is based on the property of electric currents to persist almost indefinitely in superconducting materials maintained at very low temperatures. The method has additional appeal because an appropriately constructed superconducting magnet is a high-current, low-voltage DC device especially compatible with photovoltaic array output; and its response time in accepting storage or conversion back into usable electricity is measured in milliseconds. Unfortunately, the utilization of superconductive, low temperature storage is only useful for large central power facilities and these are not yet possible because of a number of problems. Chief among these problems is the embryonic state of the technology. Because of the necessity to maintain the magnet at a low temperature (one proposal suggests a 10,000 MWh capacity magnet cooled by a pool of superfluid helium at $1.8°K$), the technical questions—of maintaining electrical insulation, the properties of structural materials used in both the magnet and its housing, and the provision for electrical connections capable of handling 100,000 amperes, all in a low temperature environment, plus the potential biological effects of the massive surrounding magnetic field—must all be addressed and resolved. The use of superconducting magnet energy storage remains a distant potential.

## STORAGE LOCATION: ANCILLARY CONSIDERATIONS

The operation of thermal and photovoltaic solar energy systems results in inherent cycles in most applications. (i.e., surplus production followed by an energy shortfall even on a daily basis). As a consequence it is usually necessary to store energy produced during surplus periods and interconnect to a non-solar energy source to cover the shortfall periods. Non-solar energy for either type of solar system presents no difficulty because of its availability and various forms. Determining the optimum storage means and location for the surplus energy produced by either system is more difficult, and especially so for photovoltaic systems.

Thermal energy is more diffuse, more easily dissipated, and more difficult to transport; qualities which make its storage at the point-of-use almost mandatory. By contrast, the nature of electrical energy—its universal utility, and its portability—results in a less intimate association between locations of the storage component and point-of-use. It is difficult to predict either the ultimately most advantageous method or the optimum location (central utility, dispersed station, or point-of-use) of storage for photovoltaic systems.

Nonetheless, simple battery systems of limited capacity will certainly be used to respond to transitory variations in load demand at the site. But most of the relatively small systems will not be able to cost-effectively install sufficient storage capacity to meet all exigencies, including several days of poor weather. Even in an application where the bulk of the load occurs during hours of reliable sunlight, and the system operates in a peaking mode (thereby successfully achieving the objective of reducing consumption of non-solar energy), energy from conventional supply will be necessary.

Locating most of the total storage capacity at central or dispersed locations holds implicit advantages. A common storage pool would free individual consumers of the operation and maintenance responsibilities and allow the use of larger, more cost-effective storage batteries.

These considerations are necessarily speculative at this point since the use of photovoltaic energy systems, large or small, has not expanded yet to the numbers needed to compel decision. In addition, there are institutional questions which must be resolved. Prominent among them is the question of payment for surplus power feedback into the network from multiple, small, consumer-owned systems. Should system owners receive a flat credit per unit of electricity returned or share in the utilities' capital and operations cost savings? The resolution of these questions and others must await the accumulation of experience as more small independent and large interconnected photovoltaic systems are put into operation.

# $\boxed{12}$    Photovoltaic Array Sizing

The accurate sizing of a photovoltaic system for a specific application involves many variables and compromises and is best carried out by a computer. Nevertheless, it is possible to perform simple estimates of sizing that can be accurate to within 10 percent of the computer generated results.

## SIZING CONSIDERATIONS

In sizing a photovoltaic array, one must consider not only the output of the photovoltaic cells but also how the electric power is to be used (AC or DC) and whether battery storage will be available. Therefore, we can distinguish between three basic types of photovoltaic systems:

1. Photovoltaic modules without either storage or connections to the utility power grid.
2. Photovoltaic modules with storage but no connection to the utility power grid.
3. Photovoltaic modules with both storage *and* connection to the utility power grid.

The first of these system types is useful in very limited applications since it can only provide power to a load when the sun is out, and even then at a very variable rate. This variability can be seen in Figure 12–1.[1] Unless one has a load which is well matched to the irradiance, this type of system is not of general usefulness.

The second type of system is generally referred to as a stand-alone system and is very commonly used in remote applications. The third type of system requires a power conditioner whose power consumption should be taken into account.

A simple sizing calculation can be considered in terms of inputs and out-

---

1. Irradiance values in photovoltaics are most conveniently expressed in watts/m². The units previously used in this handbook can be converted as follows: 1 Btu/hr-ft² = 3.154 W/m².

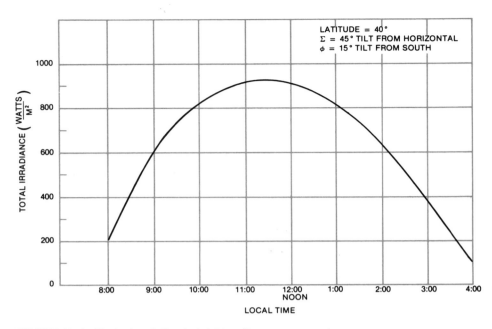

**FIGURE 12–1**   Typical variation in total irradiance over one day.

puts as shown in Figure 12–2. In other words, we must start by providing the following six inputs:

1. Nominal system voltage
2. Solar cell efficiency
3. Peak load current
4. Total average load
5. Irradiance
6. No-Sun-Drain-Period (NSDP)

The sizing calculation will, in turn, tell us how large a module we need and how many batteries are necessary for storage.

Before we consider an example, let us look at the inputs in a bit more detail. The nominal system voltage is the voltage we want from our photovol-

**FIGURE 12–2**   The sizing calculation shown schematically as a box, with inputs and outputs.

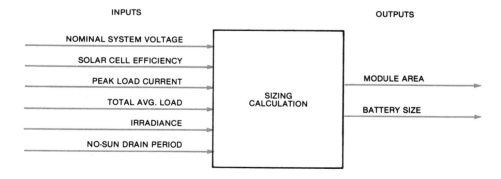

taic module. If we wish to operate devices at 12v, then that is our system voltage.[2]

The peak load current is the maximum momentary current drain of any of the devices to be operated by the photovoltaic module. The total average load is expressed in Ampere-Hours per day (A-h/day). The load for each device is determined by multiplying its ampere rating by the number of hours per day it is in operation. The total average load is the sum of these individual loads. The accuracy with which the loads are determined will strongly affect the final results. Therefore, wherever possible, actual measurements should be made of the currents and times involved. The irradiance can be obtained using the methods outlined in Chapter 2.

The No-Sun-Drain-Period is specified only if battery storage is required. It refers to the number of days during which the photovoltaic module is inoperative (due to poor weather conditions) and during which the batteries must supply the entire load with power.

## EXAMPLE OF HOW TO SIZE A PHOTOVOLTAIC POWER SYSTEM

Suppose we wish to power a transmitter/receiver which operates at 12v. On a typical day it is used to receive for 5 hours (at a current drain of 2A), to transmit for 2 hours (at a current drain of 10A), and in standby for 10 hours (at a current drain of 0.5A). The load in each of these modes is as follows:

Receive:    2A × 5 hrs/day  = 10 A-h/day
Transmit:  10A × 2 hrs/day  = 20 A-h/day
Standby:  0.5A × 10 hrs/day =  5 A-h/day

    Total load/day           35 A-h/day

Adding a safety factor of 15 percent to this gives:

    Total average load/day = 40 A-h/day

We also see that the peak current is 10A.

Next we must find the irradiance for the geographical site at which we plan to locate our photovoltaic system. This can be done in a variety of ways as explained in previous chapters. However, an approach similar to that used for thermal collectors will be used, and a detailed description can be found in Chapter 7. Table 7–9 represents a work-sheet for irradiance near Philadelphia for collector tilt angle $\Sigma = 45°$ from the horizontal and an azimuth angle $\phi = 15°$ from due south. Table 12–1 shows the same information as Table 7–9, but expressed in units more convenient for photovoltaics by use of the conversion factor 1 Btu/ft$^2$ = 3.154 Wh/m$^2$.

This table shows that the average annual irradiance is:

    3.75 kWh/m$^2$-day = 3750 Wh/m$^2$-day

---

2. However, note that if our photovoltaic modules operate at 12v, but we invert this voltage to 110v AC for running household appliances, our operating voltage is still 12v DC since that is the input we use to the power conditioner.

**TABLE 12–1**  Solar Irradiance Summary Worksheet for Photovoltaics

City: Philadelphia, PA  
Latitude: 45°  
Longitude: 75°  
Elevation: 200 ft.

Collector Tilt Angles: From horizontal, $\Sigma = 45°$  
From due south, $\phi = 15°$  
Reflection Coefficient: $C_f = 0.25$ (Grass)

Converted from Table 7–9 using 1 Btu = 3.154 $\dfrac{\text{Wh}}{\text{m}^2}$

| | Average for the Year | Jan. | Feb. | Mar. | Apr. | May | June | July | Aug. | Sept. | Oct | Nov. | Dec. |
|---|---|---|---|---|---|---|---|---|---|---|---|---|---|
| Number of days | | 31 | 28 | 31 | 30 | 31 | 30 | 31 | 31 | 30 | 31 | 30 | 31 |
| Total daily solar irradiance kWh/m² | | 5.91 | 6.68 | 7.20 | 7.22 | 6.93 | 6.70 | 6.81 | 6.98 | 6.82 | 6.43 | 5.66 | 5.30 |
| Total monthly solar irradiance kWh/m² | | 183 | 187 | 223 | 217 | 215 | 201 | 211 | 216 | 204 | 199 | 170 | 164 |
| Average percent sunshine (% ÷ 100) | | .50 | .53 | .57 | .56 | .58 | .63 | .62 | .63 | .60 | .54 | .52 | .50 |
| Average available solar energy kW/m²-mo on tilted surface | | 91.6 | 99.1 | 127 | 121 | 125 | 127 | 131 | 136 | 127 | 118 | 88.3 | 82.2 |
| Wh/m²-mo on horizontal surface | | 53.9 | 73.2 | 108 | 140 | 171 | 175 | 190 | 162 | 122 | 89.9 | 52.0 | 45.0 |
| kWh/m²-day on tilted surface | 3.75 | 2.95 | 3.54 | 4.10 | 4.04 | 4.02 | 4.22 | 4.22 | 4.39 | 4.09 | 3.79 | 2.94 | 2.65 |
| kW/h/m²-day on horizontal surface | 3.78 | 1.74 | 2.61 | 3.48 | 4.67 | 5.52 | 5.83 | 6.13 | 5.23 | 4.07 | 2.90 | 1.73 | 1.45 |

For a 12v module operating at 100 percent efficiency this represents:

$$\frac{3750}{12V}\ \text{Wh/m}^2\text{-day} = 313\ \text{A-h/m}^2\text{-day}$$

If the efficiency of the module is only 10 percent, then the output will be 31.3 A-h/m²-day. Therefore, the module area equals (40 A-h/day) ÷ (31.3 A-h/day), and 1.28 m² will supply the proper output.

A further consideration must be made after selecting a module of the right area. It must be capable of supplying the peak current of 10A. This information is obtained from the module manufacturer for the module area selected. If it cannot supply the peak current, the next larger size module must be chosen.

Another point should be noted. We have used the average annual irradiance to calculate the module area. This area will *not* be sufficient in the winter months. Therefore, if we wish to be prudent, we should use the *minimum* daily irradiance of 2.65 kWh/m²-day. Carrying through the calculation, we now require a module area of 1.81 m².

Now suppose we wish to add storage to our photovoltaic system to enable it to operate for five sunless days (NSDP = 5 days).

The battery capacity can be estimated from:

$$\text{Battery Capacity (A-h)} = \text{Total load (A-h/day)} \times \text{NSDP(days)}$$
$$= (40 \text{ A-h/day}) \times 5 \text{ days}$$
$$= 200 \text{ A-h}$$

For example, two 12v 100A-h batteries could be connected in parallel to provide the required capacity. The capacity of batteries varies strongly with temperature, and the rating at the *operating* temperature should be used.

# 13 Economic Analysis

Economic analysis can play a vital role in solar energy system planning and implementation procedures. In the planning of a new system it provides a means of determining the cost effectiveness of the various utilization techniques under consideration. With an existing system it can determine whether the performance advantages of higher priced equipment, if added to a facility, are worth the added cost.

The process is a familiar one to every consumer who determines the economic justification of a major expenditure by comparing the continuing benefits derived from it to the monthly or annual payments required by it. This comparison is the foundation of the approach employed by Richard M. Winegarner as published in the *Optical Industry and Systems Directory,* 1977, in which the justifiable costs of solar energy equipment are based upon the savings to be derived. The direct economic benefits produced by a solar energy system can be determined from the irradiance rate, efficiency, and cost of conventional energy. The cost per period or payments required to operate a solar energy system are simply a product of the amortization rate and the cost per unit area of the installed system. There are other factors, such as the beneficial effects on the environment and improvement in the nation's balance of payments which could stimulate even earlier use through governmental incentives, but this analysis considers only direct benefits. In sum, any solar energy system becomes viable whenever the value of energy produced exceeds the payments required.

## A SIMPLIFIED ANALYSIS

Since the cost level for flat plate collectors is determined from the yearly irradiance rate and the conventional utility rate, economic analysis can be performed by equating the economic benefits and the payments required:

$$I\eta_c\eta_d E = RC \tag{13-1}$$

where:  $I$ = irradiance rate, in MBtu/ft$^2$-yr

$\eta_c$ = percentage collector efficiency, in decimal form

$\eta_d$ = distribution efficiency in decimal form (accounts for storage and line losses)

$E$ = conventional energy cost, in \$/MBtu

$R$ = percentage amortization rate, in decimal form

$C$ = installed solar energy system cost, in \$/ft$^2$

**TABLE 13–1**  Irradiance Rate: Electric Rate Product
at Several U.S. Locations

| Location | $I$<br>MBtu/ft²-yr | $E$<br>$/MBtu | $I \times E$<br>$/ft²-yr |
|---|---|---|---|
| Northeast | .47 | 12 | 5.6 |
| Southern California | .65 | 8 | 5.2 |
| Tennessee | .51 | 7 | 3.6 |
| Texas | .58 | 6 | 3.5 |
| Northwest | .36 | 3 | 1.1 |

Representative values of $I$ and $E$ are shown in Table 13–1.

Solving the equation for cost per square foot:

$$C = \frac{\eta_c \eta_d I E}{R} = \$/\text{ft}^2 \tag{13-2}$$

Differentiating this equation with respect to collector efficiency yields:

$$\Delta\eta C = \frac{(\eta_1 - \eta_2)}{\eta_2} \tag{13-3}$$

With this equation we can perform marginal analysis on any system to determine whether the addition or deletion of certain components makes a system more or less economically viable.

As an example, consider the case of a residential electrical heating and cooling system augmented by a flat-plate solar-energy collection system. Assume that the addition of a heat mirror coating placed on the third cover surface of a standard flat plate collector is being considered. Will the benefits derived from the addition of the mirror coating more than offset the additional costs?

The average residential customer price for electricity in the United States in 1974 was $8.78/MBtu; suppose that addition of the heat mirror coating will produce an increase of 10 percent in collector efficiency. With a 20-year system lifetime and a money rate of 8.5 percent the amortization rate would be approximately 10 percent. Using these numbers with an average irradiance rate of 0.5 MBtu per year per square foot and a distribution efficiency of 0.85, the marginal costs can be determined as follows:

$$\Delta C = \frac{0.85 \times 0.5\text{MBtu/ft}^2\text{-yr} \times \$8.78/\text{MBtu}}{10\%/\text{yr}} \, 9.9\% = \$3.69/\text{ft}^2$$

If the addition of the coating allows the solar system to collect additional energy valued at $3.69 per square foot, the system will be more economically viable. Similar analyses can be performed to determine the economic feasibility of many controversial items, such as the number of covers, cover materials (glass or plastics), honeycombs, selective absorber surfaces, and so on. Care must be taken when analyzing a concentrator system, especially when the modification is made only to a reduced area within the system, to include the ratio of the modified or coated area to the total area in the equation as follows:

$$\Delta C = \frac{\eta_e IE}{R} \frac{A_t}{A_m} \Delta \eta_c \qquad (13\text{-}4)$$

where: $A_t$ = total area of collector

$A_m$ = reduced area modified

$\eta_e$ = electrical conversion efficiency.

The addition of a selective absorber to a trough solar collector with a concentration ratio of 12 to 1 under the same economic conditions would produce the following results, assuming an electrical conversion efficiency of 0.36:

$$\Delta C = \frac{0.36 \times 0.5\text{MBtu/ft}^2\text{-yr} \times \$8.78/\text{MBtu}}{10\%/\text{yr}} \frac{12}{\pi} \; 22.5\%$$

$$= \$13.58/\text{ft}^2$$

Thus, if a selective absorber coating can be applied to the receiver tube of a concentrating trough collector at a cost less than \$13.58 per square foot (or, stated in another way, will enable the collector to capture additional energy of equivalent value per square foot), the equipment alteration becomes economically worthwhile.

## COMPOUND INTEREST AND GROWTH OF INCOME

Since the income produced by an asset consisting of solar energy equipment is equated to the cost of energy saved from conventional sources, it is useful to consider the present value of an asset which produces an income that is increasing at a steady rate of growth.

Consider an asset which produces an income given by:

$$M_k = Me^{(k-1)r_g} \qquad (13\text{-}5)$$

Where $M_k$ is the income produced in the $k$th year (or period), $k$ is the year (or period) number and $r_g$ is the growth rate (% ÷ 100) of the income stream. This is representative of an asset which replaces some cost which is growing at a rate, $r_g$, due to inflation, shortages, or increasing production costs. The present value of the asset, assuming that it has a life of $n$ years (or periods) is:

$$PV = Me^{-r_g}\left[\frac{e^{n(r_g-r)} - 1}{1 - e^{-(r_g-r)}}\right] - e^{-nr}(S_r - S_s) \qquad (13\text{-}6)$$

Where $r$ is the nominal interest rate per year (or period), $M$ is the cost saving in the first year (or period), $e = 2.71828$, $r_g$ is the rate of growth of cost saving per year (or period), $r_g \neq r$, $n$ is the lifetime of the asset in years (or periods), and $S_r$ and $S_s$ are the replacement cost and salvage value of the asset, respectively, after $n$ years (or periods). Equation 13-6 is based on continuous compounding of interest.

If $r_g = r$, a limiting form for Equation 13-6 becomes:

$$PV = nMe^{-r} - e^{-rn}(S_r - S_s) \qquad (13\text{-}7)$$

The value of a solar energy system can be analyzed in terms of its financial value, using Equation 13-6. It is instructive to compare this present value to its cost. When and if the cost should fall below the present value, then it makes economic sense to purchase a solar energy system.

The examples that follow are used to illustrate certain characteristics of solar energy systems as an investment.

*Example 1:* Consider a solar energy system that will initially save $500 per year in conventional fuel cost. Assume that fuel costs are expected to increase at a uniform rate of 12 percent per year and that the solar system can be financed through an extension of a home mortgage loan with a nominal interest rate of 8 percent. Assume, also, that the solar system has a design life of 20 years and a salvage value equal to its replacement cost (it is possible that the salvage value could exceed the initial cost, depending on the system condition, inflation in raw materials, new technologies available, and numerous other factors). Using Equation 8–6:

$$M = 500$$
$$r_g = 0.12$$
$$r = 0.08$$
$$n = 20$$
$$S_r - S_s = 0$$

and from Equation 13–6:

$$PV = \$13,860.25$$

*Example 2:* Consider the same conditions as in Example 1, except that the salvage value is zero and the replacement cost is $10,000. Then:

$$S_r - S_s = \$10,000$$

and from Equation 13–6:

$$PV = \$11,841.29$$

*Example 3:* Consider the same conditions as in Example 1, except that the design life is 30 years instead of 20. Then $n = 30$ and from Equation 13–6:

$$PV = \$26,240.03$$

*Example 4:* Consider the same conditions as in Example 3, except that the difference between replacement cost and salvage value is $10,000 so that $S_r - S_s = -10,000$ and

$$PV = \$25,332.85$$

Note that the present value rises significantly as the design life of the solar energy system is extended, giving greater value per unit of service time as the design life is increased.

*Example 5:* Consider a severe energy crisis scenario where the growth rate of conventional fuel cost is 20 percent ($r_g = 0.20$) but where government backed incentives and subsidies provide low-interest 5 percent loans to solar energy system purchases so that $r = 0.05$. Let the initial savings be $500 per year and consider both 20- and 30-year design life, with replacement cost exceeding salvage value by $10,000:

> 20-year design life, $PV$ = $52,410.48
> 30-year design life, $PV$ = $259,368.77

Clearly, in this scenario, a solar system could be considered highly economical, especially if it has long design life.

*Example 6:* Consider an economic depression scenario where reduced use of conventional fuels and new oil and gas discoveries result in a modest 5 percent rate of growth for fuel costs ($r_g$ = 0.05) and assume interest rates are 6 percent, or $r$ = 0.06. Otherwise the same conditions as in Example 5 would prevail:

> 20-year design life, $PV$ = $5.611.05
> 30-year design life, $PV$ = $10,675.08

Examples 5 and 6 are both extreme, perhaps improbable, but they serve to illustrate the effects of compounding on interest costs and on the growth of alternative energy costs.

The decision to invest in a solar energy system on a financial basis alone rests on assumptions that extend far into the future. It is important to note that economy in costs, both initial and operating, is a function of equipment quality. Cheaper system components, particularly solar collection panels, may seem more attractive because of initial price. But they are low in price for a reason which, too often, translates into construction shortcuts that result in lower efficiencies and shortened equipment life. It is for this reason that the "disposable quality" or "planned obsolescence" characteristics of many products cannot be justified in solar energy equipment. It will not be economical unless it is built to last a long time.

Another way to understand the importance of quality in solar energy equipment is to consider that the energy expended in the production of a solar energy system (including all steps, from the mining and processing of the natural resources used to make it, through its final assembly and installation) must be more than replenished by the energy it collects during its lifetime. Otherwise, the equipment is only contributing to a net drain on total energy resources. Thus, quality and endurance are not only most important features of solar energy equipment, but also key factors to achieving the potential of superior long-term economy and contributions to improvement of the nation's energy shortage.

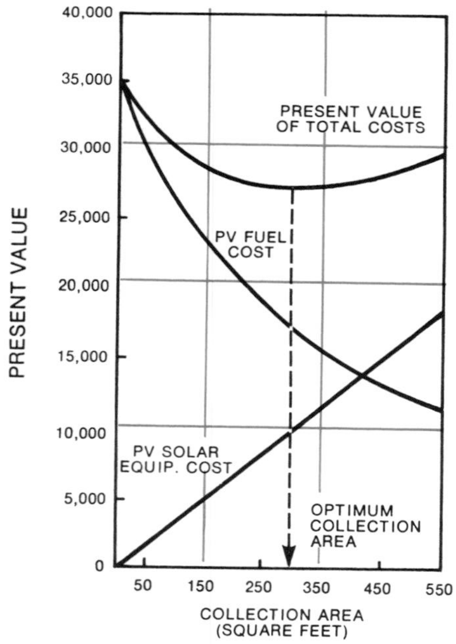

## MINIMUM COST ANALYSIS

The concept of minimizing the present value equivalent of future costs is used in determining the optimum collector area for a solar energy system. An example of this analysis is shown in Figure 13–1, where it is assumed that the present value of solar system cost is $33 per square foot of collection area. This figure portrays the sum of the initial solar energy cost and the present value of its continuing maintenance costs as a linear function of collector area. For a 1500-square-foot home in the Washington, D.C., area, the present value of the

**FIGURE 13–1** Optimization of collector area: Washington, D.C., 1500-square-foot home.

future costs of conventional fuel necessary to supplement energy requirements is shown by the declining curve on the figure. Note that the shape of this curve is sensitive to assumptions about the future cost of fuels and other alternatives for home heating. The uppermost curve represents the sum of equipment cost and fuel costs—in a phrase, the present value of total system costs. The minimum point of the total costs curve, marked by the dotted line, is used to determine the optimum collection area—optimum in the sense that the present value of total costs is minimum.

Although the example represented in Figure 13–1 applies to a specific situation with many assumptions, the results are qualitatively similar in most solar energy applications. A similar plotting for any application will show that it is most economical to design a solar energy system to provide approximately 50–80 percent of total energy demand.

Figure 13–2 illustrates the relationship between the percent of load served and the collection area for the Washington, D.C., home of 1500 square feet. It can be seen from the curve, called an f-Chart and applicable only to the example under discussion, that the acquisition of additional collection area results in a diminishing increment of improvement in the percent of load served. An iterative computer program was developed at the University of Wisconsin to produce an f-Chart for engineering specific solar energy systems; the worksheets in Chapter 7 of this handbook can be used to yield a similar evaluation by determining the relationships between percent of load served and collection area on a point-by-point basis for a wide variety of collector orientations and climatic conditions. The curve shown in Figure 13–2, however, is qualitatively similar to that which could be expected anywhere.

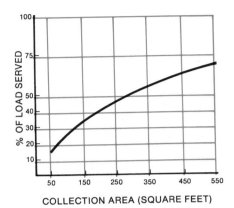

**FIGURE 13–2** Percent of annual hot water and heating load as a function of collector area: Washington, D.C., 1500-square-foot home.

# Appendices

# A Blackbody Radiation Laws

A blackbody, so called because of the resulting color, is defined as an object that absorbs all and reflects none of the radiation incident upon it. No perfect blackbody exists but the concept is important because the temperature of a blackbody can be mathematically related to the proportion between its radiant absorption and emission and, hence, a relation to the properties of real objects in actual radiative environments.

A blackbody emits radiation in accordance with Planck's Law:

$$S = \frac{C_1}{\lambda^5(e^{C_2/\lambda T} - 1)} \qquad \textbf{(A–1)}$$

where: $S$ is the spectral radiation intensity emitted from the surface of the blackbody ($W/m^2$)

$T$ is the absolute temperature of the blackbody (°K)

$\lambda$ is the wavelength of the radiation emitted (m)

$C_1$ is the first radiation constant

$C_1 = 2\pi hc^2 = 3.7418 \times 10^{-16}$ Joule-meter/second (J-m/s)

$C_2$ is the second radiation constant

$$C_2 = \frac{hc}{k} = 0.014388 \text{ meter-°K (m-°K)}$$

where:

$h = 6.6262 \times 10^{-34}$ Joule-second = Planck's constant

$c = 2.998 \times 10^8$ meters/second = the speed of light

$k = 1.3806 \times 10^{-23}$ Joules/°K = Boltzmann's constant.

Interestingly, as shown in Figure 2–1, the application of Planck's Law to a blackbody with an absolute temperature of 5750°K, near that of the solar surface, results in a theoretical curve that closely approximates the curve for the sun. Figure 2–1 also compares the theoretical characteristics of both the sun and the earth to their estimated actual emissions.

The total power emitted by a blackbody is found by integrating the spectral radiation intensity, $S$, given by Equation A–1, over all values of wavelength, $\lambda$. The result provides a relationship between the radiation intensity in units of power emitted per unit area, and the absolute temperature of a blackbody. This relationship is the Stefan-Boltzmann equation, more commonly known as the *fourth power law:*

$$I = \sigma T^4 \qquad \text{(A-2)}$$

where: $I$ is the radiation intensity, or power emitted per unit area from a blackbody at temperature $T$ (W/m² or Btu/hr-ft²)

$T$ is the absolute temperature (°K or °R)

$\sigma = \dfrac{2\pi^5 k^4}{15 h^3 c^2} = 5.6996 \times 10^{-8}$ watts/(meter²-°K⁴) is the Stefan-Boltzmann constant ($1.7132 \times 10^{-9}$ Btu/hr-ft²-°R⁴)

One convenient expression for the radiation intensity of a blackbody is obtained when $T$ is in °K and $I$ is in watts/meter²:

$$I = \left(\frac{T}{64.8}\right)^4 \qquad \text{watts/m}^2 \qquad \text{(A-3)}$$

The wavelength at which the maximum spectral radiation intensity occurs is given by Wien's Displacement Law:

$$\lambda_{\text{max}} = \frac{2.898 \times 10^{-3}}{T} \qquad \text{meters} \qquad \text{(A-4)}$$

Wien's Displacement Law is a quantitative expression to describe the displacement of peak radiation intensities toward shorter wavelengths as an object is heated. This displacement is readily seen in Figure 2–1 by comparing blackbody radiation at 5750°K and 300°K. Wien's Law also provides a basis for real objects to radiate maximum energy at different wavelengths than those at which energy is best absorbed. There are several useful effects which result from Wien's Law in the application of materials with selective absorption and reflectance of energy in different wavelength regions. One of the best known is the "greenhouse effect," which is based upon the property of glass in transmitting light but not radiant heat. Consequently, the interior of a greenhouse can be kept warm on sunny winter days as light energy is allowed to enter, but the radiant heat wavelengths are displaced by Wien's Law so that they cannot escape by direct reradiation. Another useful effect is based on selective absorption and emission by certain surfaces. These selective surfaces, discussed in detail in Chapter 4, have the property of absorbing light efficiently and converting it to heat for which emissivity is poor at the displaced wavelengths described by Wien's Law.

Absorptivity, emissivity, transmission, and reflection coefficients may be used directly in conjunction with the blackbody radiation laws to derive practical formulas for the radiation energy gained or lost by a real object in terms of its temperature. Conversely, by using the principle of equilibrium (equating energy gained with energy lost), one can determine the temperature of a real object in a relative environment.

A real object emits radiation at a rate given by:

$$I_{\text{real}} = \varepsilon \left(\frac{T}{64.8}\right)^4 \qquad \text{(A-5)}$$

where $\varepsilon$ is the emissivity of the object and $T$ is the absolute temperature of the object in °K. The spectral composition of the emitted radiation is given by:

$$S_{real} = \varepsilon S \qquad \text{(A–6)}$$

where $S$ is the blackbody spectral radiation intensity in Equation A–1 and the magnitude of maximum wavelength is given by Wien's Law if the emissivity, $\varepsilon$, is constant in the vicinity of that maximum wavelength. (Remember that $\varepsilon$ may itself be a function of $\lambda$, $T$, and the incident angle.)

An example of the use of these radiative laws is to determine the temperature of an object in a vacuum that is exposed to the sun. It will gain energy at the rate of $\alpha I$ where $\alpha$ is the absorptivity of the object and $I$ is the intensity of solar energy incident upon it. It will lose energy only by reradiating its gained energy in the form of heat to its surroundings at the same rate once its temperature has stabilized.

Therefore:

$$\alpha I = \varepsilon \left( \frac{T_{object}}{64.8} \right)^4 \qquad \text{(A–7)}$$

or

$$T_{object} = 64.8 \sqrt[4]{\frac{\alpha}{\varepsilon} I} \qquad \text{(A–8)}$$

It must be assumed that the object is not absorbing radiant heat energy from its surroundings.

# Cover Material Characterization and the Effects of Internal Reflection

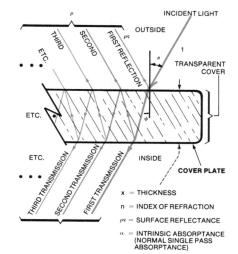

x = THICKNESS

n = INDEX OF REFRACTION

$\rho_s$ = SURFACE REFLECTANCE

$\alpha$ = INTRINSIC ABSORPTANCE (NORMAL SINGLE PASS ABSORPTANCE)

**FIGURE B–1** Internal reflections in single transparent cover.

As noted in Chapter 4, covers should be constructed of materials which possess a high transmittance, $\tau$, for solar radiation, low reflectance, $\rho$, low absorptance, $\alpha$, and, of course, should be able to withstand severe environmental hazards, such as the impact of hailstones and high winds. The reflectance, $\rho$, and transmittance, $\tau$, of a cover plate can be determined from the surface reflectance, $\rho_s$, absorptance, $\alpha$, and thickness, $x$, of the cover plates. The surface reflectance, $\rho_s$, is determined from the index of refraction, $n$, the angle of incidence, $\theta$, and the polarization of the incident radiation. For unpolarized light:

$$\rho_s = \frac{1}{2} \left( \frac{\sin^2 (\theta - \psi)}{\sin^2 (\theta + \psi)} + \frac{\tan^2 (\theta - \psi)}{\tan^2 (\theta + \psi)} \right) \tag{B–1}$$

(Fresnel's formula)

where: $\sin \psi = (\sin \theta)/n$ (Snell's Law)

determines the angle of refraction, $\psi$, where for air, $n = 1$.

The transmittance, reflectance, and absorptance are affected by internal reflections which are infinitely summed in the following closed expressions:

$$\alpha = \frac{\alpha_0 x}{\cos \psi} * \tag{B–2}$$

$$\tau = \frac{(1 - \rho_s)^2 (1 - \alpha)}{1 - (1 - \alpha)^2 \rho_s^2} \tag{B–3}$$

$$\rho = \rho_s \left[ 1 + \frac{(1 - \rho_s)^2 (1 - \alpha)^2}{(1 - \rho_s^2)(1 - \alpha)^2} \right] \tag{B–4}$$

Both the transmitted and reflected light will be slightly polarized for $\theta \neq 0$.

By ignoring the effects of polarization, the compound transmittance, $\tau_{1,2}$, reflectance, $\rho_{1,2}$, and absorptance, $\alpha_{1,2}$, of two covers with generally different properties are determined as:

* $\alpha$ may be expressed more precisely by Bouger's Law as follows:

$$\alpha = 1 - e^{\frac{\alpha_0 x}{\cos \psi}}$$

$$\tau_{1,2} = \frac{\tau_1 \tau_2}{1 - \rho_1 \rho_2} \qquad \text{(B–5)}$$

$$\rho_{1,2} = \rho_1 + \frac{\tau_1^2 \rho_2}{1 - \rho_1 \rho_2} \qquad \text{(B–6)}$$

$$\alpha_{1,2} = 1 - \tau_{1,2} - \rho_{1,2} \qquad \text{(B–7)}$$

For three dissimilar covers, the transmittance and reflectance are given by:

$$\tau_{1,2,3} = \frac{\tau_{1,2} \tau_3}{1 - \rho_{1,2} \rho_3} \qquad \text{(B–8)}$$

$$\rho_{1,2,3} = \rho_{1,2} + \frac{\tau_{1,2}^2 \rho_3}{\rho_{2,1} \rho_3} \qquad \text{(B–9)}$$

where: $\tau_{1,2,3}$ is the transmittance of three covers

$\rho_{1,2,3}$ is the reflectance of three covers

$\tau_{1,2}$  is the transmittance of two covers given by Equation B–5

$\rho_{1,2}$  is the reflectance of two covers given by Equation B–6

$\rho_{2,1}$  is the reflectance of two covers from the opposite side given by interchanging subscripts in Equation B–6 ($\rho_{1,2} \neq \rho_{2,1}$ for dissimilar covers)

$\tau_3$  is the transmittance of the third cover given by Equation B–3

$\rho_3$  is the reflectance of the third cover given by Equation B–4

Equations of the form shown in B–8 and B–9 may be extended to any number of covers by adding one cover at a time and by giving proper attention to subscript order.

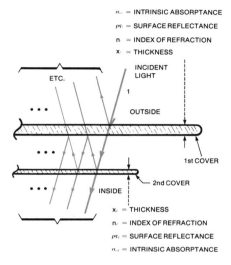

$\alpha_o$ = INTRINSIC ABSORPTANCE
$\rho_s$ = SURFACE REFLECTANCE
$n$ = INDEX OF REFRACTION
$x$ = THICKNESS

INCIDENT LIGHT
ETC.
1
OUTSIDE
1st COVER
2nd COVER
INSIDE
$x_i$ = THICKNESS
$n_i$ = INDEX OF REFRACTION
$\rho_{s_i}$ = SURFACE REFLECTANCE
$\alpha_{o_i}$ = INTRINSIC ABSORPTANCE

**FIGURE B–2**  Internal reflections in multiple transparent covers.

# C  Measurement Accuracy

## LIMITING ACCURACIES

The limiting value of measurement uncertainty attainable under laboratory conditions for basic physical measurements has been reported by the United States Bureau of Standards. A summary of these limiting accuracies is provided in Table C–1 in those ranges of interest for flat plate solar collectors.

Because most solar collector test facilities will not have available the equipment necessary to measure with these limiting accuracies, it is necessary to examine the accuracy and precision of measuring techniques which are available and in common use.

**TABLE C–1**  Limiting Accuracies of Measurement

| Parameter | Range of Interest | Technique | Measurement Uncertainty One Part In. (%) |
|---|---|---|---|
| Temperature | 200°–500°K | Platinum resistance thermometry | $10^5 = (0.001)$ |
| Liquid mass flow | 10–1000 lb/hr | Density corrected volume flow | $10^3 = (0.1)$ |
| Solar irradiance | 10–400 Btu/hr-ft² | Reference standard pyrheliometer | $5 \times 10^2 = (0.2)$ |
| Collector area | 1–$10^6$ ft² | Linear rule | $1.7 \times 10^{-7} = (6 \times 10^{-6})$ |

## TEMPERATURE

### INSTRUMENTAL ACCURACY AND PRECISION

Table C–2 lists the accuracy and precision required for temperature measurement according to ASHRAE 93-77 standards. The instrument accuracy represents the ability of the instrument to indicate the true value of the temperature. The instrument precision represents the required closeness of agreement among repeated measurements of the same temperature.

When instruments, signal conditioners, and readouts are sequentially combined, their combined error is calculated on both an absolute sum basis (for

**TABLE C–2** Temperature Measurement Requirements

| | Instrument Accuracy | Instrument Precision |
|---|---|---|
| Temperature | ±0.5°C (±0.9°F) | ±0.2°C (±0.36°F) |
| Temperature difference | ±0.1°C (±0.18°F) | ±0.1°C (±0.18°F) |

non-random error) and on an RMS basis (random error assumption). In addition to errors introduced by instrumentation, there are errors resulting from insufficient response time or the coupling of the temperature-measuring element to the object in which the temperature is being measured.

## THERMOWELLS

The accuracy of temperature measurement varies with response time in a manner that differs for each method employed. For those methods that are most commonly considered the response time is proportional to the mass and specific heat of the temperature sensing element. If the thermometer is enclosed in a thermowell, the response time of the well is generally greater (i.e., it responds more slowly) than that of the temperature sensing element within it, so the well determines the overall response time of the unit.

The use of thermowells for the measurement of fluid temperature is considered good practice for the following reasons.

1. The well provides a fixed geometry for fluid flow, no matter the type of temperature sensor placed within it. Also, sensors can be changed or moved from well to well without affecting fluid flow.
2. The well protects the temperature sensor from corrosive or chemically reactive components in the fluid which may affect the measurement or reduce the life of the sensor.
3. The well surface area provides an average over a certain region of fluid which will be representative of that region.

If fast response time is necessary, small temperature sensors can be placed directly in the fluid. Bare electrical leads in electrically conductive fluids, however, should be avoided.

Temperature sensors with a $1/e$ (2,72) response time of 30 seconds are readily available. With this response time, the sensor will represent the fluid temperature to within 0.1°F after a 20°F step change in a time of 2 minutes and 40 seconds. The first order response time of a flat plate solar collector is generally proportional to flow rate, but it will usually fall in the 1 to 10 minute range. There is a potential source of error through thermowell conduction and convection, but the error is proportional to the temperature difference between the fluid and the tubing wall so it diminishes with insulation of the tubing. Good practice leading to higher accuracy also requires thorough fluid mixing within the tubing. Two right-angle bends or staggered multi-orifice plates can be used to achieve the required fluid mixing.

**LIQUID MASS FLOW**

## BUCKET AND STOPWATCH

This technique of measuring mass flow is potentially very accurate since it uses direct measures of weight and time, each of which can be measured to reasonable accuracy. The accuracy of a manually operated stopwatch, for example, is limited primarily by variables in human response but can be regarded as within ±0.2 second with the percentage accuracy dependent upon the time period used to collect the fluid in the bucket. In practice, accuracy of 0.1 percent is achieved if a fluid mass on the order of 50 pounds is weighed. This method provides a time-averaged flow rate and is used to calibrate other flow measuring instruments in conjunction with constant head sources. An accuracy of 0.1 percent using a stopwatch would require the use of an estimated minimum of 5-minute time intervals.

## HEAD FLOWMETERS

The output reading of a head flowmeter cannot be considered as uniformly accurate since the reading is affected by small changes in the geometry of the orifice plate, nozzle, or surrounding manifolds. Errors are also magnified by the effects of wear, corrosion, deposition, and stress on the components of a head flowmeter, in addition to the instrument's sensitivity to numerous fluid parameters, such as viscosity mixing ratios, Reynolds number, and temperature. At the low flow rates characteristic of solar collectors, head flowmeters become less accurate and impose greater restrictions on flow. The instrument's time response to step changes in flow can be less than one second.

## VARIABLE AREA (ROTAMETER) FLOWMETERS

The measurement of liquid flow with a rotameter is sensitive to many of the same factors as those affecting a head flowmeter, while the direct viewing characteristic and the larger ratio of orifice perimeter to orifice area reduce the sensitivity of the instrument to changes at points along the orifice perimeter. The accuracy of a rotameter is properly expressed as a measure of the scale length so that the percentage of accuracy varies with the reading; the greatest accuracy is achieved for readings near the top of its scale. Also, the time response to a step change in flow is not the same in both directions for a rotameter. It is faster for a step increase than for a step decrease because the response time for a step decrease in flow is determined by the gravitation restoring force response time, whereas the response time for a step increase is determined by fluid mechanical response time. This will introduce a systematic error for integrated flow measurement.

# MAGNETIC FLOWMETERS

These flowmeters create no obstruction in the flow path, will measure flow in either direction, and can be used with difficult liquids. But they will operate only in an electrically conducting fluid and their accuracy is limited by the stability of the fluid's electrical conductivity.

# TURBINE FLOWMETERS

Turbine flowmeters are potentially accurate, assuming proper maintenance addressed to bearing friction and changes in the rotor due to deposits and corrosion, with good experience obtained at low flow rates of non-aqueous, self-lubricating liquids. The time interval for sampling pulses must be long to achieve high accuracy because the electrical pulses are most often discretely counted. Accuracy is thus restricted to $\pm 1$ count with percentage error from this source decreasing with an increasing time to accumulate counts.

# ACOUSTIC FLOWMETERS

The accuracy of these flowmeters is limited by the accuracy to which the acoustical properties of the liquid are known.

# ANGULAR MOMENTUM FLOWMETERS

Flowmeters using various angular momentum principles are available, offering a direct measure of mass flow. Correction factors are reduced and accuracies of better than 1 percent are possible with these instruments, according to some manufacturers' claims. Special designs using unobstructed flow paths are especially suited for biomedical instrumentation where low flow rate measurements are common.

# RADIOISOTOPE FLOWMETERS

These flowmeters, used in conjunction with dye injection techniques, are capable of accuracies in proportion to the fluid path-length available; the limit on accuracy is imposed by the dispersion and diffusion of the injected material in the fluid. Note that dispersion and diffusion rates are sensitive to many liquid properties which are themselves subject to significant error.

# THERMAL FLOWMETERS

The thermal properties of liquids are somewhat more stable and better known than are acoustic, magnetic, or other properties used in flowmeter technology. Thermal flowmeters, therefore, are sometimes regarded as more accurate by comparison, but the improvement in accuracy is not a significant one.

## POSITIVE DISPLACEMENT FLOWMETERS

The principal error in any positive displacement flowmeter is due to leakage around the moving member that is displaced during liquid flow, but there are other sources of error associated with the conversion of volume flow to mass flow, especially if air is mixed with the liquid. The time response of a positive displacement flowmeter is nearly instantaneous to step change in liquid flow, but these meters do not measure the actual rate of flow; they measure the amount of flow past a preselected time base. Thus, they are most often used to measure an accumulation of flow volume over a period of time. A true rate meter would require a differentiation of the shaft output of a positive displacement flowmeter, so the instantaneous accuracy of these meters is limited by the accuracy of the differentiation. The response time for a positive displacement flowmeter used to measure the amount of flow is essentially determined by the time period used.

The data relative to all these various measurements of liquid volume or liquid mass flow, as well as additional information on pressure drop across the instrument and required corrections, are contained in Table C–3.

**TABLE C–3**   Accuracy and Precision of Liquid Flow Techniques

| Method | Range (lb/min) | Quantity Measurement | Pressure Drop (lb/in²) | Time Response* | Corrections Required | Estimated Accuracy | | Estimated Precision | |
|---|---|---|---|---|---|---|---|---|---|
| | | | | | | % of Reading | % of Span | % of Reading | % of Span |
| Bucket & stopwatch | $10^{-6}$–$10^3$ | mass/time | <2 | >5 m | None | ±0.2 | | | ±0.1 |
| Head flowmeters | 10–$10^6$ | pressure diff. | <5 | <1 s | 5 | ±5 | | | ±0.5 |
| Variable area | $2 \times 10^{-3}$–$10^3$ | pressure diff. | <3 | <10 s | 6 | | ±2 | | ±0.25 |
| Magnetic | $10^{-3}$–$10^6$ | velocity | 0 | <1 s | 5 | | ±5 | | ±0.5 |
| Turbine | $10^{-1}$–$10^5$ | velocity | <1 | >10 s | 3 | | ±0.1 | | ±0.01 |
| Acoustic | 10–$10^6$ | velocity | 0 | <1 s | 5 | ±5 | | ±0.5 | — |
| Angular momentum | $10^{-2}$–$10^6$ | mass flow | <1 | <1 s | 2 | ±0.25 | | | ±0.01 |
| Radioisotope | $10^{-2}$–$10^6$ | velocity | 0 | <1 m | 5 | ±5 | | ±0.5 | — |
| Thermal | $10^{-1}$–$10^6$ | heat loss | 0 | <1 m | 6 | ±3 | | ±0.5 | — |
| Positive displacement: nutating disk | 16–1280 | vol. flow | <3 | <1 m | 3 | ±3 | | ±0.1 | — |
| oscillating piston | 6–120 | vol. flow | <3 | <1 m | 3 | ±0.2 | | ±0.015 | — |
| lobed impeller | 64–$1.8 \times 10^5$ | vol. flow | <3 | <1 m | 3 | ±0.2 | | ±0.015 | — |

* Within range suitable for solar collectors.

**TABLE C–4**  Classification of Accuracy of Solar Radiometers (established by World Meteorological Organization)

| | Sensitivity (MW/cm²) | Stability (%) | Temperature (%) | Selectivity (%) | Linearity (%) | Time Constant (max) | Cosine Response (%) | Azimuth Response (%) | Galvanometer | Millimeter (%) | Chronometer |
|---|---|---|---|---|---|---|---|---|---|---|---|
| Reference standard pyrheliometer | ±0.2 | ±0.2 | ±0.2 | ±1 | ±0.5 | 25 s | — | — | 0.1 unit | 0.1 | 0.1 s |
| Secondary instruments | | | | | | | | | | | |
| 1st class pyrheliometer | ±0.4 | ±1 | ±1 | ±1 | ±1 | 25 s | — | — | 0.1 unit | 0.2 | 0.3 s |
| 2nd class pyrheliometer | ±0.5 | ±2 | ±2 | ±2 | ±2 | 1 min | — | — | 0.1 unit | ±1.0 | — |
| | | | | | | | | | | Errors in recording apparatus | |
| 1st class pyranometer | ±0.1 | ±1 | ±1 | ±1 | ±1 | 25 s | ±3 | ±3 | | ±0.3 | |
| 2nd class pyranometer | ±0.5 | ±2 | ±2 | ±2 | ±2 | 1 min | ±5–7 | ±5–7 | | ±1 | |
| 3rd class pyranometer | ±1.0 | ±5 | ±5 | ±5 | ±3 | 4 min | ±10 | ±10 | | ±3 | |
| | | | | | | | | | | Errors due to wind | |
| | | | | | | | | | | % | % |
| 1st class net pyrradiometer | ±0.1 | ±1 | ±1 | ±3 | ±1 | ½ min | ±5 | ±5 | | ±0.3 | ±3 |
| 2nd class net pyrradiometer | ±0.3 | ±2 | ±2 | ±5 | ±2 | 1 min | ±10 | ±10 | | ±0.5 | ±5 |
| 3rd class net pyrradiometer | ±0.5 | ±5 | ±5 | ±10 | ±3 | 2 min | ±10 | ±10 | | ±1 | ±10 |

## SOLAR IRRADIANCE

As has been noted previously, the most accurate means of measuring solar irradiance over a period of time is the use of a continuously operating solar radiometer. The accuracy and precision of solar radiometers are compared in Table C–4, which also includes the specifications of the World Meteorological Organization (WMO) for different classifications of these instruments.

Although the limiting accuracy in measurements of the solar constant is unknown, terrestrial measurement of solar energy need not distinguish between effects that are intrinsic or secondary, as long as the energy received is locally measurable. To that end, the reference standard pyrheliometer is the most accurate instrument classification and is regarded as the standard upon which the accuracy of the others is based. The accuracy of this standard is based on calculations from first principles rather than actual calibration to the solar constant. A comparison of absolute cavity solar pyrheliometers conducted by DSET, Inc., at its test site in the Great Sonoran Desert in November of 1978 provided a measure of variation. With three different types of instruments (14 instruments in total) from ten organizations, the variation between instruments was ±0.3 percent.

## COLLECTOR AREA MEASUREMENT

Although the limiting accuracy and precision of area measurements are very high, it is important to consider these simple procedures very briefly.

STEEL TAPES

Length can be measured with steel tapes to an accuracy of one part in $10^3$ or 0.1 percent in the range of 1 meter. This results in an accuracy of 0.2 percent for rectangular areas if consistent, non-random error is assumed.

MANUFACTURING TOLERANCES

In large solar installations where the total collector area is comprised of a number of individual, discrete panels, area can be measured by multiplying one collector panel area by the number of panels. With manufacturing tolerances of $\pm\frac{1}{16}''$, the accuracy of an area measurement for one $2 \times 8$-foot collector is $\pm0.326$ percent. For $N$ such collectors, the accuracy is:

$$\pm \frac{0.325}{\sqrt{N}} \%$$ (C–1)

This error results in accuracy of more than $\pm0.1$ percent at $N = 10$ collectors. In order to maintain an accuracy of $\pm0.1$ percent in collector area measurements, it is recommended that the area of $\frac{\sqrt{N}}{3}$ of the collectors in a large array be measured with steel tapes.

## ACCURACY, TOLERANCE, PRECISION, AND SENSITIVITY

*Accuracy* refers to the degree to which a measurement reflects the actual value being measured and is usually expressed as a percentage of the absolute values in the measurement. In the measurement of temperature, however, accuracy in percentage of absolute temperature is often translated into units of temperature rather than a percentage. Thus, a thermometer accurate to $\pm0.1°C$ at $100°C$ is accurate to $\pm0.1/(273 + 100) \times 100 = \pm0.0268$ percent. This results from the fact that the temperature scale is more easily calibrated at $0°$ and $100°C$ than at its absolute zero value. For very high temperatures, accuracies return to percentages.

*Tolerance* refers to the amount within which the measured value is known to fall. It is properly expressed as an amount, not a percentage. Thus, strictly speaking, when a temperature is measured in an "accuracy" of $\pm0.1°C$, it is more proper to refer to it as being measured within a tolerance of $\pm0.1°C$.

The distinction between the accuracy and tolerance of a measurement is related to the source of error. If the source is due to uncontrollable factors in either the instrumentation or procedure, accuracy is reduced. If the source of error is a determined or controllable factor, then it contributes to a greater tolerance. When a manufacturer provides a platinum resistance thermometer with resistance specified to provide temperature measurement within a "tolerance" of $\pm0.5°C$, the accuracy can be made greater than $\pm0.5°C$ by proper calibration.

The *sensitivity* of an instrument is the change in output per unit change in input. The minimum sensitivity is the degree of precision. *Precision* is used to denote the degree to which repeated measurements can distinguish the effects of minor changes in the experimental variables.

In a measurement, $N$, consisting of component measurements $U_1$, $U_2, \ldots U_n$, each contributing a known error, $\delta U$, and related to a known relationship, $f$:

$$N = f(U_1, U_2, \ldots, U_N) \tag{C-2}$$

The absolute error, to first approximation, is given by:

$$\delta N = \left| \delta U_1 \frac{\partial f}{\partial U_1} \right| + \left| \delta U_2 \frac{\partial f}{\partial U_2} \right| + \cdots + \left| \delta U_N \frac{\partial f}{\partial U_N} \right| \tag{C-3}$$

For a flat plate solar collector the absolute error is determined as follows:

$$\eta = \frac{\dot{m} \cdot \Delta T}{I \cdot A_c} \qquad\qquad C_p = 1$$

$$\frac{\partial \eta}{\partial \dot{m}} = \frac{\Delta T}{I \cdot A_c} \qquad\qquad \frac{\partial \eta}{\partial I} = -\frac{\dot{m} \cdot \Delta T}{A_c \cdot I^2}$$

$$\frac{\partial \eta}{\partial \Delta T} = \frac{\dot{m}}{I \cdot A_c} \qquad\qquad \frac{\partial \eta}{\partial A_c} = -\frac{\dot{m} \cdot \Delta T}{I \cdot A_c^2}$$

The absolute error is:

$$\delta\eta = \left| \delta\dot{m} \frac{\Delta T}{I \cdot A_c} \right| + \left| \delta\Delta T \frac{\dot{m}}{I \cdot A_c} \right|$$
$$+ \left| \delta I \frac{\dot{m} \cdot \Delta T}{A_c \cdot I} \right| + \left| \delta A_c \frac{\dot{m} \cdot \Delta T}{I \cdot A_c} \right| \tag{C-4}$$

or:

$$\frac{\delta\eta}{\eta} = \frac{|\delta\dot{m}|}{\dot{m}} + \frac{|\delta\Delta T|}{T} + \frac{|\delta I|}{I} + \frac{|\delta A_c|}{A_c} \tag{C-5}$$

Thus, the percentage error of the efficiency is the sum of the percentage errors of mass flow, $\dot{m}$, temperature difference, $\Delta T$, solar irradiance, $I$, and collector area, $A_c$.

The absolute error represents the greatest possible error that would occur if all the measurement errors contributed to efficiency error in the same direction. If the measurement errors can be assumed to be random in both magnitude and direction, then the resultant error in efficiency can be expressed as the probable error or statistical error:

$$\delta N = \sqrt{ \left( \delta U_1 \frac{\partial f}{\partial U_1} \right)^2 + \left( \delta U_2 \frac{\partial f}{\partial U_2} \right)^2 + \cdots + \left( \delta U_N \frac{\partial f}{\partial U_N} \right)^2 } \tag{C-6}$$

The assumption of randomness implies a prior knowledge of the statistical distribution of measured values compared to some actual value. Such prior knowledge then implies that errors are truly indicators of the extent to which the actual value may differ from the measured value. This, in turn, implies that repeated measurements can serve to reduce error. Systematic or non-random elements of error cannot be treated in this manner.

The use of probable error is justified in situations where repeated measurements under identical conditions are possible and where measurements made with a sequence of instrumentation can be compared using a calibrated source, so that statistical properties can be determined and randomness can be ascertained. Unfortunately, solar experimentation cannot be repeated under identical conditions and random errors cannot be easily distinguished from systematic ones, so that one cannot always be certain that the probable error is applicable. It is for this reason that both the probable and the absolute error are included in the analysis. It is fair to say that in most cases the probable error will represent the actual error to a sufficient degree and that the absolute error represents the limit of maximum error which may be encountered if unknown sources of systematic error are present.

In the example of the calculation of collector efficiency accuracy contained in Table C–5, in which several available instrumentation choices are shown, the combined errors of instrumentation used in sequence are determined by the probable error formula because calibrated sources can be used to verify their accuracy and precision. The final determination in efficiency measurement, however, requires both the probable and absolute error, since the error introduced by the sensors may be subject to systematic as well as random factors.

In summary, Table C–5 shows the accuracy of efficiency measurement for two sets of assumptions regarding measurement accuracies that may be considered typical in practice. The data show that error increases as $\Delta T$ becomes

**TABLE C–5** Collector Efficiency Accuracy Example

Solar irradiance, $I = 300$ Btu/hr-ft$^2$, $\dfrac{\delta I}{I} = \pm 0.02$

Mass flow, $\dot{m} = 15$ lb/hr-ft$^2 \times 16$ ft$^2 = 240$ lb/hr, $\dfrac{\delta \dot{m}}{\dot{m}} = \pm 0.01$

Temperature difference, $\Delta T = T_0 - T_1 = \dfrac{\eta I A_c}{\dot{m}}$, $\dfrac{\partial \Delta T}{\Delta T} = \dfrac{1.0}{\Delta T}$ & $\dfrac{0.2}{\Delta T}$ (two cases)

Collector area, $A_c = 16$ ft$^2$, $\dfrac{\delta A_c}{A_c} = \pm 0.002$

| Solar Collector Efficiency $\eta$ | For Flow Rates Specified by ASHRAE 93-77 $\Delta T$ | Error in $\Delta T$ (Two cases) $\dfrac{\delta \Delta T}{\Delta T}$ | | Absolute (Maximum) Percent Error in Efficiency $\dfrac{\delta \eta}{\eta} \times 100$ | | Statistical (Random) Percent Error in Efficiency $\dfrac{\delta \eta}{\eta} \times 100$ | |
|---|---|---|---|---|---|---|---|
| | | $\delta \Delta T = 1.0°F$ | $\delta \Delta T = 0.2°F$ | $\delta \Delta T = 1.0°F$ | $\delta \Delta T = 0.2°F$ | $\delta \Delta T = 1.0°F$ | $\delta \Delta T = 0.2°F$ |
| 0.7 | 14°F | 0.071 | 0.014 | ±10.3 | ±4.6 | ± 7.5 | ±2.7 |
| 0.6 | 12 | 0.083 | 0.017 | ±11.5 | ±4.9 | ± 8.5 | ±2.8 |
| 0.5 | 10 | 0.100 | 0.020 | ±13.2 | ±5.2 | ±10.25 | ±3.0 |
| 0.4 | 8 | 0.125 | 0.025 | ±15.7 | ±5.7 | ±12.7 | ±3.4 |
| 0.3 | 6 | 0.167 | 0.033 | ±19.9 | ±6.5 | ±16.8 | ±4.0 |
| 0.2 | 4 | 0.250 | 0.500 | ±28.2 | ±8.2 | ±25.1 | ±5.5 |

small, and that the errors will be significant if the error in $\Delta T$ it not kept well below 1°F, and if non-random errors are present. It becomes obvious that every effort should be made to provide an accurate measurement of $\Delta T$ to eliminate sources of systematic error and that flow rates should be low enough so that $\Delta T$ will be in a range where possibilities for error are reduced.

The preceding discussion must be tempered by the discussion in pages 111 to 120 regarding the effects of wind speed, tilt angle and irradiance level. From AMETEK's test facility in Hatfield, Pennsylvania, we learned that the clear day performance we obtained was the same as that obtained at DSET, Inc., in Phoenix, Arizona, and at Lockheed Missile and Space Division in Palo Alto, California. The key factors were to have steady radiation and to correct for the tilt angle and wind speed through the equations 5–2 through 5–15.

**Properties of Materials**

**TABLE D–1**   Transparent or Translucent Cover Plate Materials

| Properties | Ordinary Float Glass | White Low Iron Glass | Acrylic Plastic (Plexiglas) | Fiberglass Reinforced Polyester | Polycarbon (Lexan) | Polyvinyl Fluoride (Tedlar) | Polyester Polyethylene (Mylar) | Fluorinated Hydrocarbons (Teflon) |
|---|---|---|---|---|---|---|---|---|
| Thickness | 1/8″ | 1/8″ | 1/8″ | 25 mils | 1/8″ | 4 mils | 5 mils | 1 mil |
| Solar transmission % at normal incidence | 85/79 | 91/88 | 89/80 | 75–80 | 73 | 88 | 80 | 96 |
| Transmission wavelengths (microns) | 0.35–0.65/ 0.2–4.0 | 0.35–0.65/ 0.2–4.0 | 0.4–1.1/ 0.2–4.0 | 0.2–4.0 | — | 0.2–4.0 | 0.2–4.0 | 0.2–4.0 |
| Infrared transmission (%)* | 2 | 2 | 2 | 3–8 | 2 | $\cong$33 | 17.8 | 58 |
| Infrared emissivity | 0.88 | 0.88 | | | | 0.59 | | 0.35 |
| Index of refraction (Air = 1.00) | −1.52 | −1.51 | 1.49–1.56 | 1.54 | 1.59 | 1.46 | 1.64–1.67 | 1.3 |
| Thermal conductivity (Btu-in/hr-ft²-°F) | 65–100 | 65–100 | 1.3 | NA | NA | NA | NA | 1.35 |
| Thermal expansion coefficient (in/in-°F) × 10⁻⁶ | 4.8 | 5–8 | 19–46 | 14–20 | 37.5 | 28 | 9–15 | 59–90 |
| Density (lb/in-ft²) (Water = 5.2) | 13 | 13 | 6 | 7.5 | 6.2 | 7 | 7 | 11 |
| Strength: Yield (psi) | 1600/6400† | 1600/6400 | 10,500 | 16,000 | 9,500 | 13,000 | 24,000 | 3,000 |
| Elastic modulus (10⁶ psi) | 10.5 | 10.5 | 0.45 | 1.1 | 0.35 | 0.26 | 0.55 | 0.07 |
| Temperature limit (°F) | 400 | 400 | 180 | 200 | 250 | 227 | 300 | 400 |
| Approximate life (years) | 500 | 500 | 10–15 | 7–20 | 10–15 | 5–10 | 4 | 15–20 |

\* In wavelengths from 3–50 microns.
† Tempered.
NA: Not Available.

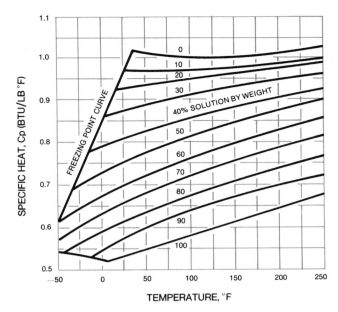

**FIGURE D–1** Properties of ethylene glycol in water solution.

**TABLE D–2** Absorber Plate Metals

| | Copper | Aluminum | Steel | Iron | Black Rubber Plastic |
|---|---|---|---|---|---|
| Thermal conductivity at 212°F (Btu-in/hr-ft²-°F) | 2616 | 1430 | 100–400 | 439 | 3–10 |
| Thermal expansion coefficient (in/in-°F) × 10⁻⁶ | 9.2 | 13.9 | 9.6 | 6.7 | — |
| Specific heat, C (Btu/lb-°F) | 0.091 | 0.21 | 0.11 | 0.10 | — |
| Density (lb/in-ft²) | 46.5 | 14.1 | 40.7 | 40.9 | 4-7 |
| Electrode potential in water (volts) | −0.340 | +2.30 | 0 | +0.441 | 0 |
| Strength: Yield (psi) | 10–20,000 | 30,000 | 50–100,000 | 30–95,000 | — |
| Elastic modulus (10⁶ psi) | 15.6 | 10 | 27–30 | 23–27 | — |

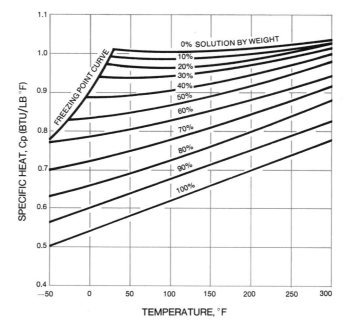

**FIGURE D–2**   Properties of propylene glycol in water solution.

**TABLE D–3**   Frame Materials

|  | Steel | Aluminum | Wood | Concrete |
|---|---|---|---|---|
| Strength: Yield (psi) | 50–100,000 | 30,000 | 8–20,000 | 500–700 |
| Elastic modulus ($10^6$ psi) | 27–30 | 10 | 1.1–2.1 | 2.5–3.5 |
| Thermal conductivity coefficient (Btu-in/hr-ft²-°F) | 100–400 | 1430 | 0.7–2.4 | 6–8.5 |
| Thermal expansion coefficient (in/in-°F) $\times 10^{-6}$ | 9.6 | 13.9 | 1.1–5.3 | 6 |
| Density (lb/in-ft²) | 40.7 | 14.1 | 2.5–4.5 | 9–13 |

## TABLE D-4 Densities and Thermal Conductivities

Thermal Conductivity (k) of Miscellaneous Substances at Room Temperature

| Material | Density at 68°F (lb/ft$^3$) | Conductivity (Btu-in/hr-ft$^2$-°F) |
|---|---|---|
| Air, still | — | 0.0169–0.215 |
| Aluminum | 168.0 | 1404–1439 |
| Asbestos board with cement | 123 | 2.7 |
| Asbestos, wool | 25.0 | 0.62 |
| Brass, red | 536.0 | 715.0 |
| Brick | | |
|   Common | 112.0 | 5.0 |
|   Face | 125.0 | 9.2 |
|   Fire | 115.0 | 6.96 |
| Bronze | 509.0 | 522.0 |
| Cabots | 3.4 | 0.25 |
| Cellulose, dry | 94.0 | 1.66 |
| Celotex (sugar cane fiber) | 13–14 | 0.34 |
| Charcoal | | |
|   Coarse | 13.2 | 0.36 |
|   6 mesh | 15.2 | 0.37 |
|   20 mesh | 19.2 | 0.39 |
| Cinders | 40–45 | 1.1 |
| Clay | | |
|   Dry | 63.0 | 3.5–4.0 |
|   Wet | 110.0 | 4.5–9.5 |
| Concrete | | |
|   Cinder | 97.0 | 4.9 |
|   Stone | 140.0 | 12.0 |
| Corkboard | 8.3 | 0.28 |
| Cornstack, insulating board | 15.0 | 0.24–0.33 |
| Cotton | 5.06 | 0.39 |
| Foamglas | 10.5 | 0.40 |
| Glass wool | 1.5 | 0.27 |
| Glass | | |
|   Common thermometer | 164.0 | 5.5 |
|   Flint | 247.0 | 5.1 |
|   Pyrex | 140.0 | 7.56 |
| Gold | 1205.0 | 2028.0 |
| Granite | 159.0 | 15.4 |
| Gypsum, solid | 78.0 | 3.0 |
| Hair felt | 13.0 | 0.26 |
| Ice | 57.5 | 15.6 |
| Iron, cast | 442.0 | 326.0 |
| Kapok | 1.0 | 0.24 |
| Lead | 710.0 | 240.0 |
| Leather, sole | 54.0 | 1.1 |
| Lime | | |
|   Mortar | 106.0 | 2.42 |
|   Slaked | 81–87 | — |

**TABLE D–4**  Densities and Thermal Conductivities (*continued*)

Thermal Conductivity (k) of Miscellaneous Substances at Room Temperature

| Material | Density at 68°F (lb/ft$^3$) | Conductivity (Btu-in/hr-ft$^2$-°F) |
|---|---|---|
| Limestone | 132.0 | 10.8 |
| Marble | 162.0 | 20.6 |
| Mineral Wool | | |
| Board | 15.0 | 0.33 |
| Fill-type | 9.4 | 0.27 |
| Nickel | 537.0 | 406.5 |
| Paper | 58.0 | 0.9 |
| Paraffin | 55.6 | 1.68 |
| Plaster | | |
| Cement | 73.8 | 8.0 |
| Gypsum | 46.2 | 3.3 |
| Redwood Bark | 5.0 | 0.26 |
| Rock Wool | 10.0 | 0.27 |
| Rubber, hard | 74.3 | 11.0 |
| Sand, dry | 94.6 | 2.23 |
| Sandstone | 143.0 | 12.6 |
| Sawdust | 8–12 | 0.41 |
| Sil-O-Cel (powered diatomaceous earth) | 10.6 | 0.31 |
| Silver | 656.0 | 2905.0 |
| Soil | | |
| Crushed quartz (4% moisture) | 100.0 | 11.5 |
| Dakota sandy loam | | |
| (4% moisture) | 110.0 | 6.5 |
| (10% moisture) | 110.0 | 13.0 |
| Fairbanks sand | | |
| (4% moisture) | 100.0 | 8.5 |
| (10% moisture) | 100.0 | 15.0 |
| Healy clay | | |
| (10% moisture) | 90.0 | 5.5 |
| (20% moisture) | 100.0 | 10.0 |
| Steel | | |
| 1% C | 487.0 | 310.0 |
| Stainless | 515.0 | 200.0 |
| Tar, bituminous | 75.0 | — |
| Water, fresh | 62.4 | 4.1 |
| Wood | | |
| Balsa | 7.3 | 0.33 |
| Fir | 34.0 | 0.8 |
| Maple | 44 | 1.2 |
| Red Oak | 48.0 | 1.1 |
| White pine | 32 | 0.78 |
| Wood fiberboard | 16.9 | 0.31 |
| Wool | 4.99 | 0.264 |

**TABLE D–5**  Solar Absorptivity and Thermal Emissivity of Commonly Used Metals

| Metal Surface | Solar Absorptivity, $\alpha$ | Solar Reflectivity*, $\rho$ | Thermal Emissivity, $\varepsilon$ |
|---|---|---|---|
| Aluminum, pure | 0.1 | 0.9 | 0.1 |
| Aluminum, anodized | 0.12–0.16 | 0.84–0.88 | 0.65 |
| Chromium | 0.4 | 0.6 | 0.2 |
| Copper, polished | 0.15 | 0.85 | 0.03 |
| Gold | 0.2–0.23 | 0.77–0.8 | 0.025–0.04 |
| Iron | 0.44 | 0.56 | 0.07–1.1 |
| Nickel | 0.36–0.43 | 0.57–0.64 | 0.1 |
| Silver, polished | 0.035 | 0.965 | 0.02 |
| Zinc | 0.5 | 0.5 | 0.05 |

* Both specular and diffused reflectivity.

**TABLE D–6**  Thermal Storage Materials

| Media | Temperature Range, °F | Melting Temperature °F | Latent Heat Btu/lb | $C_\rho$ Specific Heat Btu/lb-°F | Density lb/ft$^3$ |
|---|---|---|---|---|---|
| Water ice | 32 | 32 | 144 | 0.49 | 58 |
| Water | 90–130 | 32 | — | 1.0 | 62 |
| Steel (scrap iron) | 90–130 | — | — | 0.12 | 489 |
| Basalt (lava rock) | 90–130 | — | — | 0.20 | 184 |
| Limestone | 90–130 | — | — | 0.22 | 156 |
| Paraffin wax | 90–130 | 100 | 65 | 0.7 | 55 |
| Salt hydrates | | | | | |
| $NaSO_4 \cdot 10H_2O$ | 90–130 | 90 | 108 | 0.4 | 90 |
| $NA_2S_2O_3 \cdot 5H_2O$ | 90–130 | 120 | 90 | 0.4 | 104 |
| $NA_2HPO_4 \cdot 12H_2O$ | 90–130 | 97 | 120 | 0.4 | 94 |
| Water | 110–300 | — | — | 1.0 | 62 |
| Fire brick | 110–800 | — | — | 0.22 | 198 |
| Ceramic oxides $M_6O$ | 110–800 | — | — | 0.35 | 224 |
| Fused salts $NaNO_3$ | 110–800 | 510 | 80 | 0.38 | 140 |
| Lithium | 110–800 | 370 | 286 | $\pm 1.0$ | 33 |
| Carbon | 110–800 | 6750 | — | 0.2 | 140 |
| Lithium hydride | 1260 | 1260 | 1200 | $\pm 1.0$ | 36 |
| Sodium chloride | 1480 | 1480 | 233 | 0.21 | 135 |
| Silicon | 2605 | 2605 | 607 | — | 146 |

NOTE: By mixing fused salts, any desired melting point between 250°F and 1000°F can be attained.

# E  Shortcut DHW Sizing Method

The procedure outlined here is a method for sizing solar energy domestic hot water (DHW) systems for one- and two-family dwellings. It correlates with the f-CHART procedure developed by Beckman, Klein and Duffie at the University of Wisconsin and can be used in place of that method for measurement of average annual system performance. It is intended for new and existing buildings in the United States or Canada, where the collectors face South and are tilted at the latitude angle.

The procedure can be used to determine collector array area and storage collector size, as well as to confirm claimed system performance. Its application is limited to systems with flat-plate collectors with incident angle modifiers of approximately 0.1. Its accuracy depends upon two factors: load prediction and estimation of system efficiency.

## SIZING OF SOLAR DHW SYSTEMS FOR ONE- AND TWO-FAMILY DWELLINGS

The solar fraction of any DHW heating load depends upon several variables. The local available average irradiance and the size of the load are predetermined so that the most effective choice between collector area versus collector efficiency, heat exchange area, and storage volume can be made. A term related to available irradiance is called the location factor. The location factor is a number that has been experimentally determined so that the use of the nomograph (Fig. E–1) gives results which correspond to those obtained from f-CHART version 3.0 (a computer program of f-CHART). The following equation relates the necessary terms:

$$A = \frac{fL}{NI} \qquad \text{(E–1)}$$

where:  $A$ = gross collector area

$f$ = annual solar fraction

$L$ = Load (see below and Table E–1)

$N$ = overall system efficiency (determined from nomograph)

$I$ = average solar irradiance (see Table E–2 for your location)

Discrete values of $f$ are used to determine $N$ by means of the nomograph (Fig. E–1). By calculating several sets of the four variables, that many values for $A$ can be determined. In order to evaluate the performance of a specific area, the values of $A$ are plotted against $f$ as shown in Figures E–2, E–3 and E–4.

This procedure can be used to evaluate system changes in collector and storage tank sizes and to compare existing systems. It is also a method of determining solar fraction and consequent cost savings. A simple estimate of the total number of years for the initial cost to be recovered can be made by dividing installation cost by annual savings. See Chapter 8 for more sophisticated methods of economic analysis.

The first step is to determine the load, then by use of the nomograph, determine $N$ and then calculate $A$.

## LOAD

Load ($L$) is defined as the average rate of energy use over a typical year. It represents a power quantity expressed in Btu/hr. The average load is defined as the average rate of energy addition to the water to raise its temperature from the cold water supply temperature to the hot water use temperature. For realistic solar energy system sizing, the average rate of heat loss from the auxiliary hot water storage tank (second tank, conventional water heater) should be added and included as part of the load.

An equation for hot water load is:

$$L = \frac{w(T_h - T_c)}{2.88} + UA(\Delta T) \tag{E–2}$$

where:    $L$ = hot water load, Btu/hr

$\quad\quad\quad w$ = hot water use rate at the tap, after the mixing valve, in gallons per day

$\quad\quad\quad T_h$ = hot water use temperature after the mixing valve, °F

$\quad\quad\quad T_c$ = average annual cold water supply temperature, °F

$\quad\quad 2.88$ = factor representing the number of hours in a day divided by the weight in pounds per gallon of water (8.34)

$\quad\quad UA$ = product of overall heat transfer coefficient and surface area of the auxiliary hot water storage tank [Btu/(hr)(°F)]

$\quad\quad \Delta T$ = temperature difference between auxiliary hot water storage tank and ambient temperature outside the tank, °F.

Typical values for the variables in the equation are given in Table E–1.

As an example, the equation is solved using the middle values for each variable in Table E–1.

$$L = \frac{w(T_h - T_c)}{2.88} + UA(\Delta T)$$

$$L = \frac{70(140 - 59)}{2.88} + 5.69(81)$$

$$L = 1969 + 461$$

$$L = 2430 \text{ Btu/hr}$$

**FIGURE E–1**   DHW system efficiency nomograph.

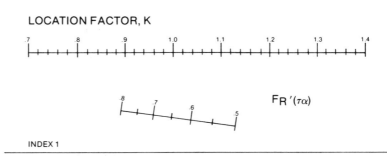

LOCATION FACTOR, K

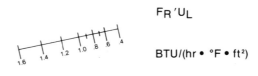

$F_R'(\tau\alpha)$

INDEX 1

$F_R'U_L$

BTU/(hr • °F • ft²)

INDEX 2

S, Gal./Ft² Collector Area

INDEX 3

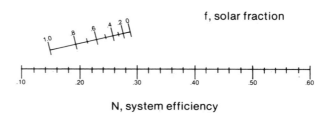

f, solar fraction

N, system efficiency

## DERIVATION OF SYSTEM EFFICIENCY EQUATION

1. The nomograph presented for DHW systems (Fig. E–1) is based on an equation that was derived from data generated by the computer program f-CHART, Version 3.0. Due to its widespread use and applicability, f-CHART was used to predict the annual thermal performance. Equations for $N$ as a function of the chosen system parameters were then developed to fit the f-CHART predictions.

2. The initial procedure for generating the DHW equation was to make f-CHART runs by varying the collector area for the simulated DHW system while holding the other program input parameters at their default values. $F_R'(\tau\alpha)$ was then varied for these same collector areas. These data were used to generate a series of $N$ (system efficiency) versus f (annual solar fraction) curves. From these curves, an equation for $N$ as a function of f and $F_R'\tau\alpha$ was developed:

$$N = [F_R'(\tau\alpha) - .20][.94 - .48f]$$

3. The same procedure was then repeated for $F_R'U_L$. This parameter was varied over a range of collector areas and a new series of $N$ versus f curves was generated. These yielded an equation for $N$ as a function of f and $F_R'U_L$:

$$N = (.65 - .227\ F_R'U_L - .3f^{1.5})$$

4. Similarly, an equation for $N$ as a function of f and the storage capacity of the preheat tank ($S$) in units of Btu/°F-ft² collector area was developed:

$$N = \left(.40 + \frac{S}{374}\right)(1 - .54\ f^{1.5})$$

5. At this point, the three equations found for $N$ were combined algebraically to form an equation for $N$ as a function of f, $F_R'(\tau\alpha)$, $F_R'U_L$, and $S$. This was done by first combining the

**TABLE E–1** Typical Values for Variables in DHW Load Equation

| Variable | Units | Low | Middle | High |
|---|---|---|---|---|
| | | | Typical Values | |
| $w$ | gal/day | 20 | 70 | 120 |
| $T_h$ | °F | 113 | 140 | 167 |
| $T_c$* | °F | 41 | 59 | 77 |
| $UA$* | Btu/hr | 2.6 | 5.69 | 7.58 |
| $\Delta T$† | °F | 50 | 81 | 90 |
| $L$ | Btu/hr | 774 | 2430 | 4432 |

* Use middle values if no data available.
† Note $\Delta T$ is *not* $T_h - T_c$: Use your best estimate for ambient outside tank temperature and assume average tank temperature is 120°F.

An alternative method of calculating hot water load, accounting more accurately for losses from the auxiliary hot water tank, is expressed by the following equation:

$$L = \frac{w(T_h - T_c)}{2.88} + 8.25(V)(x)(\Delta T) \qquad \text{(E–3)}$$

where:
$V$ = volume of water in auxiliary hot water storage tank, in gallons (typical values range from 40 to 120 gallons)

$x$ = fractional loss per hour of energy content of tank (0.015 for standard electric water heaters and 0.06 for standard gas-fired water heaters)

8.25 = factor representing the volumetric heat capacity of water at 122°F (Btu/gal·ft²)

---

equations for $F_R\tau\alpha$ and $F_R'U_L$, then combining the resulting equation with the equation for $S$. A form was assumed for the final combined equation as follows:

$$N = k_1[F_R'(\tau\alpha) - k_2](1 - k_3 F_R'U_L)(1 + k_4 S)(1 - k_5 f - k_6 f^{1.5})$$

6. The constants were determined using six equations (for the six unknowns) generated by substituting values for $N$, $F_R'(\tau\alpha)$, $F_R'U_L$, $S$, and f taken from points on the $N$ versus f curves.

7. A $K$ factor was developed for each of the 266 f-CHART cities in the following manner. Required collector area was computed, via f-CHART, for the following baseline case:

```
        f = 0.60
        w = 70 gal·day⁻¹
  Fᵣ'(τα) = 0.70
   Fᵣ'Uₗ = 0.83 Btu·hr⁻¹·°F⁻¹·ft⁻²
        S = 1.80 gal·ft⁻² (15.0 Btu·°F⁻¹·ft⁻²)
    slope = local latitude
  azimuth = 0°
 one cover
 all other variables, default values
```

f-CHART 3.0 provided A, L&I (with proper consideration for units), which were combined with f to compute $N$ using:

$$N = \frac{fL}{AI}$$

With N known, all values in the final equation are known except $K$. This equation was used to compute $K$ for each city:

$$N = K(1.27)[F_R'(\tau\alpha) - 0.20](1 - 0.4\, F_R'U_L)\left(1 + \frac{S}{18.2}\right)(1 - 0.8f - .53f^{1.5})$$

**PHILADELPHIA, PA**

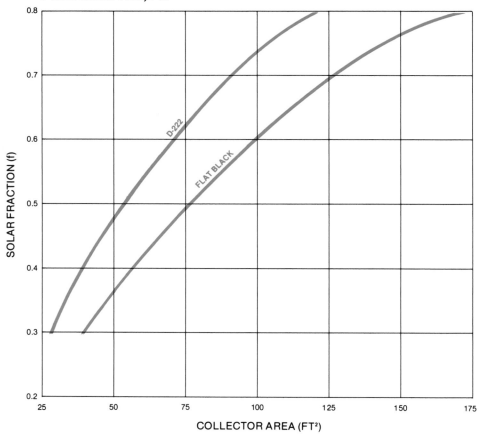

**FIGURE E–2**  Results of collector area calculation for Philadelphia.

As an example of the use of the alternative load equation, the equation is solved using the average values for each variable in Table E–1, with an electric auxiliary tank that holds 80 gallons of water.

$$L = \frac{w(T_h - T_c)}{2.88} + (8.25)(V)(x)(\Delta T)$$

$$L = \frac{70(140 - 59)}{2.88} + (8.25)(70)(0.015)(81)$$

$$L = 1969 + 702 = 2671 \text{ Btu/hr}$$

### USING THE NOMOGRAPH

The nomograph in Figure E–1 is used by simply passing straight lines through the appropriate points on the scales. The first line is drawn from the $K$ scale at the top through the $F_R'(\tau\alpha)$ scale to the horizontal line labeled Index 1. The next line is drawn from that point on Index 1 through the $F_r'U_L$ scale to Index 2. The third line is drawn similarly from that point on Index 2 through the correct point on the storage scale to Index 3, and the final line is drawn from that point on Index 3 just obtained through the f scale value to the $N$ scale.

**CHARLOTTE, NC**

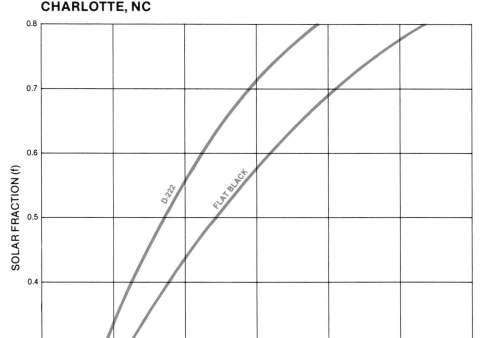

**FIGURE E–3** Collector area calculation for Charlotte.

The parameters of the nomograph are as follows:

*System Location Factor (K)*. A dimensionless constant used to account for the difference in overall system efficiencies among identical systems in different locations. *K* values for all f-CHART cities can be found in Table E–2. Each K value was selected to give zero discrepancy between f-CHART results and those from this method for base case conditions.

*f = Solar Fraction, Annual*. The ratio of the amount of energy contributed, per year, to the load by the solar energy system to the total energy required per year for the application. Units: dimensionless. The solar fraction can be expressed as a decimal between zero and one; values should be limited, however, to those between 0.3 and 0.8 for the most accurate results.

*Collector Efficiency Equation*. A linear fit through the noon efficiency points required by ASHRAE 93-77 and supplied by vendors will result in the following equation:

$$\text{Efficiency} = \eta = F_R(\tau\alpha) - F_R U_L \frac{\Delta T}{I} \qquad \textbf{(E–4)}$$

This equation is known for all certified collectors, or can be calculated by

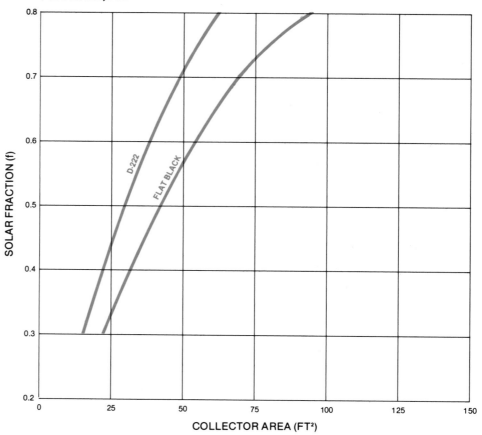

**PHOENIX, AZ**

**FIGURE E–4**  Collector area calculation for Phoenix.

regression or graphical techniques. The value $F_R(\tau\alpha)$ will be the $y$ intercept, and $F_RU_L$ the negative slope, of the collector efficiency curve (see page 246).

$F_R'(\tau\alpha)$. A dimensionless collector parameter found by multiplying the $y$-intercept on a collector efficiency curve (as defined by ASHRAE 93-77 and supplied by vendors) by the collector loop heat exchanger factor.

The collector loop heat exchanger factor is a function of a number of items normally available only to the system designer. Experience figures are acceptable for this method; practical values may be selected from the following list:

$F_R'/F_R = 1.0$  No heat exchanger (HX) present; a draindown system or some variant.

$F_R'/F_R = 0.96$  Used as a baseline case for all computer work on this standard. Properly sized heat exchanger and forced convection in both system loops (pumps) [$\geq 0.1$ ft$^2$ HX surface per ft$^2$ of collector area].

$F_R'/F_R = 0.85$  Undersized heat exchanger [$.02 - 0.1$ ft$^2$ HX surface per ft$^2$ of collector area].

$F_R'/F_R = 0.80$  Air-water heat exchanger.

$F_R'/F_R = 0.78$  Extremely small heat exchanger [$<0.02$ ft$^2$ HX surface per ft$^2$ of collector area].

**TABLE E–2** Average Solar Irradiance on Plane Facing South at Tilt Angle Equal to Latitude and System Location Factor (K) for f-CHART Cities

| City | State | N. Latitude (Degrees) | Radiation (I) (Btu/hr ft$^2$) | Location Factor (K) |
|------|-------|-----------------------|-------------------------------|---------------------|
| Birmingham | AL | 33.57 | 60.3 | 1.07 |
| Mobile | AL | 30.68 | 61.7 | 1.07 |
| Montgomery | AL | 32.30 | 62.0 | 1.08 |
| Annette | AK | 55.03 | 45.4 | .94 |
| Barrow | AK | 71.33 | 47.8 | .76 |
| Bethel | AK | 60.78 | 51.8 | 1.07 |
| Fairbanks | AK | 64.82 | 54.2 | 1.10 |
| Matanuska | AK | 61.57 | 47.5 | 1.04 |
| Page | AZ | 36.63 | 89.0 | 1.35 |
| Phoenix | AZ | 33.43 | 86.1 | 1.27 |
| Prescott | AZ | 34.65 | 84.9 | 1.31 |
| Tucson | AZ | 32.12 | 85.9 | 1.28 |
| Winslow | AZ | 35.22 | 84.4 | 1.31 |
| Yuma | AZ | 32.67 | 88.7 | 1.27 |
| Fort Smith | AR | 35.33 | 63.9 | 1.12 |
| Little Rock | AR | 34.73 | 63.5 | 1.11 |
| Arcata | CA | 40.98 | 56.2 | 1.07 |
| Bakersfield | CA | 35.42 | 79.3 | 1.22 |
| Daggett | CA | 34.87 | 85.4 | 1.27 |
| Davis | CA | 38.55 | 73.1 | 1.18 |
| El Toro | CA | 33.67 | 74.7 | 1.21 |
| Fresno | CA | 36.77 | 73.3 | 1.17 |
| Inyokern | CA | 35.65 | 104.2 | 1.38 |
| Longbeach | CA | 33.82 | 73.5 | 1.20 |
| Los Angeles | CA | 33.93 | 73.4 | 1.21 |
| Mt. Shasta | CA | 41.30 | 69.5 | 1.20 |
| Needles | CA | 34.85 | 86.8 | 1.26 |
| Oakland | CA | 37.73 | 71.0 | 1.19 |
| Pasadena | CA | 34.15 | 74.8 | 1.21 |
| Point Mugu | CA | 34.12 | 72.0 | 1.22 |
| Riverside | CA | 33.95 | 83.2 | 1.28 |
| Sacramento | CA | 38.52 | 75.0 | 1.19 |
| San Diego | CA | 32.73 | 73.5 | 1.21 |
| San Francisco | CA | 37.78 | 71.7 | 1.20 |
| San Jose | CA | 37.33 | 66.7 | 1.15 |
| Santa Maria | CA | 34.90 | 74.0 | 1.23 |
| Sunnyvale | CA | 37.42 | 73.2 | 1.20 |
| Boulder | CO | 40.00 | 66.0 | 1.20 |
| Colorado Springs | CO | 38.82 | 77.7 | 1.30 |
| Denver | CO | 39.75 | 83.0 | 1.32 |
| Eagle | CO | 39.65 | 76.5 | 1.31 |
| Grand Junction | CO | 39.12 | 79.4 | 1.29 |
| Grand Lake | CO | 40.27 | 75.6 | 1.33 |
| Pueblo | CO | 38.28 | 78.2 | 1.28 |
| Hartford | CT | 41.93 | 48.7 | .99 |
| Wilmington | DE | 39.67 | 55.7 | 1.06 |
| Washington | DC | 38.85 | 55.1 | 1.04 |
| Apalachicola | FL | 29.75 | 65.4 | 1.10 |
| Daytona Beach | FL | 29.18 | 64.9 | 1.09 |
| Gainesville | FL | 29.65 | 71.6 | 1.16 |
| Jacksonville | FL | 30.42 | 64.2 | 1.09 |
| Key West | FL | 24.55 | 73.0 | 1.13 |
| Miami | FL | 25.78 | 64.9 | 1.07 |
| Orlando | FL | 28.55 | 66.3 | 1.09 |
| Pensacola | FL | 30.47 | 68.5 | 1.14 |
| Tallahassee | FL | 30.43 | 64.1 | 1.09 |
| Tampa | FL | 27.97 | 66.3 | 1.10 |
| Atlanta | GA | 33.65 | 60.5 | 1.08 |
| Augusta | GA | 33.37 | 61.5 | 1.08 |
| Griffin | GA | 33.75 | 69.9 | 1.18 |
| Macon | GA | 32.70 | 62.0 | 1.08 |

## TABLE E–2  (Continued)

| City | State | N. Latitude (Degrees) | Radiation (I) (Btu/hr ft$^2$) | Location Factor ($K$) |
|---|---|---|---|---|
| Savannah | GA | 32.13 | 67.4 | 1.13 |
| Hilo | HI | 19.73 | 64.1 | 1.06 |
| Honolulu | HI | 21.30 | 82.8 | 1.22 |
| Boise | ID | 43.57 | 71.1 | 1.21 |
| Pocatello | ID | 42.92 | 73.2 | 1.24 |
| Idaho Falls | ID | 40.58 | 64.8 | 1.17 |
| Chicago | IL | 41.98 | 56.3 | 1.07 |
| Lemont | IL | 41.67 | 59.8 | 1.13 |
| Moline | IL | 41.45 | 57.0 | 1.10 |
| Peoria | IL | 40.67 | 62.2 | 1.15 |
| Springfield | IL | 39.83 | 60.1 | 1.11 |
| Evansville | IN | 38.85 | 57.2 | 1.06 |
| Fort Wayne | IN | 41.00 | 50.7 | 1.00 |
| Indianapolis | IN | 39.73 | 52.7 | 1.02 |
| South Bend | IN | 41.70 | 51.2 | .99 |
| Ames | IA | 42.13 | 61.6 | 1.17 |
| Des Moines | IA | 41.53 | 61.9 | 1.16 |
| Mason City | IA | 43.15 | 61.5 | 1.17 |
| Sioux City | IA | 42.40 | 62.2 | 1.16 |
| Dodge City | KS | 37.77 | 73.7 | 1.24 |
| Manhattan | KS | 39.20 | 64.6 | 1.16 |
| Topeka | KS | 39.07 | 64.8 | 1.16 |
| Wichita | KS | 37.65 | 70.4 | 1.20 |
| Covington | KY | 39.87 | 53.8 | .91 |
| Lexington | KY | 38.03 | 55.0 | 1.04 |
| Louisville | KY | 38.18 | 54.9 | 1.04 |
| Baton Rouge | LA | 30.53 | 61.0 | 1.05 |
| Lake Charles | LA | 30.22 | 60.0 | 1.04 |
| New Orleans | LA | 29.98 | 63.6 | 1.08 |
| Shreveport | LA | 32.42 | 65.1 | 1.11 |
| Caribou | ME | 46.87 | 51.1 | 1.08 |
| Portland | ME | 43.65 | 48.9 | 1.02 |
| Annapolis | MD | 38.96 | 60.7 | 1.12 |
| Baltimore | MD | 39.18 | 55.9 | 1.06 |
| Patuxent River | MD | 39.17 | 57.7 | 1.08 |
| Silver Hill | MD | 38.83 | 61.8 | 1.12 |
| Amherst | MA | 42.25 | 52.7 | 1.05 |
| Blue Hill | MA | 42.22 | 56.3 | 1.11 |
| Boston | MA | 42.37 | 51.2 | 1.01 |
| Lynn | MA | 42.47 | 53.2 | 1.02 |
| Natick | MA | 42.28 | 57.1 | 1.10 |
| Detroit | MI | 42.23 | 50.8 | 1.00 |
| East Lansing | MI | 42.73 | 55.7 | 1.08 |
| Flint | MI | 42.97 | 48.5 | .97 |
| Grand Rapids | MI | 42.90 | 51.1 | .99 |
| Lansing | MI | 42.78 | 57.1 | 1.10 |
| Sault St. Marie | MI | 46.47 | 47.8 | .99 |
| Traverse City | MI | 44.73 | 48.6 | .96 |
| Duluth | MN | 46.83 | 50.5 | 1.07 |
| International Falls | MN | 48.57 | 52.6 | 1.11 |
| Minneapolis-St. Paul | MN | 44.88 | 55.7 | 1.11 |
| Rochester | MN | 43.92 | 54.4 | 1.10 |
| St. Cloud | MN | 45.57 | 64.6 | 1.22 |
| Jackson | MS | 32.32 | 62.8 | 1.09 |
| Meridian | MS | 32.35 | 61.1 | 1.07 |
| Columbia | MO | 38.97 | 61.0 | 1.12 |
| Kansas City | MO | 39.28 | 62.3 | 1.14 |
| Springfield | MO | 37.23 | 62.4 | 1.12 |
| St. Louis | MO | 38.75 | 61.0 | 1.11 |
| Billings | MT | 45.80 | 64.9 | 1.19 |
| Dillon | MT | 45.25 | 67.6 | 1.24 |
| Glasgow | MT | 48.22 | 60.4 | 1.17 |
| Great Falls | MT | 47.48 | 62.3 | 1.17 |
| Helena | MT | 46.58 | 61.6 | 1.17 |
| Lewistown | MT | 47.05 | 60.8 | 1.17 |

## TABLE E–2  (Continued)

| City | State | N. Latitude (Degrees) | Radiation (I) (Btu/hr ft²) | Location Factor (K) |
|------|-------|----------------------|---------------------------|--------------------|
| Miles City | MT | 46.43 | 64.1 | 1.19 |
| Missoula | MT | 46.92 | 54.4 | 1.05 |
| Summit | MT | 48.32 | 56.3 | 1.14 |
| Grand Island | NE | 40.97 | 66.9 | 1.20 |
| Lincoln | NE | 40.85 | 66.1 | 1.20 |
| North Omaha | NE | 41.37 | 62.5 | 1.16 |
| North Platte | NE | 41.13 | 69.6 | 1.23 |
| Scotts Bluff | NE | 41.87 | 69.1 | 1.23 |
| Elko | NV | 40.85 | 78.2 | 1.30 |
| Ely | NV | 39.28 | 80.9 | 1.33 |
| Las Vegas | NV | 36.38 | 88.2 | 1.30 |
| Lovelock | NV | 40.07 | 87.0 | 1.34 |
| Reno | NV | 39.50 | 84.7 | 1.33 |
| Tonopah | NV | 38.07 | 88.9 | 1.35 |
| Winnemucca | NV | 40.90 | 79.1 | 1.29 |
| Concord | NH | 43.20 | 48.7 | 1.01 |
| Atlantic City | NJ | 39.45 | 64.1 | 1.16 |
| Lakehurst | NJ | 40.03 | 53.6 | 1.05 |
| Newark | NJ | 40.70 | 54.1 | 1.04 |
| Trenton | NJ | 40.22 | 62.0 | 1.15 |
| Albuquerque | NM | 35.05 | 85.9 | 1.32 |
| Farmington | NM | 36.75 | 83.9 | 1.32 |
| Roswell | NM | 33.40 | 83.7 | 1.29 |
| Zuni | NM | 35.10 | 81.8 | 1.31 |
| Albany | NY | 42.67 | 55.9 | 1.09 |
| Binghamton | NY | 42.22 | 44.2 | .91 |
| Buffalo | NY | 42.93 | 45.8 | .90 |
| Ithaca | NY | 42.45 | 53.2 | 1.05 |
| Massena | NY | 44.93 | 55.0 | 1.09 |
| New York | NY | 40.77 | 50.2 | .98 |
| Rochester | NY | 43.12 | 46.4 | .91 |
| Schenectady | NY | 42.83 | 48.7 | 1.00 |
| Syracuse | NY | 43.12 | 46.2 | .92 |
| Asheville | NC | 35.43 | 60.0 | 1.11 |
| Cape Hatteras | NC | 35.27 | 62.3 | 1.10 |
| Charlotte | NC | 35.22 | 62.7 | 1.12 |
| Cherry Point | NC | 34.90 | 63.0 | 1.11 |
| Greensboro | NC | 36.08 | 61.6 | 1.12 |
| Raleigh | NC | 35.78 | 59.1 | 1.08 |
| Raleigh-Durham | NC | 35.87 | 65.2 | 1.15 |
| Bismarck | ND | 46.78 | 61.7 | 1.19 |
| Fargo | ND | 46.90 | 58.4 | 1.16 |
| Minot | ND | 48.22 | 57.8 | 1.15 |
| Akron | OH | 40.92 | 49.7 | .97 |
| Cleveland | OH | 41.40 | 48.4 | .93 |
| Columbus | OH | 40.00 | 50.4 | .98 |
| Dayton | OH | 39.90 | 52.4 | 1.01 |
| Put-in-Bay | OH | 41.65 | 57.5 | 1.07 |
| Toledo | OH | 41.57 | 51.2 | 1.00 |
| Youngstown | OH | 41.27 | 46.3 | .91 |
| Oklahoma City | OK | 35.40 | 67.1 | 1.16 |
| Stillwater | OK | 36.15 | 69.1 | 1.18 |
| Tulsa | OK | 36.20 | 63.3 | 1.12 |
| Astoria | OR | 46.20 | 46.0 | .90 |
| Burns | OR | 43.58 | 65.8 | 1.19 |
| Corvallis | OR | 44.55 | 56.6 | 1.09 |
| Medford | OR | 42.38 | 61.2 | 1.07 |
| North Bend | OR | 43.42 | 56.5 | 1.05 |
| Pendleton | OR | 45.68 | 58.4 | 1.05 |
| Portland | OR | 45.60 | 48.3 | .90 |
| Redmond | OR | 44.27 | 65.9 | 1.18 |
| Salem | OR | 44.92 | 51.1 | .95 |
| Allentown | PA | 40.65 | 52.4 | 1.03 |
| Avoca | PA | 41.33 | 49.2 | .99 |
| Erie | PA | 42.08 | 46.6 | .91 |

## TABLE E–2 (Continued)

| City | State | N. Latitude (Degrees) | Radiation (I) (Btu/hr ft$^2$) | Location Factor ($K$) |
|---|---|---|---|---|
| Harrisburg | PA | 40.22 | 52.7 | 1.03 |
| Philadelphia | PA | 39.88 | 53.8 | 1.04 |
| Pittsburgh | PA | 40.50 | 60.3 | 1.12 |
| State College | PA | 40.80 | 55.2 | 1.07 |
| Newport | RI | 41.48 | 59.4 | 1.13 |
| Providence | RI | 41.73 | 51.5 | 1.03 |
| Charleston | SC | 32.90 | 60.5 | 1.07 |
| Columbia | SC | 33.95 | 62.7 | 1.10 |
| Granville-Spartansburg | SC | 34.90 | 68.1 | 1.16 |
| Huron | SD | 44.38 | 61.0 | 1.16 |
| Pierre | SD | 44.38 | 65.5 | 1.20 |
| Rapid City | SD | 44.15 | 65.6 | 1.20 |
| Sioux Falls | SD | 43.57 | 61.8 | 1.17 |
| Chattanooga | TN | 35.03 | 55.8 | 1.03 |
| Memphis | TN | 35.05 | 61.5 | 1.09 |
| Nashville | TN | 36.12 | 56.6 | 1.03 |
| Oak Ridge | TN | 36.02 | 60.2 | 1.09 |
| Abilene | TX | 32.42 | 70.5 | 1.17 |
| Amarillo | TX | 35.23 | 77.7 | 1.26 |
| Big Spring | TX | 32.25 | 75.0 | 1.21 |
| Brownsville | TX | 25.92 | 66.8 | 1.08 |
| Corpus Christi | TX | 27.77 | 66.4 | 1.09 |
| Dallas | TX | 32.85 | 66.1 | 1.12 |
| El Paso | TX | 31.80 | 87.3 | 1.30 |
| Fort Worth | TX | 32.83 | 65.8 | 1.12 |
| Houston | TX | 29.97 | 59.4 | 1.03 |
| Kingsville | TX | 27.52 | 65.5 | 1.08 |
| Lubbock | TX | 33.65 | 81.7 | 1.27 |
| Lufkin | TX | 31.23 | 64.0 | 1.09 |
| Midland | TX | 31.93 | 82.5 | 1.27 |
| Port Arthur | TX | 29.95 | 61.9 | 1.06 |
| San Angelo | TX | 31.37 | 70.7 | 1.16 |
| San Antonio | TX | 29.53 | 66.3 | 1.11 |
| Waco | TX | 31.62 | 65.6 | 1.11 |
| Wichita Falls | TX | 33.97 | 69.4 | 1.16 |
| Cedar City | UT | 37.70 | 82.9 | 1.32 |
| Salt Lake City | UT | 40.77 | 76.1 | 1.25 |
| Burlington | VT | 44.47 | 53.3 | 1.06 |
| Mt. Weather | VA | 39.07 | 60.6 | 1.14 |
| Norfolk | VA | 36.90 | 60.8 | 1.10 |
| Richmond | VA | 37.83 | 57.4 | 1.07 |
| Roanoke | VA | 37.32 | 58.4 | 1.09 |
| Olympia | WA | 46.97 | 45.2 | .85 |
| Prosser | WA | 46.25 | 71.4 | 1.20 |
| Pullman | WA | 46.73 | 66.5 | 1.18 |
| Richland | WA | 46.28 | 62.4 | 1.10 |
| Seattle | WA | 47.45 | 51.1 | .94 |
| Spokane | WA | 47.67 | 57.8 | 1.07 |
| Tacoma | WA | 47.25 | 47.3 | .84 |
| Whidbey Island | WA | 48.35 | 49.1 | .94 |
| Charleston | WV | 38.37 | 50.3 | .97 |
| Parkersburg | WV | 39.27 | 56.4 | 1.05 |
| Eau Claire | WI | 44.87 | 53.5 | 1.09 |
| Green Bay | WI | 44.48 | 53.7 | 1.08 |
| La Crosse | WI | 43.87 | 54.6 | 1.09 |
| Madison | WI | 43.13 | 56.0 | 1.11 |
| Milwaukee | WI | 42.95 | 55.4 | 1.09 |
| Casper | WY | 42.92 | 77.4 | 1.30 |
| Cheyenne | WY | 41.15 | 73.5 | 1.28 |
| Lander | WY | 42.80 | 84.3 | 1.35 |
| Laramie | WY | 41.30 | 74.7 | 1.31 |
| Rock Springs | WY | 41.60 | 80.3 | 1.33 |
| Sheridan | WY | 44.85 | 64.8 | 1.20 |
| Edmonton | AT | 53.57 | 61.0 | 1.22 |

**TABLE E–2** (Continued)

| City | State | N. Latitude (Degrees) | Radiation (I) (Btu/hr ft²) | Location Factor (K) |
|------|-------|------------------------|-----------------------------|----------------------|
| Lethbridge | AT | 49.63 | 66.8 | 1.24 |
| Vancouver | BC | 48.98 | 44.2 | .83 |
| Churchill | MA | 58.75 | 60.4 | 1.12 |
| Winnipeg | MA | 49.90 | 64.5 | 1.25 |
| Moncton | NB | 46.12 | 50.3 | 1.06 |
| St. Johns | NF | 47.52 | 42.7 | .92 |
| Aklavik | NW | 68.23 | 48.2 | .98 |
| Kapuskasing | OT | 49.42 | 51.1 | 1.12 |
| Ottawa | OT | 45.45 | 59.6 | 1.17 |
| Toronto | OT | 43.67 | 52.7 | 1.05 |
| Montreal | QU | 45.50 | 52.3 | 1.05 |

$F_R'U_L$. A collector parameter in units of Btu/(°F)(ft²)(hr), found by multiplying the collector loop heat exchanger factor by the negative slope of the collector efficiency curve referred to above.

*Storage Capacity (S)*. The capacity of the DHW preheat tank per unit collector area in units of gal/ft²; 1.82 is a typical value for DHW systems. Use the tank volume below the heating element for single tank systems.

The purpose of using the nomograph is to determine overall system efficiency (N). The value of N should be between 0.2 and 0.55 for most practical cases. Once N is determined for a given f, A can be calculated and the f vs. A graph constructed. Figures E–2, E–3 and E–4 show the results of calculating the area needed with L = 2430 Btu/day for two collectors. One is a black chrome, single-glazed AMETEK Model D-222 and the other is identical except for a flat black paint coating. Let us now proceed through the necessary steps that were necessary to construct these plots.

The linear fit equations to the ASHRAE tests are:

$$\text{AMETEK D-222} \quad \eta = .76 \quad - \quad .75 \quad \Delta T/I \qquad \textbf{(E–5)}$$
$$\text{Black Paint} \quad \eta = .7897 - 1.3953 \ \Delta T/I \qquad \textbf{(E–6)}$$

If "good practice" is used in selecting the heat exchange area, the correction factor is 0.96 (from page 250) and the following results are obtained:

| Parameter | AMETEK D-222 Black Chrome | Flat Black Paint |
|-----------|----------------------------|-------------------|
| $F_R'(\tau\alpha)$ | .725 | .758 |
| $F_R'U_L$ | .72 | 1.34 |

As shown in Table E–2:

| | | |
|---|---|---|
| Philadelphia | $I = 53.8$ and $K = 1.04$ |
| Charlotte | $I = 62.7$ and $K = 1.12$ |
| Phoenix | $I = 86.7$ and $K = 1.27$ |

From equation E–1:

$$\text{Philadelphia} \quad A = \frac{f}{N} \times \frac{L}{I} = \frac{f}{N} \times \frac{2430}{53.8} = \frac{f}{N} \times 45.17$$

$$\text{Charlotte} \quad A = \frac{f}{N} \times \frac{2430}{62.7} = \frac{f}{N} \times 38.76$$

$$\text{Phoenix} \quad A = \frac{f}{N} \times \frac{2430}{86.1} = \frac{f}{N} \times 28.22$$

The ratio of storage volume to collector area will be set at 2.0 for this example.

## CALCULATION OF AREA (A)

For each city and for each collector, start at the appropriate Location Factor $(K)$ number (found in Table E–2) and draw a straight line between that point through the appropriate $F_R'(\tau\alpha)$ to Index 1 (in Figure E–1). From that intersection point, draw a line through the appropriate $F_R'U_L$ to Index 2. From that intersection point, draw a line through $S = 2$ gal/ft$^2$ to Index 3. From that intersection, draw lines through f = .3, .4, .5, .6, .7 and .8 to determine $N$. With this information $A$ can be calculated using Equation E–1.

With $A$ computed, we can construct Table E–3 and use the data to plot the graphs shown in Figures E–2, E–3 and E–4.

**TABLE E–3**  Data for Plotting, Collector Area in Three Locations

| | AMETEK D-222 Black Chrome | | | Flat Black Paint | | |
|---|---|---|---|---|---|---|
| | f | N | A | f | N | A |
| Philadelphia | .3 | .475 | 28 | .3 | .32 | 42 |
| | .4 | .45 | 40 | .4 | .302 | 60 |
| $A = \frac{f}{N} \times 45.17$ | .5 | .415 | 54 | .5 | .28 | 81 |
| | .6 | .38 | 71 | .6 | .25 | 108 |
| | .7 | .34 | 93 | .7 | .23 | 137 |
| | .8 | .3 | 120 | .8 | .20 | 181 |
| Charlotte | .3 | .505 | 23 | .3 | .345 | 34 |
| | .4 | .48 | 32 | .4 | .325 | 48 |
| $A = \frac{f}{N} \times 38.756$ | .5 | .45 | 43 | .5 | .30 | 65 |
| | .6 | .41 | 57 | .6 | .275 | 85 |
| | .7 | .37 | 73 | .7 | .245 | 111 |
| | .8 | .32 | 97 | .8 | .215 | 144 |
| Phoenix | .3 | .57 | 15 | .3 | .39 | 22 |
| | .4 | .53 | 21 | .4 | .365 | 31 |
| $A = \frac{F}{N} \times 28.22$ | .5 | .5 | 28 | .5 | .34 | 42 |
| | .6 | .45 | 38 | .6 | .31 | 55 |
| | .7 | .41 | 48 | .7 | .28 | 71 |
| | .8 | .355 | 64 | .8 | .24 | 94 |

## ANALYSIS OF F-A GRAPHS

First, calculate the cost of hot water per year using the example of 2430 Btu/hr.

$$\frac{2430 \text{ Btu/hr} \times 24 \text{ hr/day} \times 365 \text{ day/yr}}{3413 \text{ Btu/kWh}} \times \$.095/\text{kWh} = \$597.50$$

If the installed cost of a two-collector (52.3 ft$^2$) AMETEK black chrome system is $3500 and the tax credit is 40%, then the net cost is $2100. Compare that with a $3300 black paint system whose net cost is $980 after the tax credit.

Using Figures E–2 through E–4, we ascertain the following:

|  | Solar Fraction (f) | Yearly Savings ($) | Years to Pay Back |
|---|---|---|---|
| **Philadelphia** | | | |
| Black Chrome | .5 | 296.25 | 7.09 |
| Black Paint | .38 | 225.15 | 8.79 |
| **Charlotte** | | | |
| Black Chrome | .57 | 337.73 | 6.22 |
| Black Paint | .46 | 272.55 | 7.26 |
| **Phoenix** | | | |
| Black Chrome | .75 | 444.38 | 4.73 |
| Black Paint | .6 | 355.50 | 5.57 |

The AMETEK D-222 black chrome collector system saves $71.10 more each year than the black paint collector in Philadelphia; $65.18 more each year in Charlotte; and $88.88 more each year in Phoenix with the above assumptions.

Adding another collector to the Philadelphia system at a retail cost of $500 ($300 after the tax credit) gives a total area of 78.45 ft$^2$. The solar fraction would be raised to 0.64 from 0.5 and the differential would be 0.14 times $592.50, or $82.95 per year. Then, $300 ÷ $83.95/yr = 3.6 years, which is clearly a good investment. The overall system payback is reduced to 6.33 years.

## SUMMARY

This shortcut method gives the consumer a tool with which to compare collector systems and to calculate performance on an average basis. "On average" is the key phrase here, since performance during the summer will be higher than during the winter. To verify data on your specific application under local conditions, use either procedures outlined here, or the f-CHART computer program, which prints out expected month-by-month performance. Most system manufacturers have access to this program and can model month-by-month performance, as well as different orientations. The shortcut method is restricted to use of a "tilt at latitude" and "due south" orientation. Two examples of the value of examining month-by-month data are discussed here.

1. In the Philadelphia area, there are many tankless water heaters. These oil-fired units could be shut down for the spring, summer and fall months when only hot water is needed. Additionally, the winter months are often cloudy, with a low solar fraction. Better economic performance can be enjoyed by a lower-than-latitude collector tilt. To prevent condensation, a low level of heat must be added to the shut-down oil-fired heater.

2. In Phoenix, the summers are so warm and the solar input so high that collectors should usually be tilted higher than the latitude angle for best average annual performance.

These observations can only be made by studying month-by-month performance for a specific locality.

From a multi-month simultaneous test of an AMETEK two-collector and three-collector domestic hot water system, it was found that f-CHART will accurately predict system performance to within a few percentage points. To obtain this correlation, the *actual* average irradiance had to be substituted for the multi-year *average* values that are used in the f-CHART procedure. (This data was presented by J. Nelson of AMETEK, Inc., at the ASHRAE meeting in Cincinnati on June 30, 1981. The data was taken at AMETEK's test facility in Hatfield, Pennsylvania.)

# Bibliography

**The Widening Energy Gap and Alternative Energy Sources**

Cheney, Eric S. "U.S. Energy Resources: Limits and Future Outlook." *American Scientist* 62 (January–February 1974): 14.

Hottel, H. C., and Howard, J. B. *New Energy Technology: Some Facts and Assessments.* Cambridge: The MIT Press, second printing 1972.

Reed, C. B. *Fuels, Minerals, and Human Survival.* Ann Arbor: Ann Arbor Science, 1975.

Rose, David J. "Energy Policy in the U.S." *Scientific American* 230, no. 1 (January 1974):20.

Starr, Chauncey. "Energy and Power." *Scientific American* 225, no. 3 (September 1971):37.

Thirring, Hans. *Energy for Man: From Windmills to Nuclear Power,* New York: Harper & Row, 1976.

**The Prospects for Solar Energy**

Bradford, P. V. "Investing in Solar Energy." Securities Research Division of Merrill Lynch, New York, NY, 1975.

Butt, Sheldon H. "Solar Heating and Cooling, A Developing Market." Presented before the Second Annual Meeting of the Solar Energy Industries Association in Washington, DC, 1975. Published by the SEIA.

Duffie, John A., and Beckman, William A. "Solar Heating and Cooling." *Science* 191, no. 4223 (16 January 1976):143.

Kaplan, Gadi. "Planning Solar's Future." *IEEE Spectrum* (June 1976):55.

Pollard, William G. "The Long Range Prospects for Solar Energy." *American Scientist* 64 (July–August 1976):424.

———. "The Long Range Prospects for Solar-Derived Fuels." *American Scientist* 64 (September–October 1976):509.

Stoll, Richard D. "Solar Collector Manufacturing Activity January Through June 1976." Nuclear Energy Analysis Division, Office of Coal, Nuclear & Electric Power Analysis, Federal Energy Administration, Room 207, Old Post Office Building, Washington, DC 20461.

**The Sun**

Flammarion, G. Camille, ed. *Flammarion Book of Astronomy.* New York: Simon & Schuster, 1964.

Lindsay, Sally, ed. "The Turbulent Sun." *Natural History* (November 1976): 54–81.

**The Solar Spectrum and the Solar Constant**

Coulson, Kinsell L. *Solar and Terrestrial Radiation.* New York: Academic Press, 1975.

Duffie, John A., and Beckman, William A. *Solar Energy Thermal Processes.* New York: John Wiley & Sons, 1974.

Kreider, Jan F., and Kreith, Frank. *Solar Heating and Cooling.* New York: McGraw Hill, 1975.

## Blackbody Radiation Laws

Coulson, Kinsell L. *Solar and Terrestrial Radiation*. New York: Academic Press, 1975.

Duffie, John A., and Beckman, William A. *Solar Energy Thermal Processes*. New York: John Wiley & Sons, 1974.

## Apparent Motion of the Sun

*The American Ephemeris*. Published each forthcoming year by the U.S. Government Printing Office, Washington, DC.

American Society of Heating, Refrigerating and Air Conditioning Engineers, Inc. *Handbook of Fundamentals*. 1981. 345 E. 47th Street, New York, NY 10017.

Brinkworth, B. J. *Solar Energy for Man*. Somerset, NJ: John Wiley & Sons, 1972.

Kreider, Jan F., and Kreith, Frank. *Solar Heating and Cooling*. New York: McGraw Hill, 1975 (pp. 45–56).

## Estimation of Solar Irradiance

American Society of Heating, Refrigerating, and Air Conditioning Engineers, Inc. *Handbook of Fundamentals*. 1981. 345 E. 47th Street, New York, NY 10017.

Yellott, John I., ed. "Solar Energy Utilization for Heating and Cooling." Chapter 58, 1978 edition, *Applications Handbook*. American Society of Heating, Refrigerating and Air Conditioning Engineers, Inc., 345 E. 47th Street, New York, NY 10017. Also reprinted as a National Science Foundation Report by the U.S. Government Printing Office, pamphlet NSF 74–41, stock number 038-000-00188-4.

## Shadow Sizes

Waugh, Albert E. *Sundials, Their Theory and Construction*. New York: Dover Publications, 1973.

## Measurements of Solar Irradiance

Bennett, Iven. "Monthly Maps of Mean Daily Insolation for the United States." *Solar Energy* 9, no. 3 (July–September 1965):145.

Coulson, Kinsell L. *Solar and Terrestrial Radiation*. New York: Academic Press, 1975.

Klein, S. A. "Calculation of Flat Plate Utilizability." *Solar Energy* 21:393–402.

———. "Calculation of Monthly Average Insolation on Tilted Surfaces." *Solar Energy* 19:325–329.

Liu, B. Y. H., and Jordan, R. C. "Daily Insolation of Surfaces Tilted toward the Equator." *ASHRAE Transactions* 526–541 (1962).

———. "The Long Term Average Performance of Flat Plate Solar Energy Collectors." *Solar Energy* 7:53–74.

Lof, G. O. G., Duffie, J. A., and Smith, C. O. "World Distribution of Solar Radiation." July 1966. Report no. 21, Solar Energy Laboratory, University of Wisconsin, Engineering Experiment Station.

Quinlan, Frank. "Availability of Solar Radiation Data in the United States." Paper PH-79-8, No. 3, *ASHRAE Transactions*, Volume 85, Part 1.

## Climatic Variables

*Climatic Atlas of the United States*. Reprinted by the National Oceanic and Atmospheric Administration, U.S. Department of Commerce. Available from the National Climatic Center, Federal Building, Asheville, NC 28801.

*Engineering Design Manual for Outdoor Advertising Structures*. Published by the Outdoor Advertising Association of America, Inc., 625 Madison Ave., New York, NY 10022. (See section on wind loads.)

*Statistical Abstract of the United States*. Published annually by the U.S. Department of Commerce, Bureau of the Census. Available from the U.S. Government Printing Office, Section G, Geography and Environment.

## Diffuse and Reflected Components of Solar Insolation

American Society of Heating, Refrigerating and Air Conditioning Engineers, Inc. *Applications Handbook*. 1981. 345 E. 47th Street, New York, NY 10017.

Kreider, J. F., and Keith, F. *Solar Heating and Cooling*. New York: McGraw-Hill, 1975 (p. 77).

## System Approximations

"Solar Energy for Space Heating and Hot Water." 1976. SE-101. Published by the Energy Research and Development Administration, Washington, DC.

## Flat Plate Collectors

Bowen, John C. "The Problems of Treating $U_L$ as a Constant." *Proceedings of the 1980 Annual Meeting of the International Solar Energy Society, Inc.* Reprints available from ISES, 205B McDowell Hall, Univ. Delaware, Newark, DE 19711.

Hottel, H. C., and Woertz, B. B. "The Performance of Flat-Plate Solar Heat Collectors." *Transactions of the American Society of Mechanical Engineers* 64 (1942):91.

Leckie, Jim, et al., *Other Homes and Garbage*. San Francisco: Sierra Club Books, 1975.

### Selective Surfaces

Baumeister, Philip. "A Comparison of Solar Photothermal Coatings." *Proceedings of the Society of Photo-Optical Instrumentation Engineers* 85 (1976):47–61.

McDonald, Glen E. "Selective Coating for Solar Panels." National Technical Information Service Report N76-15603, December 1975. Obtained from NTIS, 5285 Port Royal Road, Springfield, VA 12161.

Schreyer, J. M., et al. "Selective Absorptivity of Carbon Coatings." Report Y/DA-6701, July 1976. Oak Ridge Y-12 Plant, Oak Ridge, TN. Available from the Union Carbide Corporation Nuclear Division or from the Energy Research and Development Administration.

Tabor, H. "Selective Surfaces for Solar Collectors." Chapter IV of *Low Temperature Engineering Application of Solar Energy*. American Society of Heating, Refrigerating and Air Conditioning Engineers, Inc., 345 E. 97th Street, New York, NY 10017.

## Concentrating Collectors

Nelson, D. T., Evans, D. L., and Bansal, R. K. "Linear Fresnel Lens Concentrators." *Solar Energy* 17, no. 5-C (1975):285.

Rabl, A. "Comparison of Solar Concentrators." *Solar Energy* 18 (1976):93.

Winston, R. "Light Collection Within the Framework of Geometrical Optics." *Journal of the Optical Society of America* 60 (1970):245.

## Photovoltaic Collectors

Chalmers, Bruce. "The Photovoltaic Generation of Electricity." *Scientific American* 235, no. 4 (October 1976):34.

## Measurement Techniques and Procedures

Considine, Douglas M. *Process Instruments and Controls Handbook*. New York: McGraw-Hill, 1974.

Duebelin, Ernest O. *Measurement Systems—Applications and Design*. New York: McGraw-Hill, 1975.

"Methods of Testing for Rating Solar Collectors Based on Thermal Performance." NBSIR 74-635, December 1974. Interim Report prepared for the National Science Foundation by the National Bureau of Standards.

"Method of Testing Solar Collectors Based on Thermal Performance." ASHRAE Proposed Standard 93 P, June 1, 1976.

**General References**

Agosto, William N. "Microwave Power: A Far-Out System." *IEEE Spectrum* (May 1976):48.

Agranoff, Joan, ed. *Modern Plastics Encyclopedia 1975–1976*. New York: McGraw-Hill, 1976.

Balcomb, J. Douglas, and Hedstrom, James C. "Sizing Collectors for Space Heating." *Sunworld* no. 1 (July 1976):25. Published by the International Solar Energy Society, P.O. Box 26, Highett, Victoria 3190 Australia.

Baumeister, Theodore, ed. *Mark's Standard Handbook for Mechanical Engineers*, 7th ed. New York: McGraw-Hill, 1967; 8th ed., 1978.

Considine, Douglas M. *Process Instruments and Controls Handbook*. New York: McGraw-Hill, 1974.

Duffie, John, and Beckman, William. *Solar Energy Thermal Processes*. New York: John Wiley & Sons, 1974.

Gunther, Raymond C. *Refrigeration, Air Conditioning, and Cold Storage*. Chilton Books, 1969.

Hicks, Tyler G., ed. *Standard Handbook of Engineering Calculations*. New York: McGraw-Hill, 1972.

Kays, W. M., and London, A. L. *Compact Heat Exchangers*. 2d ed. New York: McGraw-Hill, 1964.

Klein, S. A., Beckman, W. A., and Duffie, J. A. "Design Procedure for Solar Heating Systems." *Solar Energy* 18, no. 2 1976:113.

Kreider, Jan F., and Kreith, Frank. *Solar Heating and Cooling*. New York: McGraw-Hill, 1975.

*Materials Engineering*, Vol 81. Materials Selector, mid-September 1975. Published by Reinhold Publishing Company, 600 Summer Street, Stamford, CT 06904.

Meinel, Aden B., and Meinel, Marjorie P. *Applied Solar Energy*. Reading, Mass.: Addison-Wesley, 1976.

Peny, Robert H., and Chilton, Cecil H. *Chemical Engineer's Handbook*. 5th ed. New York: McGraw-Hill, 1973.

Ratzel, A. C., and Bannerot, R. B. "Optimal Material Selection for Flat Plate Solar Energy Collectors Utilizing Commercially Available Materials." Presented at the 16th National Heat Transfer Conference, St. Louis, August 8–11 1976. Available from the American Institute of Chemical Engineers, 345 E. 47th Street, New York, NY 10017.

Tabor, H. "Radiation, Convection, and Conduction Coefficients in Solar Collectors." *Bulletin of the Research Council of Israel* 6C:155, 1958.

Williams, J. Richard. *Solar Energy Technology and Applications*. Ann Arbor: Ann Arbor Science, 1977.

Weast, Robert C., ed. *Handbook of Chemistry and Physics*. CRC Press, 18901 Cranwood Parkway, Cleveland, Ohio 44128.

# Index

Page numbers in *italics* refer to illustrations. Page numbers followed by t refer to tables.